I0503127

Steady-State and Transient Models of Groundwater Flow and Advective Transport, Eastern Snake River Plain Aquifer, Idaho National Laboratory and Vicinity, Idaho

By Daniel J. Ackerman, Joseph P. Rousseau, Gordon W. Rattray, and Jason C. Fisher

Prepared in cooperation with the U.S. Department of Energy

Scientific Investigations Report 2010–5123

U.S. Department of the Interior
U.S. Geological Survey

U.S. Department of the Interior
KEN SALAZAR, Secretary

U.S. Geological Survey
Marcia K. McNutt, Director

U.S. Geological Survey, Reston, Virginia: 2010

For more information on the USGS—the Federal source for science about the Earth, its natural and living resources, natural hazards, and the environment, visit http://www.usgs.gov or call 1-888-ASK-USGS

For an overview of USGS information products, including maps, imagery, and publications, visit http://www.usgs.gov/pubprod

To order this and other USGS information products, visit http://store.usgs.gov

Suggested citation:
Ackerman, D.J., Rousseau, J.P., Rattray, G.W., and Fisher, J.C., 2010, Steady-state and transient models of groundwater flow and advective transport, Eastern Snake River Plain aquifer, Idaho National Laboratory and vicinity, Idaho: U.S. Geological Survey Scientific Investigations Report 2010-5123, 220 p.

Contents

Contents—Continued

Figures

Figures—Continued

Figures—Continued

Figures—Continued

Tables

Tables—Continued

Conversion Factors, Datums, and Abbreviations and Acronyms

Conversion Factors

Multiply	By	To obtain
Length		
inch (in.)	2.54	centimeter (cm)
foot (ft)	0.3048	meter (m)
mile (mi)	1.609	kilometer (km)
Area		
acre	4,047	square meter (m^2)
square foot (ft^2)	0.09290	square meter (m^2)
square mile (mi^2)	2.590	square kilometer (km^2)
Volume		
gallon (gal)	3.785	liter (L)
acre-foot (acre-ft)	1,233	cubic meter (m^3)
Flow rate		
foot per second (ft/s)	0.3048	meter per second (m/s)
foot per day (ft/d)	0.3048	meter per day (m/d)
foot per year (ft/yr)	0.3048	meter per year (m/yr)
cubic foot per second (ft^3/s)	0.02832	cubic meter per second (m^3/s)
cubic foot per second per square mile [(ft^3/s)/mi^2]	0.01093	cubic meter per second per square kilometer [(m^3/s)/km^2]
Radioactivity		
curie (Ci)	37,000	megabecquerel (MBq)
curie per year (Ci/yr)	37,000	megabecquerel per year (MBq/yr)
picocurie per milliliter (pCi/mL)	0.037	becquerel per milliliter (Bq/mL)
picocurie per liter (pCi/L)	0.037	becquerel per liter (Bq/L)
Hydraulic conductivity		
foot per day (ft/d)	0.3048	meter per day (m/d)
Hydraulic gradient		
foot per mile (ft/mi)	0.1894	meter per kilometer (m/km)

Temperature in degrees Celsius (°C) may be converted to degrees Fahrenheit (°F) as follows:

$$°F=(1.8×°C)+32.$$

Concentrations of chemical constituents in water are given either in milligrams per liter (mg/L) or micrograms per liter (μg/L).

Datums

Vertical coordinate information is referenced to the National Geodetic Vertical Datum of 1988 (NGVD 29).

Horizontal coordinate information is referenced to the North American Datum of 1983 (NAD 83).

Altitude, as used in this report, refers to distance above the vertical datum.

The projection Albers used in the North American Datum of 1927, a central meridian of 113°W., standard parallels of 42° 50′ N. and 44° 10′ N., a false easting of 656,166.67 ft, and the latitude of projection's origin was 41° 30′ N.

Abbreviations and Acronyms

Abbreviation or Acronym	Definition
ANL	Argonne National Laboratory (now MFC)
CFA	Central Facilities Area
CPP	Chemical Processing Plant (changed to ICPP, then INTEC)
CU	composite unit
DOE	U.S. Department of Energy
EBR	Experimental Breeder Reactor
ESRP	Eastern Snake River Plain
HC	hydraulic conductivity
ICPP	Idaho Chemical Processing Plant (formerly CPP, now INTEC)
INL	Idaho National Laboratory
INTEC	Idaho Nuclear Technology and Engineering Center (formerly ICPP)
LSIT	large-scale infiltration test
MFC	Materials and Fuels Complex (formerly ANL)
NRF	Naval Reactors Facility
RASA	Regional Aquifer System Analysis
RTC	Reactor Technology Complex (formerly TRA)
RWMC	Radioactive Waste Management Complex
SDA	Subsurface Disposal Area
SS	specific storage
SY	specific yield
TAN	Test Area North
TRA	Test Reactor Area (now RTC)
USGS	U.S. Geological Survey
VANI	ratio of horizontal to vertical hydraulic conductivity

Steady-State and Transient Models of Groundwater Flow and Advective Transport, Eastern Snake River Plain Aquifer, Idaho National Laboratory and Vicinity, Idaho

By Daniel J. Ackerman, Joseph P. Rousseau, Gordon W. Rattray, and Jason C. Fisher

Abstract

Three-dimensional steady-state and transient models of groundwater flow and advective transport in the eastern Snake River Plain aquifer were developed by the U.S. Geological Survey in cooperation with the U.S. Department of Energy. The steady-state and transient flow models cover an area of 1,940 square miles that includes most of the 890 square miles of the Idaho National Laboratory (INL). A 50-year history of waste disposal at the INL has resulted in measurable concentrations of waste contaminants in the eastern Snake River Plain aquifer. Model results can be used in numerical simulations to evaluate the movement of contaminants in the aquifer.

Saturated flow in the eastern Snake River Plain aquifer was simulated using the MODFLOW-2000 groundwater flow model. Steady-state flow was simulated to represent conditions in 1980 with average streamflow infiltration from 1966–80 for the Big Lost River, the major variable inflow to the system. The transient flow model simulates groundwater flow between 1980 and 1995, a period that included a 5-year wet cycle (1982–86) followed by an 8-year dry cycle (1987–94). Specified flows into or out of the active model grid define the conditions on all boundaries except the southwest (outflow) boundary, which is simulated with head-dependent flow. In the transient flow model, streamflow infiltration was the major stress, and was variable in time and location. The models were calibrated by adjusting aquifer hydraulic properties to match simulated and observed heads or head differences using the parameter-estimation program incorporated in MODFLOW-2000. Various summary, regression, and inferential statistics, in addition to comparisons of model properties and simulated head to measured properties and head, were used to evaluate the model calibration.

Model parameters estimated for the steady-state calibration included hydraulic conductivity for seven of nine hydrogeologic zones and a global value of vertical anisotropy. Parameters estimated for the transient calibration included specific yield for five of the seven hydrogeologic zones. The zones represent five rock units and parts of four rock units with abundant interbedded sediment. All estimates of hydraulic conductivity were nearly within 2 orders of magnitude of the maximum expected value in a range that exceeds 6 orders of magnitude. The estimate of vertical anisotropy was larger than the maximum expected value. All estimates of specific yield and their confidence intervals were within the ranges of values expected for aquifers, the range of values for porosity of basalt, and other estimates of specific yield for basalt.

The steady-state model reasonably simulated the observed water-table altitude, orientation, and gradients. Simulation of transient flow conditions accurately reproduced observed changes in the flow system resulting from episodic infiltration from the Big Lost River and facilitated understanding and visualization of the relative importance of historical differences in infiltration in time and space. As described in a conceptual model, the numerical model simulations demonstrate that flow is (1) dominantly horizontal through interflow zones in basalt and vertical anisotropy resulting from contrasts in hydraulic conductivity of various types of basalt and the interbedded sediments, (2) temporally variable due to streamflow infiltration from the Big Lost River, and (3) moving downward downgradient of the INL.

The numerical models were reparameterized, recalibrated, and analyzed to evaluate alternative conceptualizations or implementations of the conceptual model. The analysis of the reparameterized models revealed that little improvement in the model could come from alternative descriptions of sediment content, simulated aquifer thickness, streamflow infiltration, and vertical head distribution on the downgradient boundary. Of the alternative estimates of flow to or from the aquifer, only a 20 percent decrease in the largest inflow, the northeast boundary underflow, resulted in a recalibrated parameter value just outside the confidence interval of the base-case calibrated value.

Particle-tracking calculations using the particle-tracking program MODPATH were used to evaluate (1) how simulated groundwater flow paths and travel times differ between the steady-state and transient flow models, (2) how wet- and dry-climate cycles affect groundwater flow paths and travel times, and (3) how well model-derived groundwater flow directions and velocities compare to independently derived estimates. Particle tracking also was used to simulate the growth of tritium (^{3}H) plumes originating at the Idaho Nuclear Technology and Engineering Center and the Reactor Technology Complex over a 16-year period under steady-state and transient flow conditions (1953–68). The shape, dimensions, and areal extent of the ^{3}H plumes were compared to a map of the plumes for 1968 from ^{3}H releases at the Idaho Nuclear Technology and Engineering Center and the Reactor Technology Complex beginning in 1952.

Collectively, the particle-tracking simulations indicate that average linear groundwater velocities, based on estimates of porosity, and flow paths are influenced by two primary factors: (1) the dynamic character of the water table and (2) the large contrasts in the hydraulic properties of the media, primarily hydraulic conductivity. The simulated growth and decay of groundwater mounds as much as 34 ft above the steady-state water table beneath the Big Lost River spreading areas, sinks, and playas, and to a lesser extent beneath the Big Lost River channel lead to non-uniform changes in the altitude of the water table throughout the model area. These changes affect the orientation and magnitude of water-table gradients and affect groundwater flow directions and velocities to a greater or lesser degree depending on the magnitude, duration, and proximity of the transient stress. Simulation results also indicate that temporal changes in the local hydraulic gradient can account for some of the observed dispersion of contaminants in the aquifer near the major sources of contamination at the INTEC and the RTC and perhaps most observed dispersion several miles downgradient of these facilities. The distance downgradient of the INTEC that simulated particle plumes were able to reasonably reproduce the shape and dimensions of the 1968 ^{3}H plume extended only to the boundary of zones of abundant sediment, about 4 miles downgradient of the INTEC. This boundary encompasses the entire area represented by the 1968 25,000 picocuries/liter ^{3}H isopleths. Particle plumes simulated beyond this boundary were narrow and long, and did not reasonably reproduce the shape, dimensions, or position of the leading edge of the ^{3}H plume as shown in earlier reports; however, as noted in an assessment of the interpreted plume, few data were available in 1968 to characterize its true areal extent and shape.

Introduction

The Idaho National Laboratory (INL) was established by the U.S. Atomic Energy Commission, now the U.S. Department of Energy (DOE), in 1949 to build, operate, and test nuclear reactors. The scope of work at the INL increased from the 1950s through the 1970s to include other nuclear-research programs, the reprocessing of spent nuclear fuel, and the storage and disposal of radioactive waste. A history of more than 50 years of waste disposal associated with nuclear-reactor research and nuclear-fuel reprocessing at the INL has resulted in measurable concentrations of contaminants in the eastern Snake River Plain (ESRP) aquifer beneath the INL.

The INL covers an area of about 890 mi^2 and overlies the west-central part of the ESRP in southeast Idaho (fig. 1). The underlying ESRP aquifer, designated as a sole source aquifer by the U.S. Environmental Protection Agency (1991), is a major source of water for agricultural, industrial, and domestic use in southeast Idaho. Wastewater disposal sites at the Test Area North (TAN), Naval Reactors Facility (NRF), Reactor Technology Complex (RTC, formerly known as Test Reactor Area, or TRA), and Idaho Nuclear Technology and Engineering Center (INTEC, formerly known as the Idaho Chemical Processing Plant, or ICPP) (fig. 1) have been primary sources of radioactive and chemical waste contaminants in water from the ESRP aquifer. These wastewater disposal sites have, in the past, included lined evaporation ponds, unlined infiltration ponds and ditches, drain fields, and injection wells. Waste materials buried in shallow pits and trenches within the Subsurface Disposal Area (SDA) at the Radioactive Waste Management Complex (RWMC) have been sources of contaminants in groundwater (table 1).

The presence of contaminants in the ESRP aquifer (fig. 2) has led to public concern about the quality of water in the aquifer and about how contaminated groundwater might affect the region and its economy. The DOE needs to understand thoroughly the movement and fate of these contaminants in the subsurface to minimize health risks and to plan effectively for any necessary remediation. To meet these needs, the U.S. Geological Survey (USGS) developed computer-based numerical models to simulate the movement of groundwater and the advective transport of contaminants in the aquifer under steady-state and transient flow conditions.

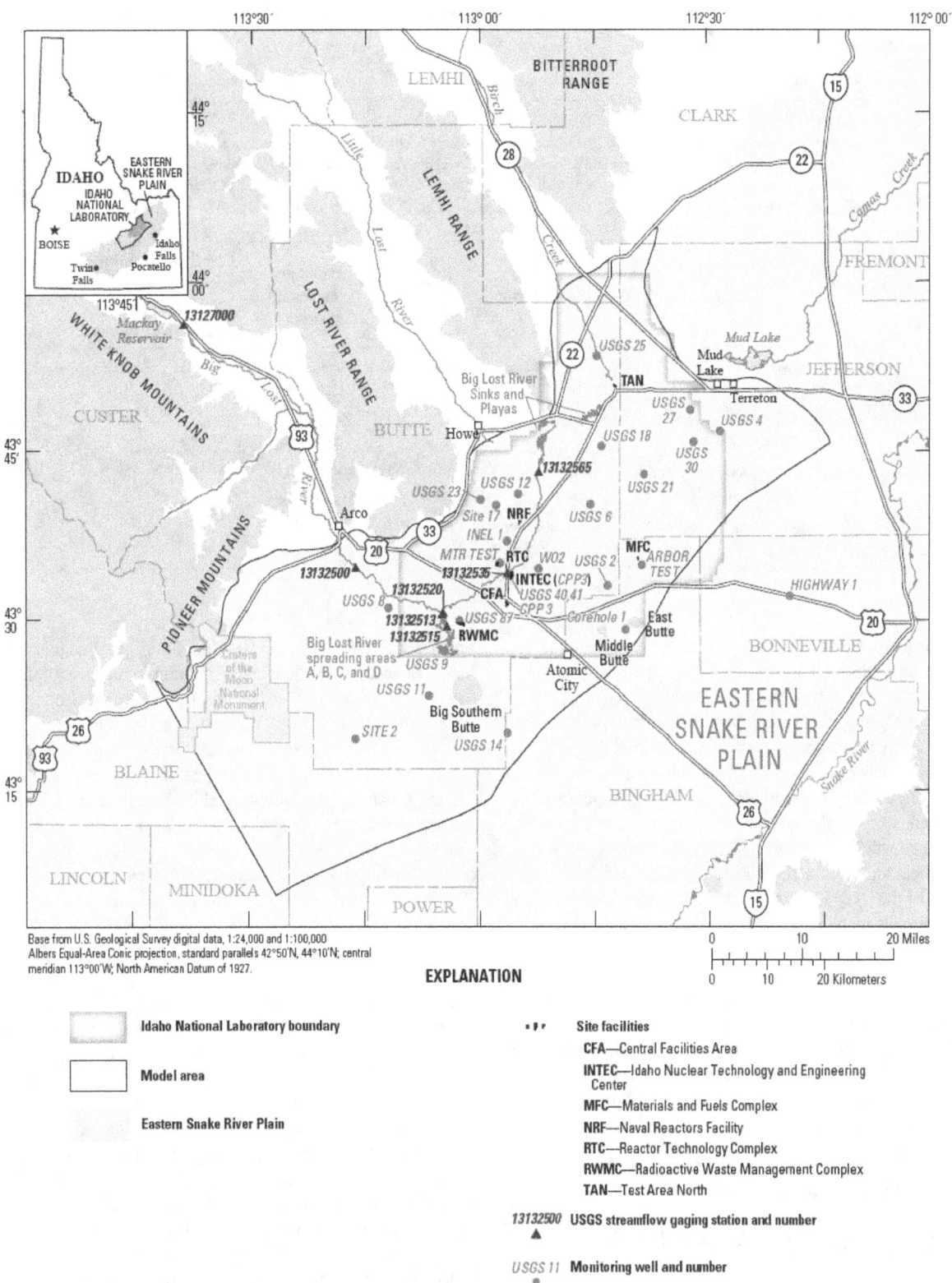

Figure 1.   Location of the Idaho National Laboratory, the model area, selected facilities, wells, boreholes, streamflow-gaging stations, and Craters of the Moon National Monument, Idaho.

Table 1.   Summary of surface and subsurface wastewater disposal at selected facilities, Idaho National Laboratory, Idaho.

[**Estimated disposal quantities:** Estimated disposal quantities for Reactor Technology Complex, Test Reactor Area, and Idaho Nuclear Technology and Engineering Center from French and others, 1999 Estimated disposal quantities for the Radioactive Waste Management Complex from Mann and Knobel, 1987 and Becker and others, 1998 **Contaminant concentrations:** from Mann and Beasley, 1994; Bukowski and others, 1998; and Bartholomay and others, 2000 **Abbreviations**: MCLs, U S Environmental Protection Agency maximum contaminant levels for drinking water; Ci, curies; Mgal, million gallons; ^{137}Cs, cesium-137; ^{3}H, tritium; ^{90}Sr, strontium-90; TCE, trichloroethene; PCE, tetrachloroethene; DCE, cis- and trans-1,2-dichloroethene; Cr, chromium, ^{129}I, iodine-129; CCl_4, carbon tetrachloride]

Facility	Disposal site	Years of disposal	Estimated disposal quantities	Contaminants with concentrations that have exceeded MCLs in groundwater
Test Area North	Well, pond	1953 to 1993	61 Ci 717 Mgal	^{137}Cs, ^{3}H, ^{90}Sr, TCE, PCE, DCE
Reactor Technology Complex	Well, ponds	1952 to 1998	53,879 Ci 5,180 Mgal	^{3}H, Cr
Idaho Nuclear Technology and Engineering Center	Well, ponds	1952 to 1998	22,254 Ci 19,165 Mgal	^{3}H, ^{90}Sr, ^{129}I
Radioactive Waste Management Complex	Excavation pits, trenches	1952 to 1970	1,532,600 Ci 0.09 Mgal	CCl_4

Groundwater flow and contaminant transport in the aquifer beneath the INL were simulated by Robertson (1974), and then evaluated and resimulated by Goode and Konikow (1990a, b). The calibrated numerical flow-transport and advective-transport models described in this report build on the work of earlier models, substantially more information than was available to earlier investigators, and a refined conceptual model of flow (Ackerman and others, 2006) beneath the INL and vicinity that implies downward movement and deeper circulation of contaminants that migrate offsite of the INL. New information available for developing these models includes improved understanding of the hydrogeology beneath the INL and of the amount and timing of transient recharge to the aquifer beneath the INL, including a near record amount of streamflow infiltration from the Big Lost River in 1983 and 1984, and a period with little or no infiltration between 1987 and 1994. Additionally, more than 30 years of water-level data for numerical-model calibration and measurements of contaminant concentrations in the aquifer are available.

Purpose and Scope

This report presents the results of simulations and analyses of three-dimensional (3-D) steady-state and transient groundwater flow models of the ESRP aquifer beneath the INL and vicinity. These models can be used to simulate contaminant movement in the aquifer to (1) determine long-term risks associated with contaminants that are in the aquifer currently or might be in the future from additional, slow releases of residual contamination in the unsaturated zone or from wastes that are left buried in shallow pits and trenches; (2) identify the best locations for monitoring contaminant movement in the aquifer; (3) evaluate the effect of future groundwater usage at the INL on water availability in the ESRP aquifer; and (4) evaluate risks to the aquifer associated with the operation of future nuclear research facilities.

The steady-state and transient flow models were constructed using the USGS modular, three-dimensional, finite-difference groundwater flow model, MODFLOW-2000 (Harbaugh and others, 2000). Steady-state flow was simulated to represent conditions in 1980 with average streamflow infiltration from 1966–80 for the Big Lost River, which is the main variable inflow to the system. The transient flow model simulates groundwater flow that occurred between 1980 and 1995, a period that included a 5-year wet cycle (1982–86) followed by an 8-year dry cycle (1987–94). Development of the numerical models involved (1) formulation and evaluation of various spatial and temporal discretization schemata; (2) specification of boundary and initial conditions; (3) weighting of calibration heads and head differences; (4) calibration; (5) specification of convergence and mass-balance criteria; (6) sensitivity analyses for evaluation of various conceptual assumptions, approximations, and simplifications; and (7) particle-tracking simulations to visualize 3-D flow and to assess model-derived groundwater flow paths, travel times, velocities, and source areas for consistency with other independent lines of evidence.

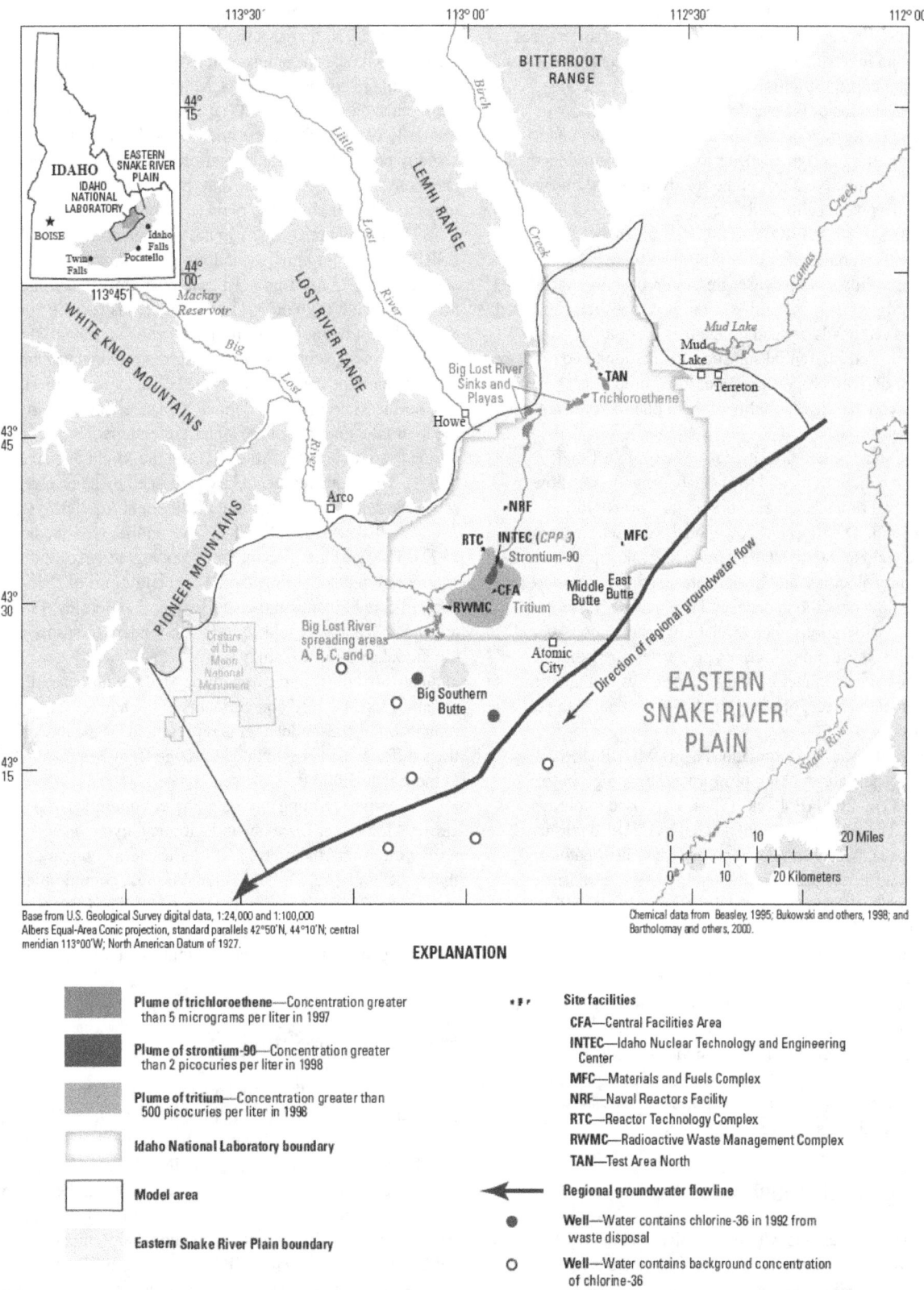

Base from U.S. Geological Survey digital data, 1:24,000 and 1:100,000
Albers Equal-Area Conic projection, standard parallels 42°50'N, 44°10'N; central
meridian 113°00'W; North American Datum of 1927.

Chemical data from Beasley, 1995; Bukowski and others, 1998; and
Bartholomay and others, 2000.

EXPLANATION

Plume of trichloroethene—Concentration greater than 5 micrograms per liter in 1997

Plume of strontium-90—Concentration greater than 2 picocuries per liter in 1998

Plume of tritium—Concentration greater than 500 picocuries per liter in 1998

Idaho National Laboratory boundary

Model area

Eastern Snake River Plain boundary

• ฯ ɾ   Site facilities

CFA—Central Facilities Area

INTEC—Idaho Nuclear Technology and Engineering Center

MFC—Materials and Fuels Complex

NRF—Naval Reactors Facility

RTC—Reactor Technology Complex

RWMC—Radioactive Waste Management Complex

TAN—Test Area North

Regional groundwater flowline

● Well—Water contains chlorine-36 in 1992 from waste disposal

○ Well—Water contains background concentration of chlorine-36

Figure 2.   Locations of documented contaminant plumes in the eastern Snake River Plain aquifer that exceed or have exceeded maximum contaminant levels for drinking water as defined by the U.S. Environmental Protection Agency, Idaho National Laboratory and vicinity, Idaho. Contaminant concentrations exceeding maximum contaminant levels for drinking water are summarized in table 1.

Models were calibrated by adjusting aquifer properties to match simulated and observed heads or head differences by using an inverse modeling approach. The "trial-and-error" calibration approach initially was used to calibrate the simplest versions of the steady-state model, primarily to test the overall soundness of the conceptual model and to select important design elements to include in subsequent models. Complex versions of the steady-state and transient flow models were calibrated using the inverse modeling capabilities of MODFLOW–2000 (Hill and others, 2000). Optimized estimates of aquifer properties were calculated using a non-linear statistical regression procedure that adjusts aquifer properties to minimize differences between simulated and observed heads or head differences. Various summary, regression, and inferential statistics, in addition to comparisons of model-derived properties to measured properties, were used to evaluate the rationality of the model-derived aquifer properties.

The steady-state flow model was used to estimate hydraulic conductivity and to evaluate groundwater flow paths, travel times, and velocities under flow conditions during 1980. The transient flow model was used to estimate specific yield and to evaluate groundwater flow paths, travel times, and velocities, and to simulate advective transport for variable flow conditions between 1980 and 1995. Because of scaling considerations, data availability, and computational constraints, these model-derived properties represent regional-scale estimates of aquifer properties that approximate the integrated effects of many smaller-scale features and processes.

Particle-tracking simulations using MODPATH, the USGS particle-tracking post-processing program for MODFLOW–2000 (Pollock, 1994), were used to produce maps of 3-D flow in the aquifer. MODPATH also was used to evaluate (1) how simulated groundwater flow paths and travel times differ between the steady-state and transient flow models, (2) how wet- and dry-climate cycles affect groundwater flow paths and travel times, (3) how episodic streamflow infiltration affects advective transport, and (4) how well the simulated groundwater flow directions and average linear velocities represent actual conditions in that part of the aquifer most affected by contamination.

Additional information on individual wells, boreholes, and streamflow-gaging stations used in this study are found in appendix tables A1 and A2. Additional maps showing the locations of these sites are in appendix B.

Description of Model Area

The model area is within the ESRP, a relatively flat topographic depression about 200-mi long and 50- to 70-mi wide surrounded by mountains on three sides and rimmed by the Snake River near the edges of its eastern and southern boundaries and by canyons as deep as 550-ft along and near

its southwestern boundary. The altitude of the plain rises from about 2,900 ft near King Hill in the southwest to more than 6,200 ft near the southwestern extent of the Yellowstone Plateau in the northeast (fig. 3). Mountains surrounding the plain reach altitudes of 12,000 ft. The surface of the plain is primarily composed of loess and olivine basalt with many visually prominent volcanic landforms, such as cinder and lava cones, shield volcanoes, and rhyolite domes that rise as much as 2,500 ft above the plain (Lindholm, 1996, p. 5).

The model area (fig. 3), in the west-central part of the ESRP, covers 1,940 mi^2, extends 35 mi from northwest to southeast and 75 mi from northeast to southwest, and includes most of the INL (890 mi^2). The model area is bounded on the northwest by mountain fronts and valleys tributary to the plain and on the southeast by an inferred regional groundwater flowline near the central axis of the ESRP (figs. 2 and 3). The northeast boundary is defined by a steep increase in the hydraulic gradient of the water table upgradient of the northeastern boundary of the INL and the Mud Lake area (fig. 3). The southwest boundary is defined by a northwest to southeast gridline from the USGS Regional Aquifer System Analysis (RASA) model of the ESRP aquifer (Garabedian, 1992). This gridline is about 25 mi downgradient of the southwestern extent of measured concentrations of INL-derived contaminants in the aquifer (fig. 2) (Beasley, 1995, appendix 2) and was selected as the boundary to ensure that the model area for this study would be sufficiently large to simulate contaminant movement beyond the farthest known extent of INL-derived contaminants in the aquifer. The vertical dimension of the model area is represented by the thickness of the aquifer, which is estimated to range from less than 600 ft to more than 3,000 ft.

The small farming and ranching communities of Arco, Howe, Mud Lake, Terreton, and Atomic City are near the northwestern, northeastern, and southeastern boundaries of the model area (fig. 2). Populations of these communities range from 25 to slightly more than 1,000. The Craters of the Moon National Monument (fig. 1) is a remote and largely undeveloped 83-mi^2 area near the southwestern extent of the model area.

Hydrogeologic Setting

The climate in the model area is semiarid. Mean annual precipitation at the INL is about 0.7 ft (Goodell, 1988, fig. 5; Clawson and others, 1989, tables D-1 and D-2). About 0.1 to 0.3 ft of snow accumulates at the INL during winter, and peak streamflow from snowmelt and runoff from the adjacent mountains typically occurs in late spring and early summer. Between 1950 and 1988, the mean annual air temperature at the Central Facilities Area (CFA; fig. 2) was 30°F; the coldest and warmest mean monthly temperatures were measured in January and during July–August, respectively (Clawson and others, 1989, table B-3).

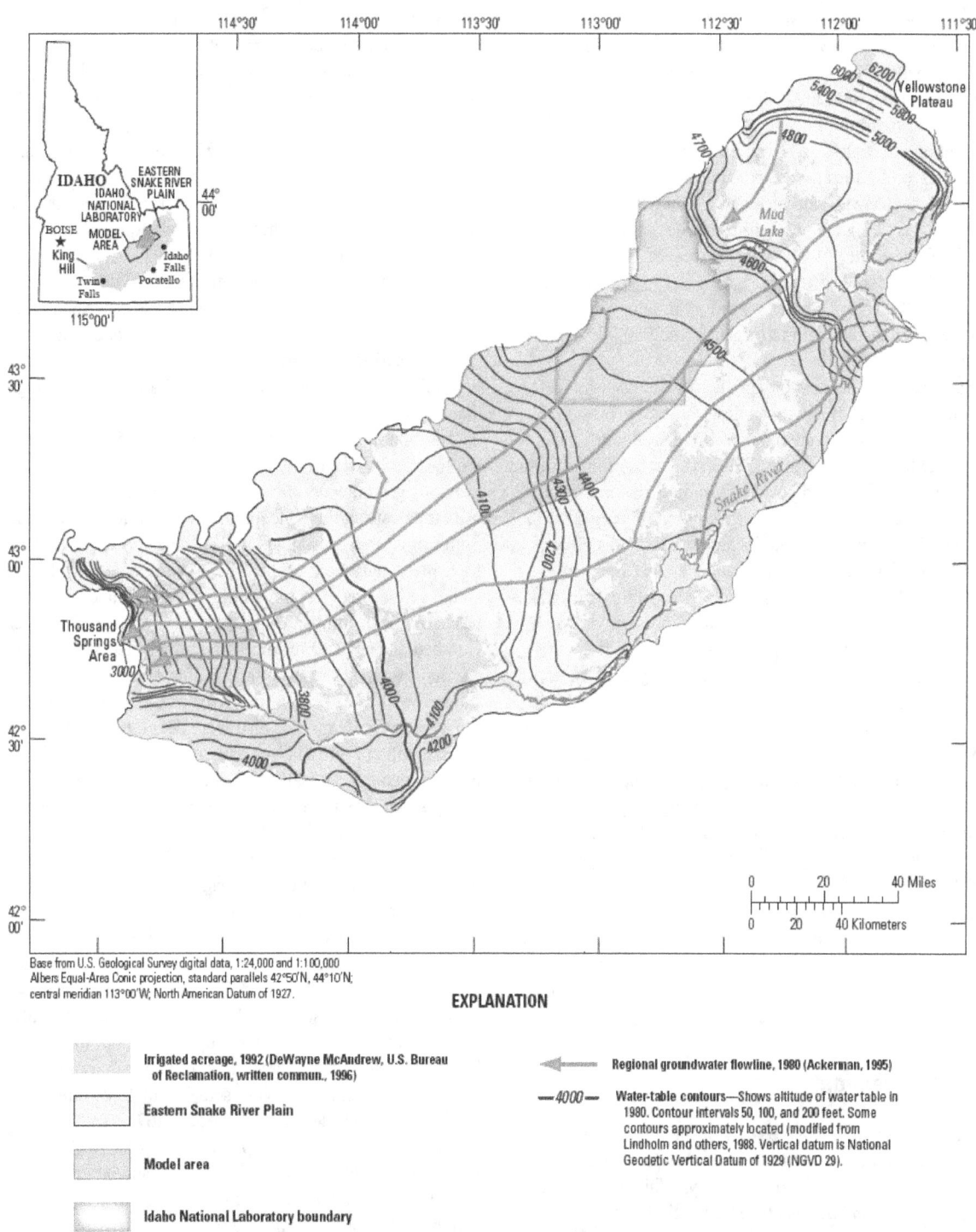

Base from U.S. Geological Survey digital data, 1:24,000 and 1:100,000
Albers Equal-Area Conic projection, standard parallels 42°50'N, 44°10'N;
central meridian 113°00'W; North American Datum of 1927.

EXPLANATION

Irrigated acreage, 1992 (DeWayne McAndrew, U.S. Bureau of Reclamation, written commun., 1996)

Eastern Snake River Plain

Model area

Idaho National Laboratory boundary

Regional groundwater flowline, 1980 (Ackerman, 1995)

—4000— Water-table contours—Shows altitude of water table in 1980. Contour intervals 50, 100, and 200 feet. Some contours approximately located (modified from Lindholm and others, 1988. Vertical datum is National Geodetic Vertical Datum of 1929 (NGVD 29).

Figure 3. Direction of regional groundwater flow, discharge areas, and irrigated acreage, eastern Snake River Plain, Idaho.

Groundwater in the ESRP aquifer flows to the southwest and discharges primarily through large springs and seeps along the Snake River in the Thousand Springs area along the southwestern margin of the ESRP (fig. 3). Recharge to the ESRP aquifer is from infiltration of surface water diverted for irrigation, underflow from tributary drainages, stream and canal losses, and infiltration of precipitation (Garabedian, 1992, p. 11, table 15). Land that is irrigated with groundwater on the ESRP is along the southeastern and southern margins of the plain, from north of Idaho Falls to west of Twin Falls, and in the Mud Lake area northeast of the INL (fig. 3). In 1980, about 1,760,000 acre-ft of water was withdrawn from about 4,000 wells across the plain to irrigate about 930,000 acres (Garabedian, 1992, p. 19-21).

Surface-water flow onto the INL and vicinity is intermittent, and nearly all streamflow infiltrates through the thick unsaturated zone and recharges the ESRP aquifer. Depth to the water table ranges from 200 ft in the northern part of the model area to 1,000 ft in the southern part. Streams tributary to the ESRP near the INL originate in mountain ranges north and west of the model area and include the Big Lost River, the Little Lost River, Birch Creek, and Camas Creek (fig. 1). The Big Lost River, on the west-central side of the ESRP, is the most significant surface-water feature in the model area. The Big Lost River drainage basin drains an area of about 1,410 mi[2] upstream of the USGS streamflow-gaging station (13132500) near Arco (fig. 1). Water from the river is stored in Mackay Reservoir (capacity 44,370 acre-ft, Bennett, 1990) about 40 mi upstream of the INL (fig. 1). In most years, much of the water in the river downstream of the reservoir is diverted for irrigation or infiltrates through the riverbed before reaching Arco. To prevent flooding of downstream facilities, a large percentage of the flow from the Big Lost River downstream of Arco is diverted to a series of four interconnected spreading basins near the southwest boundary of the INL (fig. 1). During years with sufficient precipitation, the Big Lost River flows onto the broad, undulating ESRP near Arco and terminates in the Big Lost River Sinks and a series of three interconnected playas east of Howe (fig. 1).

Previous Investigations

The most comprehensive study of the ESRP aquifer was completed as part of the USGS Regional Aquifer System Analysis (RASA) program. In that study, Whitehead (1986, 1992), Lindholm and others (1988), and Lindholm (1996) described the regional hydrogeologic framework, hydrologic properties, and geologic controls on groundwater movement. Kjelstrom (1986, 1995) described streamflow gains and losses, and compiled regional groundwater budgets. Garabedian

(1986, 1992) synthesized hydrologic data and constructed a regional groundwater flow model of the entire ESRP aquifer (10,800 mi[2]). These studies provided the regional context for constructing the groundwater flow models described in this report.

Anderson and Liszewski (1997) combined information from geophysical logs (table A1; Anderson, Ackerman, and others, 1996), rock cores, and outcrops to describe the stratigraphy of the unsaturated zone and aquifer beneath the INL and vicinity. The work of Anderson and Liszewski (1997) provides the basis for defining the hydrogeologic units used in the groundwater flow models described in this report. Ackerman (1991) and Anderson and others (1999) described the distribution of hydraulic properties in the ESRP aquifer and the geologic features controlling that distribution. Bennett (1990) estimated the amount, distribution, and timing of recharge to the ESRP aquifer at the INL from streamflow infiltration in the Big Lost River. Previous groundwater models also provided information used to construct the conceptual and numerical flow models. Spinazola (1994) simulated groundwater flow in the Mud Lake area upgradient of the INL. Ackerman (1995) developed an advective-transport model of regional groundwater flow based on the groundwater flow model of the ESRP aquifer by Garabedian (1992). McCarthy and others (1995) developed a groundwater flow model for the area of the INL to support groundwater remediation for specific INL facilities. Several INL facility-scale models also were developed to support risk assessments for remedial investigation feasibility studies, including coupled unsaturated- and saturated-zone flow and transport models for TAN (Schaffer-Perini, 1993) and the SDA (Magnuson and Sondrup, 1998).

The current modeling effort also builds on the groundwater flow and contaminant transport modeling results of Robertson (1974) and Goode and Konikow (1990b). Robertson (1974) simulated two-dimensional steady-state and transient groundwater flow and contaminant transport. Robertson's steady-state model, with no recharge from the Big Lost River, successfully reproduced the unusually wide contaminant plumes in the upper 200 ft of the aquifer; however, a larger transverse ($\alpha_T = 450$ ft) than longitudinal ($\alpha_L = 300$ ft) dispersivity was used. The α_L to α_T ratio derived from Robertson's steady-state model, 0.67, is smaller than the ratios of 24 other published pairs of α_L and α_T, which ranged from 1 to 53 with a mean of 8.8 (Gelhar and others, 1992, table 1). This small α_L to α_T ratio, and the large simulated values for α_L and α_T, indicated that the steady-state model of Robertson (1974) did not adequately represent some of the geologic and hydrologic features governing contaminant transport. Dispersivity values were not reported for the transient flow simulations.

Goode and Konikow (1990b) recognized that groundwater flow directions and velocities varied temporally in the shallow flow field, partly caused by recharge from the Big Lost River near contaminated groundwater at the INL, and that the large model-derived α_T of Robertson (1974) could have accounted for the effects of these variations. If these transient variations in the flowfield are ignored or undefined, they can result in larger estimates for α_T, and smaller α_L to α_T ratios to account for advective transport caused by transient influences. Using the steady-state and transient flow models of Robertson (1974), but including the magnitude, timing, and spatial distribution of recharge from the Big Lost River, Goode and Konikow (1990b) evaluated the effect of transient groundwater flow at the INL on estimates of α_L and α_T. Although the simulations of Goode and Konikow (1990b) resulted in α_T values that were smaller than values for α_L, their results were inconclusive and they were unable to determine the effect that transient variations in the flowfield had on dispersivities.

Ackerman and others (2006) developed a conceptual model of groundwater flow and contaminant transport for the area simulated in this study. This conceptual model provides a qualitative description of groundwater flow, emphasizing those geologic and hydrologic features that most strongly affect contaminant movement in the aquifer, and forms the basis of the numerical models described in this report.

Conceptual Model

The conceptual model of Ackerman and others (2006), upon which the numerical models described in this report are based, is a simplified description of the aquifer system. The conceptual model includes (1) descriptions of the hydrogeologic framework and hydraulic properties of the media; (2) descriptions of the spatial and temporal characteristics of the model boundaries; (3) estimates of inflows, outflows, and fluxes across model boundaries; (4) descriptions of the approaches used to estimate the components of the steady-state water budget and an assessment of the uncertainties associated with those estimates; (5) descriptions of groundwater flow paths, flow velocities, and the possible effects of transient influences on groundwater flow directions; (6) descriptions of the stratigraphic, structural, and hydrologic controls on groundwater flow; and (7) an assessment of how groundwater flow may affect contaminant movement in the aquifer.

Primarily because of data limitations, scaling considerations, and computational constraints, the conceptual model relies on simplifications that involve the grouping of individual basalt flows and basalt-flow groups, temporal averaging of aquifer inflows and outflows, spatial averaging of inflows and outflows across model boundaries, and an implicit representation of flow through the unsaturated zone to the water-table boundary. These simplifications largely preclude simulating the effects of small-scale physical features,

short-term hydrologic processes, and non-uniform (or uneven) distribution of inflows and outflows across model boundaries, but are appropriate for modeling the transport of conservative, nonreactive contaminants for geographically large areas.

Data limitations included (1) uneven spatial distribution of wells and boreholes, areally and vertically (fig. 4), (2) sparse well and borehole coverage in areas remote from the main facilities on the INL (fig. 4), (3) scaling-compatibility issues involving the application of small-scale (laboratory) measurements to a large-scale study, (4) uncertainties arising from partial well and borehole penetrations and different well completions that complicate interpretations of water-level, water-chemistry, and hydraulic-conductivity measurements, and (5) discontinuous or nonexistent hydrologic records.

Most wells in the model area penetrate only the upper 200 ft of the aquifer. Depths of 114 monitoring wells that were used to evaluate transmissivity and hydraulic conductivity range from 21 to 991 ft below the water table. Only 13 of these wells penetrate more than 300 ft of the aquifer, and the average penetration of all wells is 182 ft. These and other monitoring wells are completed as open holes or holes with one or more perforated intervals or wells screens.

Hydrogeologic Framework

The ESRP is underlain by a thick sequence of Tertiary and Quaternary volcanic rocks and sedimentary interbeds that may be more than 10,000 ft thick (Whitehead, 1992, pl. 3). Stratigraphic data from 10 deep boreholes indicate that the base of the aquifer in the western part of the INL ranges from 815 to 1,710 ft below land surface and likely coincides with the top of a thick and widespread layer of clay, silt, sand, and altered basalt of Tertiary age (1.7 to 2.2 Ma) (Anderson and Liszewski, 1997, p. 11). Surface-based electrical-resistivity surveys and drillhole data (fig. 4A) indicate the aquifer base in the model area ranges from about 700 to 4,800 ft below land surface and the thickness of the aquifer may exceed 2,500 ft in the eastern half of the INL and 4,000 ft in the southeastern part of the model area (Whitehead, 1986, sheet 2; 1992, pl. 6; Ackerman and others, 2006, p. 25 and fig. 14).

Quaternary basalt flows make up about 85 percent of the unsaturated zone and aquifer; sediments, andesite flows, and rhyolite domes of Quaternary age make up the remaining 15 percent. The database compiled by Anderson and Liszewski (1997) indicates that the unsaturated zone and aquifer are composed of at least 178 basalt-flow groups, 103 sedimentary interbeds, 6 andesite-flow groups, and 4 rhyolite domes. Stratigraphic units were combined by Anderson and Liszewski (1997, p. 14), based on similar age, into 14 composite stratigraphic units each made up of 5 to 90 stratigraphic subunits of similar age. Surficial mapping and correlation of subsurface stratigraphic units among numerous outcrops and geophysical logs and cores from 333 boreholes (fig. 4B) at and near the INL indicate that many of these stratigraphic units are continuous across large parts of the model area.

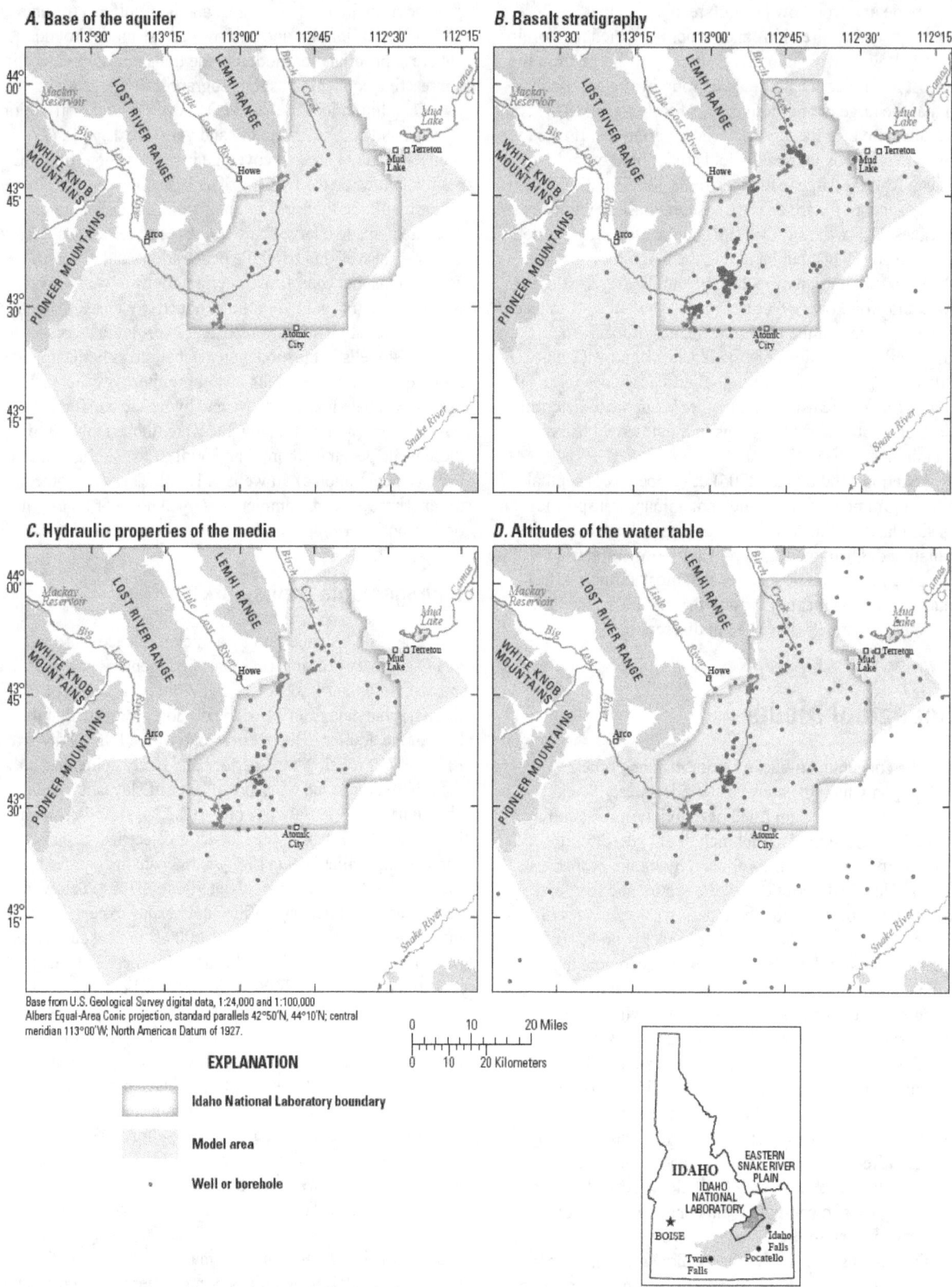

Figure 4. Locations of wells and boreholes used to define (*A*) base of the aquifer, (*B*) basalt stratigraphy, (*C*) hydraulic properties of the media, and (*D*) altitudes of the water table, Idaho National Laboratory and vicinity, Idaho. Additional well and borehole maps are in appendix B.

Hydrogeologic Units

In the conceptual model, the fractured basalts, interflow zones, and interbedded sediments of the ESRP aquifer are represented as an equivalent porous media with non-uniform properties, and are grouped into three primary hydrogeologic units (table 2 and fig. 5), each with different hydraulic properties (table 3). In this representation of the aquifer the small-scale heterogeneities and anisotropies of individual basalt flows, basalt flow groups, and interbedded sediments are not preserved, and the resulting hydraulic conductivity of each hydrogeologic unit reflects the aggregate lithology, thickness, and number of basalt flows, interflow zones, and sedimentary interbeds in each hydrogeologic unit (tables 2 and 3).

The three hydrogeologic units used to represent the complex basalt stratigraphy of the aquifer are referred to as (1) hydrogeologic unit 1, younger rocks that are 440,000–650,000 years old, (2) hydrogeologic unit 2, younger rocks that are 650,000–800,000 years old, and (3) hydrogeologic unit 3, intermediate-age rocks that are 800,000–1,800,000 years old (table 2). Younger rocks form the uppermost part of the aquifer in much of the model area and intermediate-age rocks make up the largest volume of the aquifer in the model area. Altered rocks that are older than about 1,800,000 years underlie the aquifer. Other rocks of hydrologic significance in the model area are rhyolite domes that penetrate the younger and intermediate-age rocks.

Hydrogeologic Unit 1

Hydrogeologic unit 1 is composed of many thin, densely fractured basalt flows and interbedded sediment simulated as 0 to about 500 ft thick. Hydrogeologic unit 1 is present at the water table beneath most of the INL, but is absent in the aquifer in the northwestern and southwestern parts of the model area (fig. 5).

Results of single-well aquifer tests in 67 wells with perforated or open intervals only in hydrogeologic unit 1 indicate that the hydraulic conductivity of these rocks ranges from about 0.01 to 24,000 ft/d (table 3). Almost two-thirds of these estimates are larger than 100 ft/d and about one-third are larger than 1,000 ft/d; the large values reflect the effects of numerous interconnected interflow zones associated with the many thin flows in this unit (table 3). Porosity measurements of 1,504 core samples from hydrogeologic unit 1 and the overlying unsaturated zone at and near the RWMC ranged from 0.01 to 0.43 with 90 percent of the reported values, representing densely fractured basalt, between 0.05 and 0.27 (table 3). In the RASA transient model simulations, Garabedian (1992, p. 44-46) used an average specific yield of 0.05 for all types of basalt and intermediate composition volcanic rocks, such as rhyolite and andesite, and 0.20 for all types of sediment in the upper 200 ft of the regional aquifer system. Ackerman (1995, p. 10) used a calibrated effective porosity of 0.21 to simulate advective transport in the uppermost 200 ft of the ESRP aquifer.

Table 2. Composite unit stratigraphy used to define hydrogeologic units and older rocks, Idaho National Laboratory and vicinity, Idaho.

[**Abbreviation:** CU, composite unit; ft, foot; >, greater than]

Hydrogeologic unit or rock	Age (thousands of years before present)	Composite unit designation	Thickness (ft)	Number of basalt flow groups	Number of andesite flow groups	Number of sediment interbeds	Number of penetrating boreholes
Hydrogeologic unit 1—Younger rocks consisting of thin, densely fractured basalt and interbedded sediment	440–650	CU 4	0–482	9	4	11	172
		CU 5	0–329	3		6	143
		CU 6	0–347	5		8	118
Hydrogeologic unit 2—Younger rocks consisting of massive, less densely fractured basalt and interbedded sediment	650–800	CU 7	0–409	7		10	70
Hydrogeologic unit 3—Intermediate-age rocks consisting of slightly altered fractured basalt and interbedded sediment	800–1,800	CU 8 to 14	0–4,031	41	1	26	86
Older rocks consisting of intensely altered fractured basalt, rhyolitic ash-flow tuffs, and interbedded sediment—underlies base of the aquifer	> 1,800						13

A.

Base from U.S. Geological Survey digital data, 1:24,000 and 1:100,000
Albers Equal-Area Conic projection, standard parallels 42°50'N, 44°10'N; central
meridian 113°00'W; North American Datum of 1927.

EXPLANATION

Hydrogeologic units at the water table

1—Younger rocks consisting of densely fractured basalt and interbedded sediment

2—Younger rocks consisting of massive, less densely fractured basalt and interbedded sediment.

3—Intermediate-age rocks consisting of slightly altered fractured basalt and interbedded sediment

Silicic rocks—Includes rhyolite domes Big Southern Butte (A), Middle Butte (B), and East Butte (C). Small unnamed dome and Cedar Butte andesite not shown

Idaho National Laboratory boundary

——— Model area boundary

Site facilities

CFA—Central Facilities Area

INTEC—Idaho Nuclear Technology and Engineering Center

MFC—Materials and Fuels Complex

NRF—Naval Reactors Facility

RTC—Reactor Technology Complex

RWMC—Radioactive Waste Management Complex

TAN—Test Area North

A———*A'* Trace of section

Figure 5. Distribution of hydrogeologic units at the (*A*) water table and (*B*) cross section along line of section *A–A'*, Idaho National Laboratory and vicinity, Idaho. Trace of section is shown in figure 5A.

B.

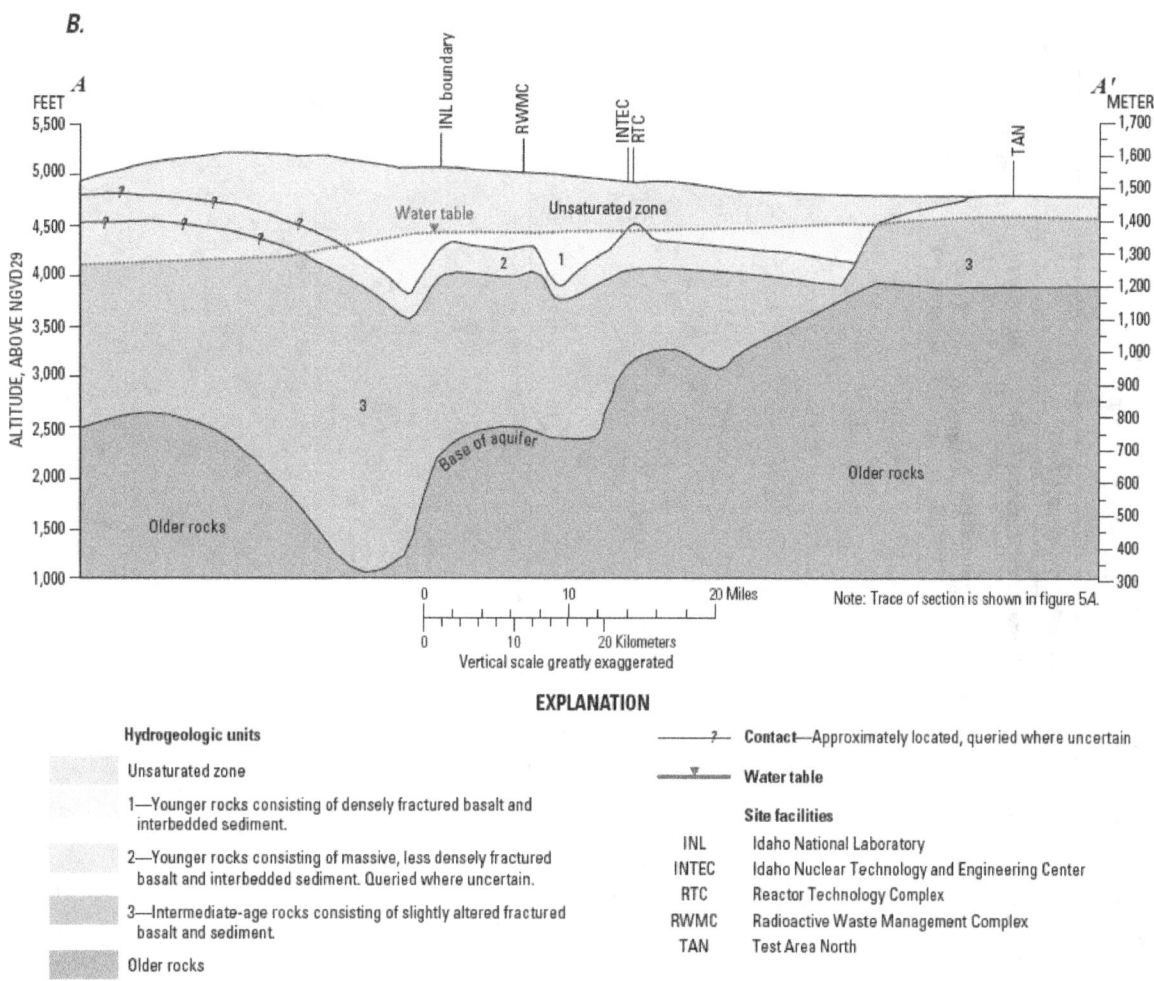

EXPLANATION

Hydrogeologic units

Unsaturated zone

1—Younger rocks consisting of densely fractured basalt and interbedded sediment.

2—Younger rocks consisting of massive, less densely fractured basalt and interbedded sediment. Queried where uncertain.

3—Intermediate-age rocks consisting of slightly altered fractured basalt and sediment.

Older rocks

——?—— Contact—Approximately located, queried where uncertain

——▽—— Water table

Site facilities

INL   Idaho National Laboratory
INTEC Idaho Nuclear Technology and Engineering Center
RTC   Reactor Technology Complex
RWMC   Radioactive Waste Management Complex
TAN   Test Area North

Figure 5.—Continued.

Table 3. Ranges of hydraulic conductivities and porosities for hydrogeologic units and older rocks and summary statistics of basalt flow thickness within hydrogeologic units, Idaho National Laboratory and vicinity, Idaho.

[**Hydraulic conductivity:** Conductivities were estimated from aquifer tests in wells having perforated or open intervals only within the respective hydrogeologic unit or rock. Modified from Ackerman and others (2006, tables 2 and 3) and Perkins (2008, table 2). **Abbreviations** ft, foot; ft/d, foot per day; NA, not applicable; w, well; c, core; 100s, hundreds; <, less than]

Hydrogeologic unit or rock	Number of basalt flows	Basalt flow thickness (ft)				Property and sample data					
		Minimum	Median	Mean	Maximum	Hydraulic conductivity (ft/d)			Number of wells (w) or core samples (c)	Range or median porosity	Number of core samples
						Smallest	Geometric midpoint	Largest			
Hydrogeologic unit 1—Younger rocks consisting of thin, densely fractured basalt and interbedded sediment	1,095	2	18	20	54	0.01	15	24,000	67 (w)	0.05 to 0.27	>1,000
Hydrogeologic unit 2—Younger rocks consisting of massive, less densely fractured basalt and interbedded sediment	212	4	28	29	78	6.5	95	1,400	4 (w)	0.11	100s
Hydrogeologic unit 3— Intermediate-age rocks consisting of slightly altered fractured basalt and interbedded sediment	441	2	22	23	107	.32	88	24,000	14 (w)	0.05 to 0.08	20
Older rocks consisting of intensely altered fractured basalt, rhyolitic ash-flow tuffs, and interbedded sediment—underlies base of the aquifer	NA	NA	NA	NA	NA	.002	.008	.03	1 (w)	<0.09 to 0.19	19
Sediment	NA	NA	NA	NA	NA	5.9×10^{-7}	.038	240	97 (c)	0.35 to 0.63	51

Hydrogeologic Unit 2

Hydrogeologic unit 2 is composed of massive, less densely fractured basalt flows and interbedded sediment that are 0–409 ft thick and average 266 ft in the 15 boreholes that penetrate the full thickness of this unit in the model area. Hydrogeologic unit 2 includes one of the thickest and most extensive basalt flow groups (basalt flow group I as defined by Anderson and Liszewski, 1997) beneath the INL and is important for interpretations of contaminant movement in the aquifer. Basalt-flow group I and an overlying layer of clay and silt (designated as the H-I interbed) underlie all but the northern and extreme southeastern parts of the INL (Anderson and others, 1997, p. 19). Hydrogeologic unit 2 underlies most of the INL, intersects the water table south and west of the INL, and is absent in the aquifer in the northwestern and southwestern parts of the model area (fig. 5).

The hydraulic conductivity and bulk porosity of hydrogeologic unit 2 generally are less than that of hydrogeologic unit 1 because of the thick, massive interiors and fewer interconnected interflow zones associated with the basalt flows of this unit (table 3). Hydraulic conductivity values range from 6.5 to 1,400 ft/d for single-well aquifer tests in four wells with perforated intervals only in hydrogeologic unit 2 (table 3). The porosity of the massive basalt of hydrogeologic unit 2 probably is within the lower end of the range estimated for porosity of the densely fractured basalt of hydrogeologic unit 1, 0.05 to 0.27. Knutson and others (1992, p. 4–21) reported a median porosity of 0.11 for hundreds of nonvesicular basalt cores from the unsaturated zone and aquifer at and near the RWMC (table 3), a value that may approximate the porosity of massive basalt because it does not include the effects of more vesicular basalts and interflow zones.

Hydrogeologic Unit 3

Hydrogeologic unit 3 is composed of slightly altered, fractured basalt and interbedded sediment and constitutes the full thickness of the aquifer in the northwestern and southwestern parts of the model area (fig. 5). Hydrogeologic unit 3 is simulated as 0–4,031 ft thick, and thickens substantially from west to east and from northeast to southwest in the model area (figs. 5B and 6B). The interpreted distribution of older rocks that underlie hydrogeologic unit 3 and that form the base of the aquifer indicates changes of more than 2,000 ft in the saturated thickness of hydrogeologic unit 3 from northeast to southwest across the central part of the model area (fig. 5B) that may be related to differential subsidence and uplift.

Results of single-well aquifer tests in 14 wells with perforated or open intervals only in hydrogeologic unit 3 indicate that the hydraulic conductivity ranges from about 0.32 to 24,000 ft/d (table 3). However, a comparison of 24 hydraulic conductivity estimates of hydrogeologic

units 1 and 2 (undifferentiated) near the INTEC with 68 hydraulic conductivity estimates of hydrogeologic unit 3 near TAN indicate that the average hydraulic conductivity of hydrogeologic unit 3 near TAN is about 1 order of magnitude smaller than hydrogeologic units 1 and 2 near the INTEC (John Welhan, Idaho State Geological Survey, written commun., 1999). The porosity of hydrogeologic unit 3 probably is within the lower end of the range of that estimated for the densely fractured basalt of hydrogeologic unit 1, 0.05 to 0.27. Median values for 10 nonvesicular and 10 vesicular basalt cores at TAN were about 0.05 and 0.08, respectively (table 3; Allan Wylie, Idaho Water Resources Research Institute, written commun., 2000). Median values of porosity for hundreds of nonvesicular and vesicular basalt cores from hydrogeologic units 1 and 2 (undifferentiated) at and near the RWMC were 0.11 and 0.22, respectively (Knutson and others, 1992, p. 4–21). Comparison of these porosity values indicates that porosities of hydrogeologic unit 3 may be smaller than porosities of hydrogeologic units 1 and 2.

Older Rocks that Underlie the Base of the Aquifer

The base of the aquifer is interpreted as the contact between hydrogeologic unit 3, which is present throughout most of the model area, or hydrogeologic unit 2 where unit 3 is absent, and the underlying Tertiary-age rocks (fig. 5). This contact is characterized by an abrupt downward increase in altered basalt and interbedded sediment (Mann, 1986, p. 4–5; Whitehead, 1992, p. 10; Anderson and Liszewski, 1997, p. 11; Morse and McCurry, 1997, p. 6–7). Based on electrical-resistivity surveys (Whitehead, 1986, sheet 2), the estimated altitude and depth to the base of the aquifer ranges from about 80 to 4,200 ft above NGVD of 1929 (fig. 6A) and from about 700 to 4,800 ft below land surface, respectively, resulting in an aquifer thickness that ranges from about 200 to 4,000 ft (fig. 6B) (Ackerman, 2006, p. 25). Direct evidence indicating the location of the base of the aquifer was limited to data from 13 deep wells and boreholes (fig. 4A). In seven wells and coreholes this contact also coincides with an increase in the alteration of basalts, an abrupt increase in temperature gradient interpreted to represent a change from convective-dominated to conductive-dominated heat flow, and a decrease in hydraulic conductivity and porosity (Mann, 1986, p. 21; Morse and McCurry, 1997, p. 6–7; Welhan and Wylie, 1997, p. 99; Morse and McCurry, 2002, p. 222; Mazurek and others, 2004). Collectively these factors indicate that water circulation is limited in the older rocks.

Few estimates of hydraulic conductivity and porosity are available for these older rocks. Hydraulic conductivities ranging from 0.002 to 0.03 ft/d (table 3) were calculated from four aquifer tests (Mann, 1986, p. 21) in borehole INEL1, about 4 mi north of the INTEC (fig. 1). Porosity values range from less than 0.09 to 0.19 (table 3).

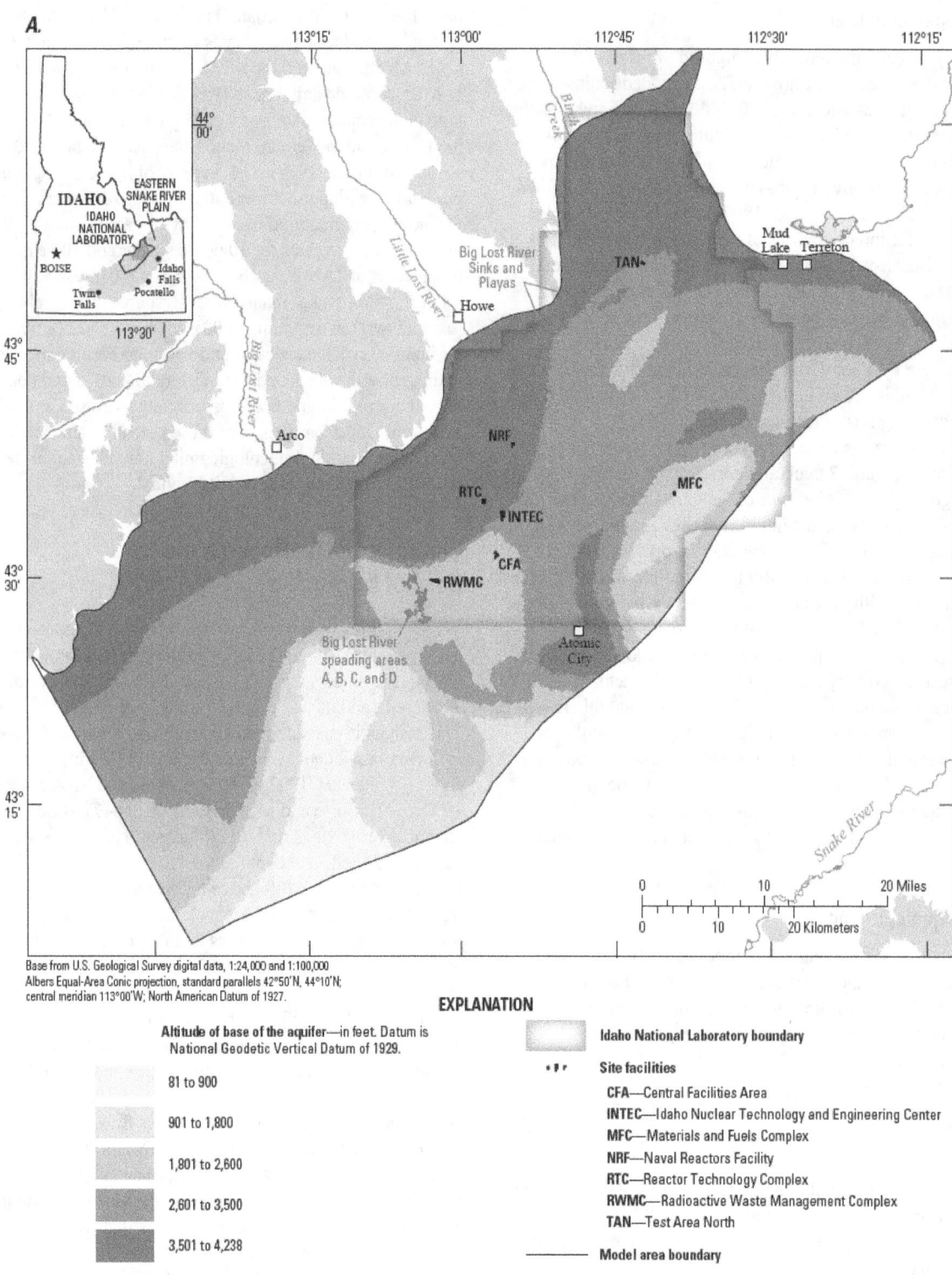

Figure 6.   Aquifer (*A*) altitude and (*B*) thickness as defined by borehole data and surface-based electrical-resistivity surveys, Idaho National Laboratory and vicinity, Idaho.

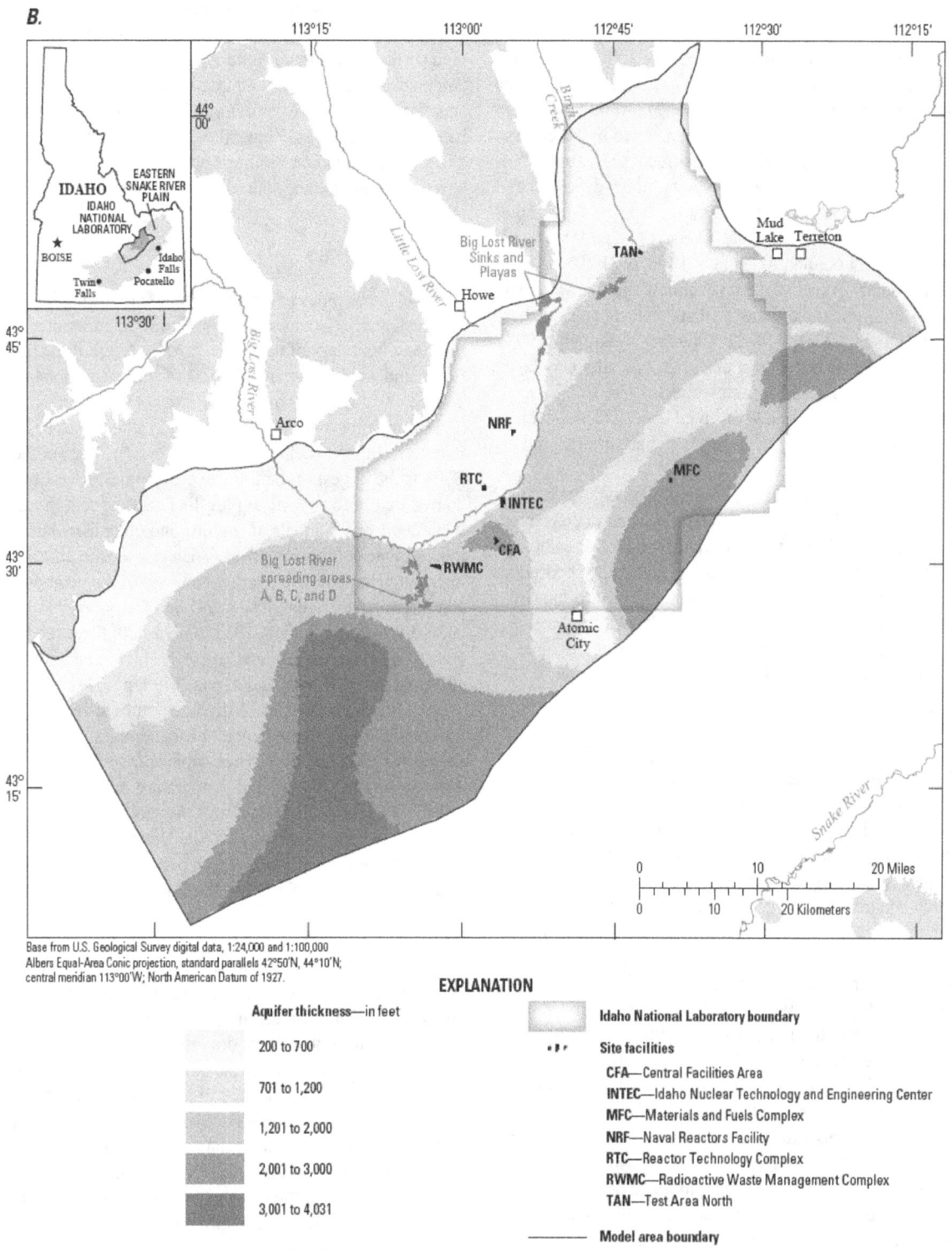

B.

EXPLANATION

Aquifer thickness—in feet

200 to 700

701 to 1,200

1,201 to 2,000

2,001 to 3,000

3,001 to 4,031

Idaho National Laboratory boundary

Site facilities

CFA—Central Facilities Area
INTEC—Idaho Nuclear Technology and Engineering Center
MFC—Materials and Fuels Complex
NRF—Naval Reactors Facility
RTC—Reactor Technology Complex
RWMC—Radioactive Waste Management Complex
TAN—Test Area North

Model area boundary

Figure 6.—Continued.

Sediment Interbeds

Basalts in the model area are intebedded with sediment throughout the unsaturated zone, hydrogeologic units 1, 2, and 3, and the underlying Tertiary rocks (fig. 7) and compose from less than 1 to 50 percent of the rocks penetrated by boreholes in the model area (fig. 4B). Four depositional processes contributed sediment to the area: (1) windblown deposits of clay, silt, and fine sand are present in thin deposits throughout most of the model area (Nace and others, 1975, p. 35; Anderson, Liszewski, and Ackerman, 1996, p. 3); (2) fluvial deposits of sandy gravel in stream channels and finer-grained clayey silt in terminal playas at the distal ends of river systems (Nace and others, 1975, p. 19–27; Gianniny and others, 1997, p. 31); (3) lacustrine deposits of thick clay and silty clay layers in and near Mud Lake (Stearns and others, 1939, p. 17, 39; Spinazola, 1994, p. 10); and (4) alluvial deposits along and near the mouths of tributary valleys and adjacent mountain fronts.

Sediment interbeds in the three hydrogeologic units are mostly fine grained, ranging from very fine sand to clay-sized (Perkins, 2008), and tend to reduce aquifer bulk hydraulic conductivity and increase bulk porosity. Laboratory hydraulic conductivity measurements from 97 core samples (primarily from the unsaturated zone at and near the RWMC and the INTEC), described as having a silt loam to gravely texture, ranged from 5.9×10^{-7} to 240 ft/d (table 3). Laboratory porosity measurements from these same core samples ranged from 0.35 to 0.63 (table 3).

The amount of sediment penetrated by individual boreholes ranges from less than 5 percent of the stratigraphic column to more than 50 percent near the terminus of the Big Lost and Little Lost Rivers in an inferred depositional center known informally as the Big Lost Trough (Anderson, Ackerman, and others, 1996; Geslin and others, 1997; Gianniny and others, 1997). The boundaries of the Big Lost Trough in this report are expanded beyond that of Gianniny and others (1997, p. 31) to include the coalescing sedimentary deposits along and near the channel, floodplain, sinks, and playas of the Big Lost River, Little Lost River, Birch Creek, Camas Creek, and Mud Lake (fig. 7). Within the Big Lost Trough, sediments grade from fluvial, sandy gravel within stream channels to finer-grained clayey silt within terminal playas at the distal ends of the river systems that historically have drained into this structural depression.

Two interpretations of the Big Lost Trough are shown in figure 7 to account for differences in methods, data, and interpolation techniques used to evaluate the distribution of sediment. One area, interpreted from borehole logs (Whitehead, 1992, pl. 5), shows where sediment was estimated to compose from 100 to 999 ft of the stratigraphic section (including older rocks) (fig. 7A). A second area, where sediment was estimated to constitute more than 11 percent of the stratigraphic section (excluding older rocks) (fig. 7B), was based on interpretations from the geologic map of the

INL (Kuntz and others, 1994) and natural-gamma logs from wells drilled to depths of more than 300 ft below land surface (Anderson, Ackerman, and Liszewski, 1996; Anderson, Ackerman and others, 1996). Water-table contours, superimposed over this structural feature, show the relation between hydraulic gradients, flexures in water-table contours, and areas of abundant sediment accumulation defined by boreholes with greater than 11 percent sediment.

Rhyolite Domes

Rhyolite domes (exposed as buttes) in the model area are clustered in two areas (fig. 5); three domes are near the southeast boundary of the model area—Middle Butte, East Butte, and an older unnamed dome of limited surficial extent between Middle Butte and East Butte not shown in figure 5 (Kuntz and others, 1994). The other rhyolite dome, Big Southern Butte is south of the INL boundary. Big Southern Butte is the largest of the four rhyolite domes. Rhyolite domes are uplifted, vertical plug-like masses, consisting of thick flows and blankets of rhyolite and other associated intermediate composition rocks such as andesite, that are interpreted to penetrate a large thickness of the younger rocks and intermediate-age rocks (Kuntz and Dalrymple, 1979, p. 30-34; Spear and King, 1982, p. 396-400; Kuntz and others, 1994; Anderson and Liszewski, 1997; Hughes and others, 1999, fig. 16; McCurry and others, 1999, p. 170-174).

Hydraulic properties of rhyolite domes have not been measured. Based on the general characteristics of these less vesicular rocks, however, water-table contours (Bartholomay and others, 1997, fig. 9), and temperature data from well Corehole 1 (fig. 1) that penetrates the saturated part of the unnamed dome between Middle Butte and East Butte, rhyolite domes probably have low permeability and may have hydraulic properties similar to those of the massive basalt of hydrogeologic unit 2. A temperature log of Corehole 1 (Morse and McCurry, 2002, fig. 2) indicates that most of the unnamed dome is in hydraulic contact with the cold water of the aquifer. Inferred water-table contour deflections around the unamed dome and Middle and East Buttes (fig. 7); however, probably indicate that the hydraulic conductivity of these rocks is smaller than that of the adjacent fractured basalt, hydrogeologic unit 1. The lack of a similar deflection of contours near Big Southern Butte probably indicates that this dome is fractured at depth or may lie above the water table. If the hydraulic conductivity and porosity of the rhyolite domes are similar to that of the massive basalt of hydrogeologic unit 2, then the conductivity probably ranges from about 6.5 to 1,400 ft/d and porosity probably is within the lower end of the 0.05 to 0.27 range of porosity estimated for hydrogeologic unit 1 (table 3). In the RASA study, a value of 0.65 ft/d for hydraulic conductivity and a value of 0.05 for specific yield were used for silicic volcanic rocks (Garabedian, 1992, p. 44-46, tables 19 and 20).

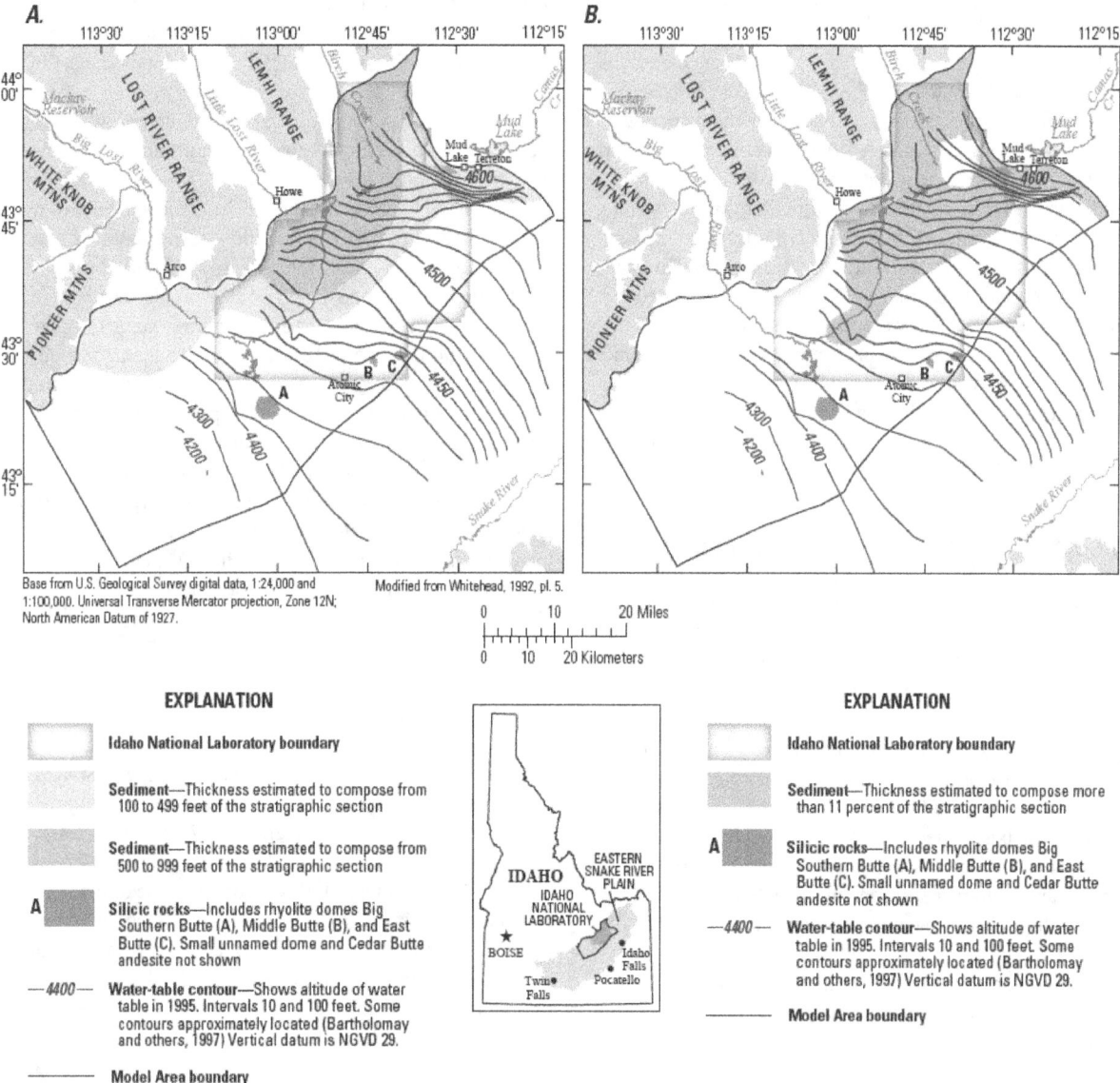

Figure 7.   Water-table contours and areas of abundant sediment in the model area, Idaho National Laboratory and vicinity, Idaho. Areas of abundant sediment were interpreted from (*A*) borehole logs and (*B*) geologic maps and natural-gamma logs.

Volcanic Rift Zones, Vents, and Vent Corridors

Volcanic rift zones are areas of increased concentrations of volcanic activity on the ESRP. Vent corridors are narrow zones of aligned vents, dikes, and fissures in and near volcanic rift zones (Anderson and others, 1999; Ackerman and others, 2006, fig. 7). Fissures and tension cracks associated with these features may provide local conduits for flow, and deep-seated dikes that feed these fissures may impede groundwater flow. Although these structural features are identified in the conceptual model (Ackerman and others, 2006, p. 11), their effects on groundwater flow are largely unknown. These features may produce preferential flow; however, the available geochemical and hydrologic information do not provide evidence for preferential flow and therefore these features are not represented in the numerical models and are not discussed further.

Conceptual Model Boundaries and Fluxes

Three physical and three artificial boundaries define the model area. The physical boundaries are the water table, the northwest mountain-front, and the base of the aquifer (fig. 8). The artificial boundaries are the northeast regional-underflow, the southeast flowline, and the southwest regional-underflow (fig. 8). In the conceptual model, inflow to the aquifer occurs across the water table, northwest mountain-front, northeast regional-underflow, and base of the aquifer boundaries, and from waste-disposal injection wells operated in the past. Outflow occurs across the southwest regional-underflow boundary and from irrigation and industrial production wells. No flow is exchanged across the southeast boundary as a consequence of the boundaries definition as a flowline. Although the conceptual model characterizes the base of the aquifer as an inflow boundary, few data are available to characterize the direction, quantity, and distribution of flow across this boundary.

Inflows and outflows across model boundaries are constant in a steady-state groundwater flow model, but vary in a transient groundwater flow model. In a steady-state model, the assumption is that the groundwater system is in approximate equilibrium. Inflows into the model area equal outflows resulting in no changes in aquifer storage. A transient groundwater flow model represents a dynamic system with changes in inflows, outflows, and aquifer storage. Both model types are presented in this report, thus inflows and outflows across boundaries are discussed in terms of a time-averaged snapshot from 1980, and in terms of dynamic fluxes between 1981 and 1995. More than 50 years of water-level observations in more than 100 wells (with variable lengths of record; fig. 4D) indicate that the aquifer is never truly at steady state, but the available data indicate that the steady-state assumption is most closely approximated by water-table conditions in 1980 (Ackerman and others, 2006, p. 36, table 6).

Inflows and outflows, in cubic feet per second, across model boundaries are summarized in a groundwater budget that was developed for the conceptual model for 1980 (fig. 9; table 4). The 157 ft^3/s residual for the estimated budget was about 7 percent of inflow with inflows exceeding outflow. Budget uncertainty, using the 5-year (1976–80) stress period of the transient 3-D RASA flow model as a basis of comparison, was ±7 percent of the average value of the total inflow (2,394 ft^3/s) and ±10 percent of the average value of the total outflow (2,316 ft^3/s) for the RASA and conceptual model (Ackerman and others, 2006, p. 38). Differences in water-budget estimates between the conceptual model and the RASA models were attributed to (1) coarse discretization of time and space for the RASA model, (2) interpolation errors resulting from inflow and outflow estimates that are assumed to be uniformly distributed within the RASA model grid cells, (3) differences in hydrogeologic (or professional) judgment, and (4) differences in accounting methodology (Ackerman and others, 2006, p. 36).

Inflow Boundaries

Estimated inflow to the model area during 1980 was 2,239 ft^3/s (table 4). Of the inflow components for the conceptual model water budget, about 7 percent enters at the water table, about 4 percent is from change in storage, about 2 percent is from upward flow across the base of the aquifer, and about 86 percent enters as underflow along the northwest mountain-front and northeast regional-underflow boundaries.

In the numerical models, inflow from storage is required to be zero for the steady-state model and varies temporally in the transient model. Like the conceptual model, underflow across the northwest mountain front between tributary valleys and flow across the southeast boundary were assumed to be zero. The underflow across the base of the aquifer from underlying rocks was treated as insignificant and assumed to be zero in the numerical models.

Water-Table Boundary

The water-table boundary is represented by the contact between the unsaturated zone and saturated zone. Depth to the water table varies from 200 ft in the northern part of the model area to 1,000 ft in the southern part. The altitude of the water table decreases from about 4,600 ft near the northeast boundary to about 4,100 ft near the southwest boundary of the model area (fig. 9).

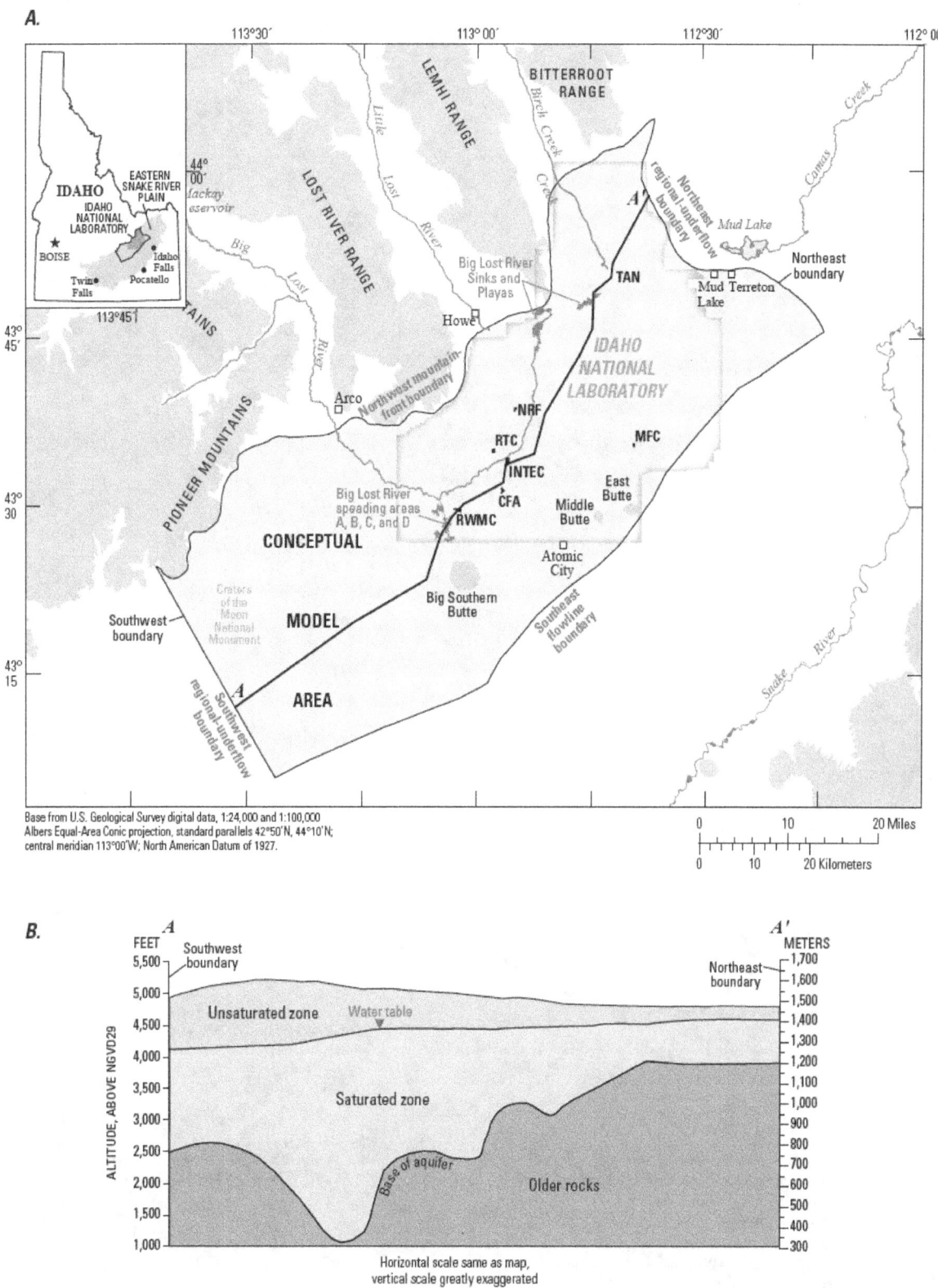

Figure 8. Location of the water table, northwest mountain-front, northeast regional-underflow, southwest regional-underflow, southeast flowline, and base of the aquifer boundaries, Idaho National Laboratory and vicinity, Idaho.

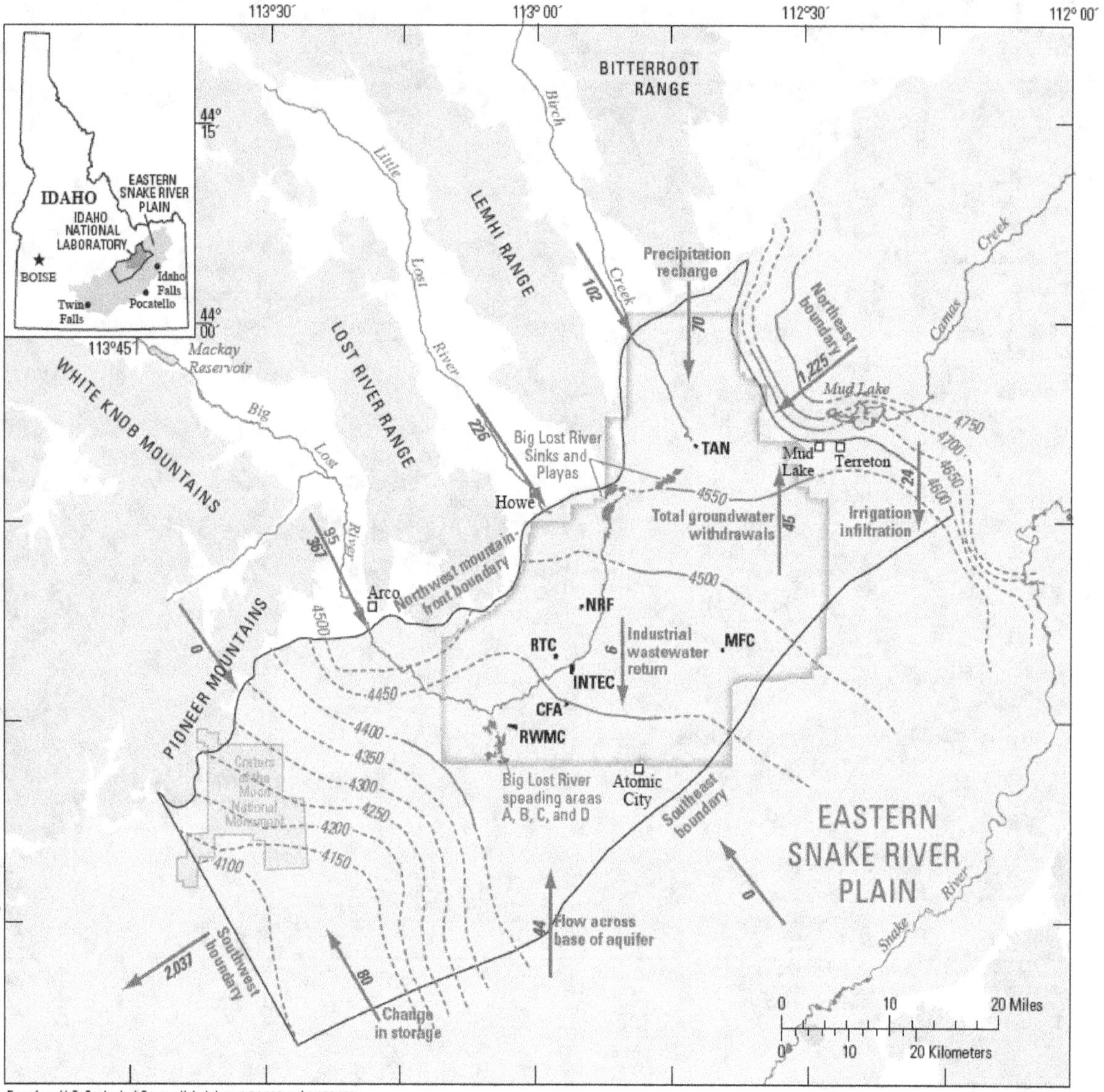

Base from U.S. Geological Survey digital data, 1:24,000 and 1:100,000
Albers Equal-Area Conic projection, standard parallels 42°50'N, 44°10'N;
central meridian 113°00'W; North American Datum of 1927.

EXPLANATION

Idaho National Laboratory boundary

Model area

Eastern Snake River Plain boundary

◄ ▪ ► Site facilities

   CFA—Central Facilities Area
   INTEC—Idaho Nuclear Technology and Engineering Center
   MFC—Materials and Fuels Complex
   NRF—Naval Reactors Facility
   RTC—Reactor Technology Complex
   RWMC—Radioactive Waste Management Complex
   TAN—Test Area North

$\frac{95}{367}$ → **Volumetric flux**—Value is rounded to nearest 1 cubic foot per second. For the Big Lost River, the number above the arrow in red (95) represents streamflow-infiltration recharge, and the number below the arrow in dark red (367) represents underflow

— 4200 — **Water-table contour**—Shows altitude of water table in March 1980. Interval 50 feet. Dashed where approximately located. Datum is NGVD 29. Modified from Linholm and others, 1988.

Figure 9.   Generalized 1980 groundwater budget components for the model area, Idaho National Laboratory and vicinity, Idaho.

Table 4. Summary of conceptual model boundary characteristics, estimated 1980 inflows and outflows, and water budget, Idaho National Laboratory and vicinity, Idaho.

[Steady-state model input values rounded to nearest tenth **Abbreviation**: ft^3/s, cubic foot per second; NA, not applicable; <, less than]

Inflow boundaries	Conceptual model				Steady-state model input	
	Boundary characteristics		Flow (ft^3/s)	Percentage of total flow	Numerical	
	Spatial	Temporal			Spatial boundary characteristics	Flow (ft^3/s)
Water-table boundary						
Precipitation recharge	Uniform	Constant	70	3	Uniform	70.0
Streamflow infiltration						
Big Lost River	Nonuniform	Variable	95	4	Nonuniform	[1] 125.3
Little Lost River	Nonuniform	Variable	0	0	Nonuniform	[2] 3.0
Birch Creek	Nonuniform	Variable	0	0	Nonuniform	[3] 20.0
Streamflow-infiltration subtotal			95	4		148.3
Irrigation infiltration	Nonuniform	Variable	24	1	Nonuniform	21.6
Industrial wastewater return	Nonuniform	Variable	6	<1	Nonuniform	5.9
Northwest mountain-front boundary	Nonuniform	Constant				
Big Lost River underflow	Uniform	Constant	367	16	Nonuniform	[4] 361.0
Little Lost River underflow	Uniform	Constant	226	10	Nonuniform	[5] 223.0
Birch Creek underflow	Uniform	Constant	102	5	Nonuniform	[6] 62.0
Mountain-front underflow	Nonuniform	Constant	0	0	Uniform	.0
Northwest mountain-front subtotal			695			646.0
Northeast regional-underflow boundary	Nonuniform	Constant	1,225	55	Nonuniform	1,225.0
Southeast-flowline boundary	Uniform	Constant	0	0	Uniform	.0
Change in storage	Uniform	Variable	80	4		NA
Base of the aquifer	Uniform	Constant	44	2	Uniform	.0
Total inflow			2,239	100		2,116.7
Outflow boundaries						
Southwest regional-underflow boundary	Nonuniform	Variable	2,037	98	Nonuniform	NA
Groundwater withdrawals						
Irrigation	Nonuniform	Variable	37	2	Nonuniform	37.2
Industrial	Nonuniform	Variable	8	<1	Nonuniform	7.6
Total outflow			2,082	100		
Budget residual			157			
Percentage difference between inflow and outflow				7		

[1] Mean annual discharge for 1966–80

[2] Simulated average flow

[3] Simulated diversion returns

[4] Basin yield minus mean annual discharge for 1966–90

[5] Basin yield minus simulated streamflow infiltration

[6] Basin yield minus diversion of streamflow outside the model area

In the conceptual and steady-state models, flow through the unsaturated zone to the water table is treated implicitly as spatially nonuniform, time-averaged net infiltration recharge to the aquifer that is assumed to occur at a constant rate, whereas in the transient model, flow across the water table from some components varies in space and time. Components of the water budget that contribute water to the aquifer across the water-table boundary that are spatially uniform and temporally constant include precipitation recharge; spatially nonuniform and temporally (in the transient model) variable contributors include streamflow infiltration, and industrial wastewater and irrigation return flow. Typical unsaturated-zone flow processes such as moisture depletion from evaporation and evapotranspiration, downward percolation and lateral redistribution, alternate wetting and drying, and perching are not treated as distinct flow processes.

Precipitation Recharge

Precipitation recharge is characterized as spatially uniform and temporally constant in the steady-state and transient flow models (table 4). The effect of precipitation recharge in the model area has not been discernable in water-level measurements. In areas where precipitation accumulates as runoff into small, closed basins, numerous sedimentary interbeds in the thick unsaturated zone (more than 40 interbeds have been identified [Ackerman and others, 2006, fig. 11]) intercept and laterally redistribute downward percolation (Nimmo and others, 2002). This redistribution tends to offset the effects of local runoff accumulation and probably maintains an approximately uniform and constant rate of recharge to the aquifer. Estimates of precipitation recharge range from 0.01 ft/yr (Cecil and others, 1992, p. 713) to as much as 0.08 ft/yr (Kjelstrom, 1995, p. 11). In the conceptual model water budget, 5 percent of the mean annual precipitation was used to establish a maximum recharge rate of 0.04 ft/yr. Volumetrically, precipitation recharge across the water-table boundary is 70 ft^3/s and represents 3 percent of the total inflow estimate (table 4). When distributed across the model area, precipitation recharge represents a flux of 0.04 (ft^3/s)/mi^2. This small areal flux indicates that the effect of precipitation recharge on groundwater flow directions and velocities is probably small.

Streamflow infiltration

In the conceptual steady-state model, streamflow infiltration is characterized as spatially nonuniform and temporally constant, and in the conceptual transient model as spatially nonuniform and temporally variable (table 4). Streamflow onto the INL is episodic and nearly all streamflow infiltrates through the thick, unsaturated zone and recharges the ESRP aquifer. Three streams potentially contribute infiltration to the water table—Big Lost River, Little Lost

River, and Birch Creek—but only recharge from the Big Lost River is considered a significant source of infiltration (table 4; fig. 9). Estimates of recharge for the Big Lost River were based on measured discharge (fig. 10) and estimated infiltration rates in the Big Lost River Sinks, Playas, channel, and spreading areas (Bennett, 1990). Infiltration from the Little Lost River and Birch Creek is small because most streamflow either infiltrates or is diverted upstream of the INL boundary. Although flow from the Big Lost River also is diverted upstream of the INL boundary, during years with sufficient precipitation, flow from the Big Lost River enters the ESRP near Arco and terminates on the INL in a series of three interconnected playas southwest of TAN. From 1946 to 2003, the annual mean discharge of the Big Lost River near Arco (streamflow-gaging station 13132500, fig. 10) was 95.3 ft^3/s (Brennan and others, 2005, p. 196).

Streamflow fluctuates greatly in response to short-term wet and dry climate cycles that typically last from 3 to 8 years. During dry climate cycles, streamflow in the Big Lost River near Arco frequently is zero. In 1983 and 1984, streamflow infiltration in the Big Lost River channel, spreading areas, sinks, and playas at the INL accounted for more than 20 percent of the inflow component of the steady-state water budget and caused water-level rises exceeding 10 ft locally (figs. 11 and 12). Flux across the water-table boundary from streamflow infiltration in the Big Lost River channel, spreading areas, sinks, and playas averaged about 20 (ft^3/s)/mi^2 from 1966 through 1980 and was almost 200 (ft^3/s)/mi^2 at the spreading areas in 1984. Streamflow infiltration fluxes are 2 and 3 orders of magnitude larger than flux from precipitation recharge. Because of its large flux and proximity to known sources and areas of contamination in the aquifer, streamflow infiltration from the Big Lost River is considered more important to understanding contaminant transport than the other components of inflow across the water-table boundary.

Streamflow infiltration was apportioned based on differences in measured discharge between upstream and downstream streamflow-gaging stations and allocated based on river mileage and surface area enclosed within the perimeter of each surface-water impoundment. The simulated distribution of streamflow infiltration is described in the Numerical Model section of this report under the heading Boundary Conditions.

Irrigation and Industrial Water Return Flows

Infiltration of agricultural irrigation water in the northern part of the study area, infiltration of landscape irrigation at facilities, and disposal of industrial waste to ponds and ditches at facilities at the INL were characterized as spatially nonuniform and temporally variable. Refined estimates of the boundaries and types of infiltration areas used for model simulations resulted in slightly different rates of irrigation

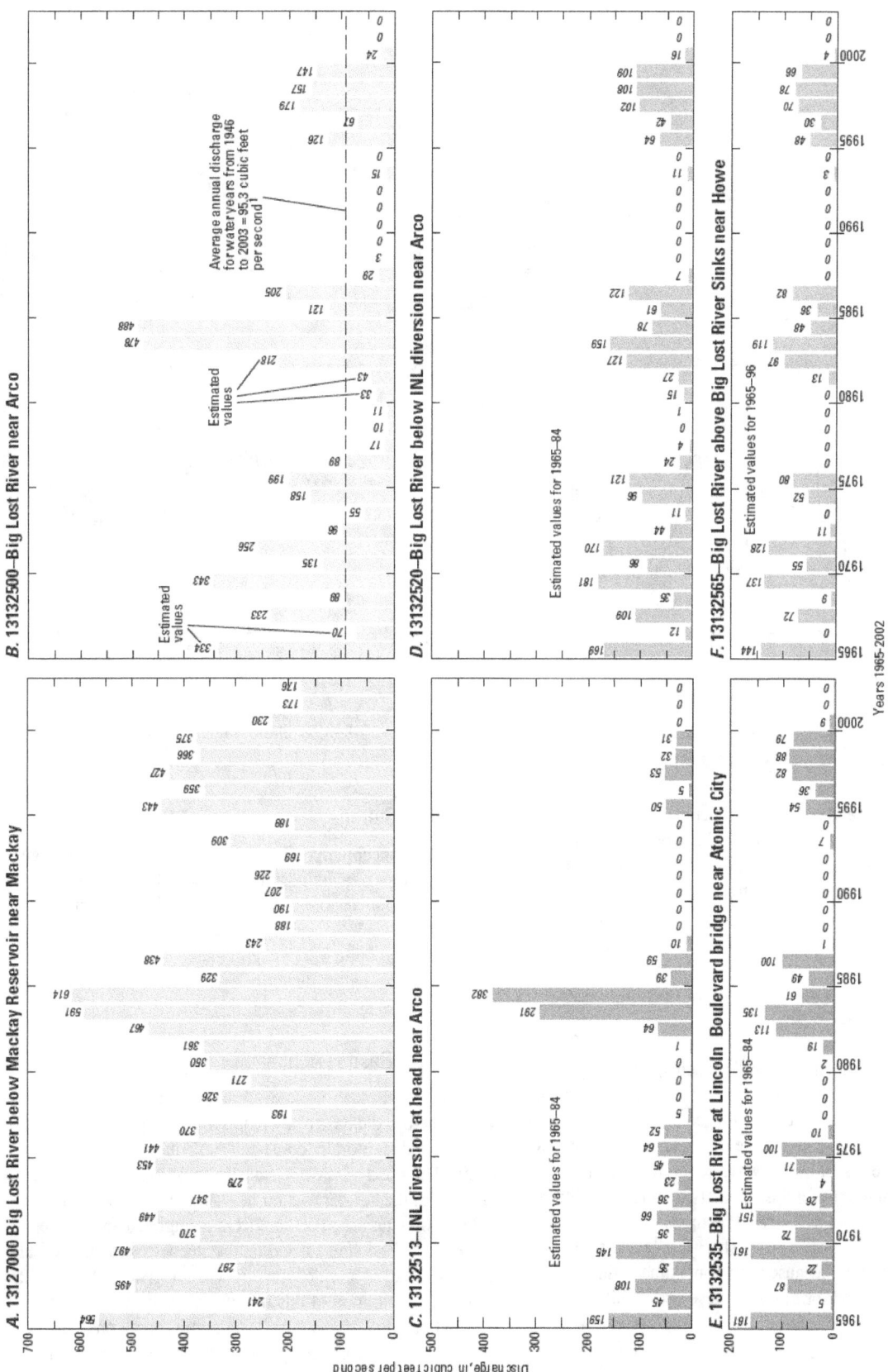

Figure 10. Annual mean discharge at streamflow-gaging stations on the Big Lost River, Idaho, for 1965–2002. (*A*) Big Lost River below Mackay Reservoir near Mackay (13127000), (*B*) Big Lost River near Arco (13132500), (*C*) INL diversion at head near Arco (13132513), (*D*) Big Lost River below INL diversion near Arco (13132520), (*E*) Big Lost River at Lincoln Boulevard bridge near Atomic City (13132535), and (*F*) Big Lost River above the Big Lost River Sinks near Howe (13132565). Locations of streamflow-gaging stations are shown in figure 1. Calendar year and water year discharge data for these six streamflow-gaging stations can be accessed at http://idaho.usgs.gov/.

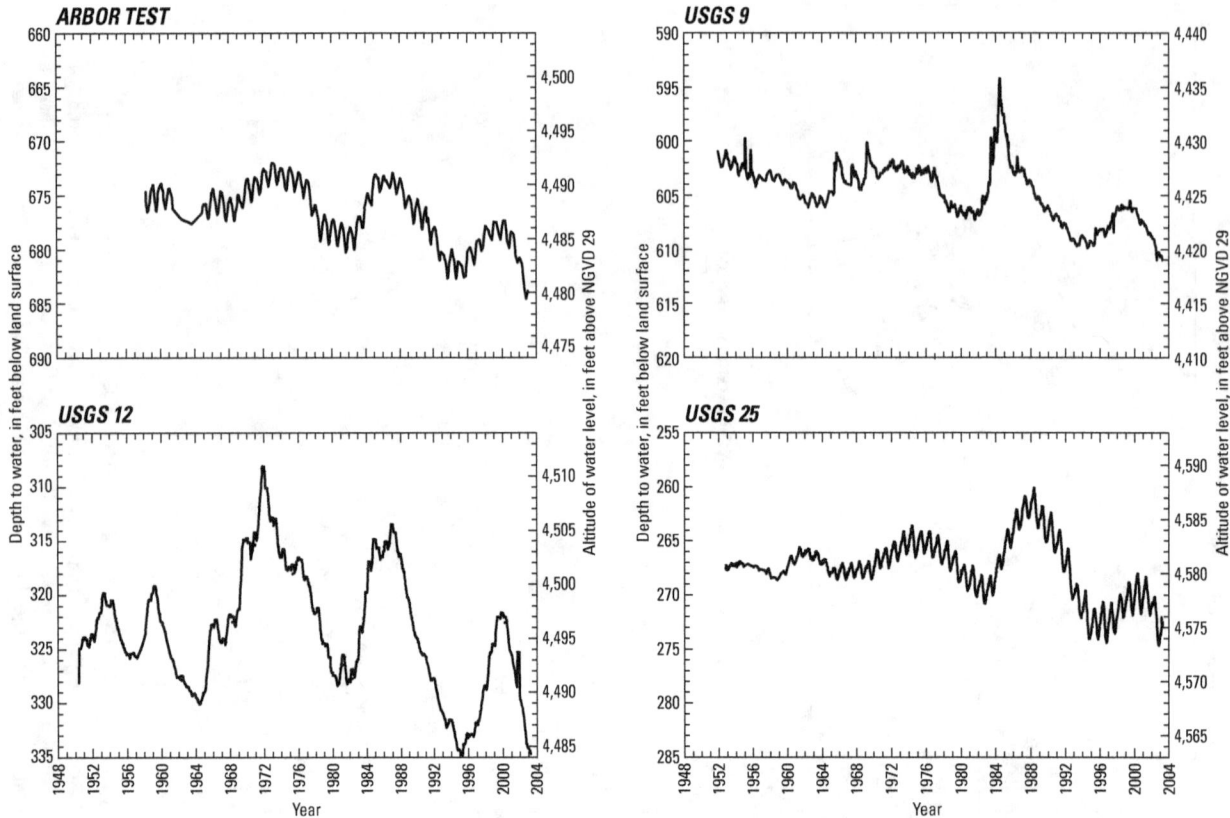

Figure 11. Water levels in wells Arbor Test, USGS 12, USGS 9, and USGS 25 for 1948–2003, Idaho National Laboratory and vicinity, Idaho. Locations of wells are shown in figures 1 and 12.

infiltration than those given in the conceptual model (table 4). Irrigation infiltration rates were applied uniformly within individual irrigation areas and were simulated as constant but were applied only during the stress period representing the irrigation season, May through August of each year in the transient model. Infiltration rates of 2.55×10^{-8} and 4.66×10^{-9} ft/s for surface-water and groundwater irrigation areas, respectively, were used and are equivalent to those of Spinazola (1994, p. 32-33, fig. 38).

Industrial wastewater return flows at the INL in 1980 were 6 ft^3/s, and included wastewater discharged in disposal ponds or ditches where it infiltrated into the unsaturated zone and the aquifer and into injection wells completed in the aquifer (table 4; fig. 9). The volume of wastewater discharged into infiltration ponds and wells varied over the years in response to the amount of industrial activity at INL facilities and changes in wastewater-disposal methods. Returns at the RTC and INTEC could have major short-term effects on contaminant migration because most contaminants in the aquifer beneath the INL originate at these two facilities and more than 50 percent of industrial wastewater returns were injected or infiltrated into the aquifer at these two facilities.

Water from Storage

In the conceptual model, releases of water from storage accounted for 4 percent of inflow in the 1980 water budget and were based on measured changes in water levels from October 1979 to October 1980 and RASA derived storage coefficients ranging from 0.05 to 0.2 (Ackerman and others, 2006, p. 2). The net gain to the aquifer from a decrease in storage during this period was 80 ft^3/s (table 4; fig. 9).

Water From Base of Aquifer

The conceptual-model water budget includes inflow of 44 ft^3/s across the base of the aquifer characterized as spatially uniform and temporally constant (Ackerman and others, 2006, p. 29). The steady-state and transient models, however, treat the base of the aquifer as a no flow boundary because this inflow estimate, based on data from a single 10,365-ft-deep corehole, INEL1 (fig. 1), and on the steepest vertical hydraulic gradients (0.071) as calculated by Mann (1986, table 2, p. 22), is small. Distribution of this inflow across the model area resulted in a flux of about 0.02 (ft^3/s)/mi^2, about one-half the flux estimate for precipitation recharge, 0.04 (ft^3/s)/mi^2.

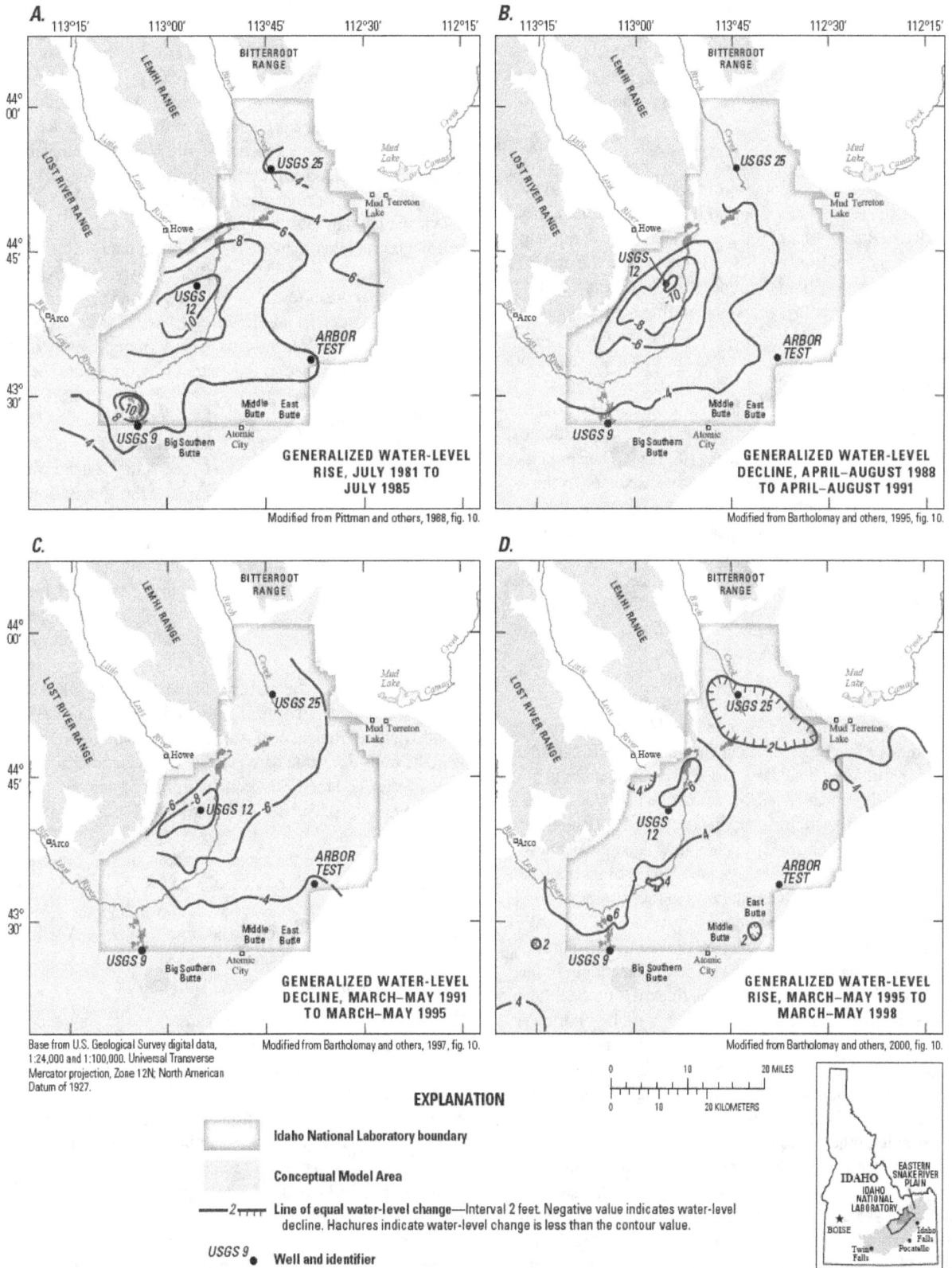

Figure 12. Water-level changes for (*A*) July 1981 to July 1985, (*B*) April–August 1988 to April–August 1991, (*C*) March–May 1991 to March–May 1995, and (*D*) March–May 1995 to March–May 1998, Idaho National Laboratory and vicinity, Idaho.

Additionally, the location of this boundary is from several hundred to more than 1,000 ft beneath areas of known contamination in the aquifer and, because of this, inflow across this boundary probably has minimal influence on the advective movement of contaminants in the aquifer.

Northwest Mountain-Front Boundary

The northwest mountain-front boundary is defined by the edge of the mountain fronts of the Pioneer Mountains, Lost River Range, Lemhi Range, and Bitterroot Range and the mouths of the Big Lost River, Little Lost River, and Birch Creek valleys (fig. 9). In the conceptual steady-state and transient models, inflow across the northwest mountain-front boundary is represented as spatially nonuniform and temporally constant. Inflows across this boundary represent 31 percent of the total inflow to the model area (table 4; fig. 9) and are limited to tributary valley underflow from the alluvial aquifers of the Big Lost River (367 ft^3/s), Little Lost River (226 ft^3/s), and Birch Creek (102 ft^3/s). The amount of underflow from each stream valley is calculated as the basin yield minus simulated streamflow infiltration. Basin yield is defined as the annual mean water yield to the Snake River Plain from mountainous areas bordering the plain for water years 1934–80 (Kjelstrom, 1986, sheet 2) and includes groundwater underflow and surface-water flow onto the plain. Inflow across the intervening mountain fronts is assumed to be zero and these are treated as no-flow sections along the northwest mountain-front boundary (fig. 9).

Water-table altitudes in the alluvial aquifers near the boundaries of the model area are several hundred feet above the altitude of the water table in the wells of the ESRP aquifer near the northwest boundary. These large differences in altitude indicate that underflow across the interface between the tributary valley alluvial aquifers and the ESRP aquifer probably is a combination of lateral flow and downward percolation in perched alluvial aquifers where these perched aquifers interfinger with basalt flows in the unsaturated zone of the ESRP aquifer and as lateral flow where the alluvial aquifer interfingers with basalt flows in the ESRP aquifer.

The estimated width of the aquifer at the mouth of the Big Lost River valley is 6.1 mi. The thickness of the alluvial aquifer in the Big Lost River valley at a well 4 mi northwest of Arco and near the mouth of the valley is as much as 760 ft (Crosthwaite and others, 1970, p. 73, fig. 25). Interbedded basalt and sediment at the mouth of the Big Lost River valley contain several vertically discrete water-bearing zones. The water table of the shallowest zone is 900 ft above that of the ESRP aquifer 6 mi south of Arco in well Weaver and Lowe (table A1; fig. B1). The less permeable parts of the sediment and basalt sequence probably cause some lateral movement of groundwater in perched aquifers to flow onto the Snake River Plain and percolate downward to recharge the aquifer (Crosthwaite and others, 1970, p. 71-75).

The width of the aquifer at the mouth of the Little Lost River valley was estimated at 5.7 mi. The thickness of the alluvial aquifer in the Little Lost River valley is at least 312 ft about 12 mi northwest of Howe (fig. 1) (Mundorff and others, 1963, table 6), and may be thicker. Interbedded basalt and alluvium near the mouth of the Little Lost River valley may cause some groundwater to flow onto the plain as perched aquifers above the ESRP aquifer. Cascading water enters the casing in some wells near the mouth of the valley. The water table in the alluvium immediately east of Howe is nearly 200 ft higher than the water table in the ESRP aquifer only 1 mile to the south (Mundorff and others, 1963, p. 23) indicating that water levels in wells may stand at progressively lower levels as deeper aquifers are penetrated (Mundorff and others, 1963, p. 23-25). Most underflow from the Little Lost River valley is assumed to flow laterally at depth into the ESRP aquifer.

The width of the aquifer at the mouth of the Birch Creek valley was estimated at 5.2 mi. The thickness of the alluvial aquifer at the mouth of the Birch Creek valley is unknown, but based on the estimated thicknesses of the alluvial aquifers in the Big Lost River and the Little Lost River valleys, may be more than several hundred feet thick. The water table in the ESRP aquifer at well ANP 7, about 8 mi southeast of the northwest boundary and near the mouth of the Birch Creek valley, is at an altitude of about 4,580 ft. This water table is about 1,400 ft lower than the water table at the Wagoner Ranch well (table A1; fig. B2) about 11 mi upstream of where Birch Creek crosses the northwest model boundary. It is not known if a relatively uniform gradient is between wells ANP 7 and Wagoner Ranch, or if a steep decrease in the water table is in the transition zone between the alluvial aquifer in Birch Creek valley and the ESRP aquifer similar to that observed near the mouths of the Big Lost River and Little Lost River valleys.

Underflow from the tributary valleys was apportioned and distributed on the basis of the estimated width and thickness of the alluvial aquifers and is described in the Numerical Model section of this report under the heading Boundary Conditions.

Northeast Regional-Underflow Boundary

The northeast regional-underflow boundary coincides with a steep increase in the hydraulic gradient near the northeastern boundary of the INL (fig. 9). The hydraulic gradient along the trace of this boundary ranges from 27 to 60 ft/mi, averages about 36 ft/mi, and reflects changes in aquifer transmissivity where basalt interfingers with less permeable sediments. Inflow across this 500–800 ft thick boundary was estimated at 1,225 ft^3/s by applying Darcy's Law, a hydraulic conductivity of 140 ft/d (a small value to reflect the presence of the less permeable sediments), an average hydraulic gradient of 36 ft/mi, a boundary length of 35 mi, and a saturated thickness of 600 ft (Ackerman and others, 2006, p. 35). This inflow estimate represents 55 percent of the total inflow to the model area (table 4) and is in close agreement with the value of 1,207 ft³/s derived from the RASA model. In the conceptual steady-state and transient models, underflow along the northeast regional-underflow boundary is represented as spatially nonuniform and temporally constant. Underflow across the northeast boundary was distributed on the basis of estimated variations in the thickness and hydraulic conductivity of the aquifer along the northeast boundary and is described in the Numerical Model section of this report under the heading Boundary Conditions.

Outflow Boundaries

Total outflow from the model area is 2,082 ft³/s (table 4). This estimate is 18 percent (469 ft³/s) less than that derived from the RASA model (2,551 ft³/s), and 7 percent (157 ft³/s) less than the conceptual-model inflow estimate (table 4). Of the outflow components, about 98 percent is underflow across the southwest boundary to adjacent parts of the aquifer and about 2 percent is from groundwater withdrawals for irrigation and industrial water use.

Southwest Regional-Underflow Boundary

In the steady-state and transient models, outflow across the southwest boundary is represented as spatially nonuniform and in the transient model as temporally variable. The length and shape of this boundary are not well documented. Surface-based electrical-resistivity surveys (Whitehead, 1986, sheet 2) indicate that the aquifer thickness along this boundary ranges from about 500 to 3,000 ft (fig. 6B). The variable flow across this boundary reflects the character of large, episodic inflows from Big Lost River streamflow infiltration. Head definition for the southwest boundary of the aquifer was limited to water-level measurements from six wells that are from 3 to 10 mi from the southwest boundary (fig. 4D); consequently, outflow estimates across this boundary perhaps are the least reliable of all the water-budget estimates. Outflow across the southwest boundary was estimated at 2,037 ft³/s (table 4; fig. 9) using Darcy's Law and data described in Ackerman and others (2006, p. 35), and is similar to the value, 2,000 ft³/s, summed for underflow across this boundary from the flow-net analyses by Mundorff and others (1964, pl. 4).

Groundwater Withdrawals

In the steady-state and transient models, groundwater withdrawals from the aquifer are industrial withdrawals at onsite facilities and irrigation withdrawals for use offsite (fig. 9), and are represented as spatially nonuniform and, in the transient model, temporally variable. For the steady-state and transient models, irrigation withdrawals are constant, but are applied seasonally in the transient model. Groundwater withdrawals in the model area in 1980 were estimated to be

45 ft³/s (Ackerman and others, 2006, p. 34 and table 6). Slight differences in subtotals for well withdrawals between the conceptual and numerical models are due to refined estimates of irrigation and industrial withdrawals. Local effects of industrial withdrawals were not expected to have long-term effects on aquifer response beneath the central part of the INL; however, withdrawals and returns at the RTC and INTEC could have major short-term effects on contaminant migration because most contaminants in the aquifer beneath the INL originate at these two facilities. These facilities also account for more than 50 percent of industrial groundwater withdrawal at the INL. Groundwater used for irrigation in the model area is withdrawn from the aquifer primarily near Mud Lake, but smaller quantities also are withdrawn near Howe (Ackerman and others, 2006, p. 34).

No-Flow Boundaries

In the steady-state and transient flow models, the base of the aquifer and the southeast flowline boundaries are designated as no-flow boundaries (fig. 9). The boundary at the base of the aquifer has been described as an inflow boundary conceptually, but treated as a no-flow boundary in the numerical models.

The southeast-flowline boundary, near the central axis of the ESRP, is represented by a southwest-trending flowline, projected vertically through the thickness of the aquifer (800–4,000 ft) (fig. 6B), and consequently flow across this boundary, conceptualized as spatially uniform and temporally constant, is presumed to be zero. Water-table maps constructed for the ESRP aquifer for 1928–30, 1956–58, and 1980 (Stearns and others, 1938; Mundorff and others, 1964; and Lindholm and others, 1988) indicate that changes in water levels along and east of the southeast boundary are spatially uniform. Consequently, flow directions along most of the southeast boundary of the model area are relatively stable, and flowlines constructed for this area are assumed to be temporally constant (Ackerman and others, 2006, p. 35).

Conceptualization of Groundwater Flow and Contaminant Transport

Water-table contours indicate that the subregional and regional directions of groundwater flow in the model area are from northeast to southwest (fig. 13A). The volume of groundwater flow increases progressively in a direction downgradient of the northeast boundary. This increased flow is the result of tributary-valley underflows along the northwest mountain-front boundary (695 ft³/s), precipitation, irrigation, industrial wastewater, and streamflow infiltration across the water-table boundary (195 ft³/s). Together these additions account for about 44 percent of the outflow across the southwest boundary; the remainder of the outflow originates from underflow along the northeast boundary (1,225 ft³/s).

A.

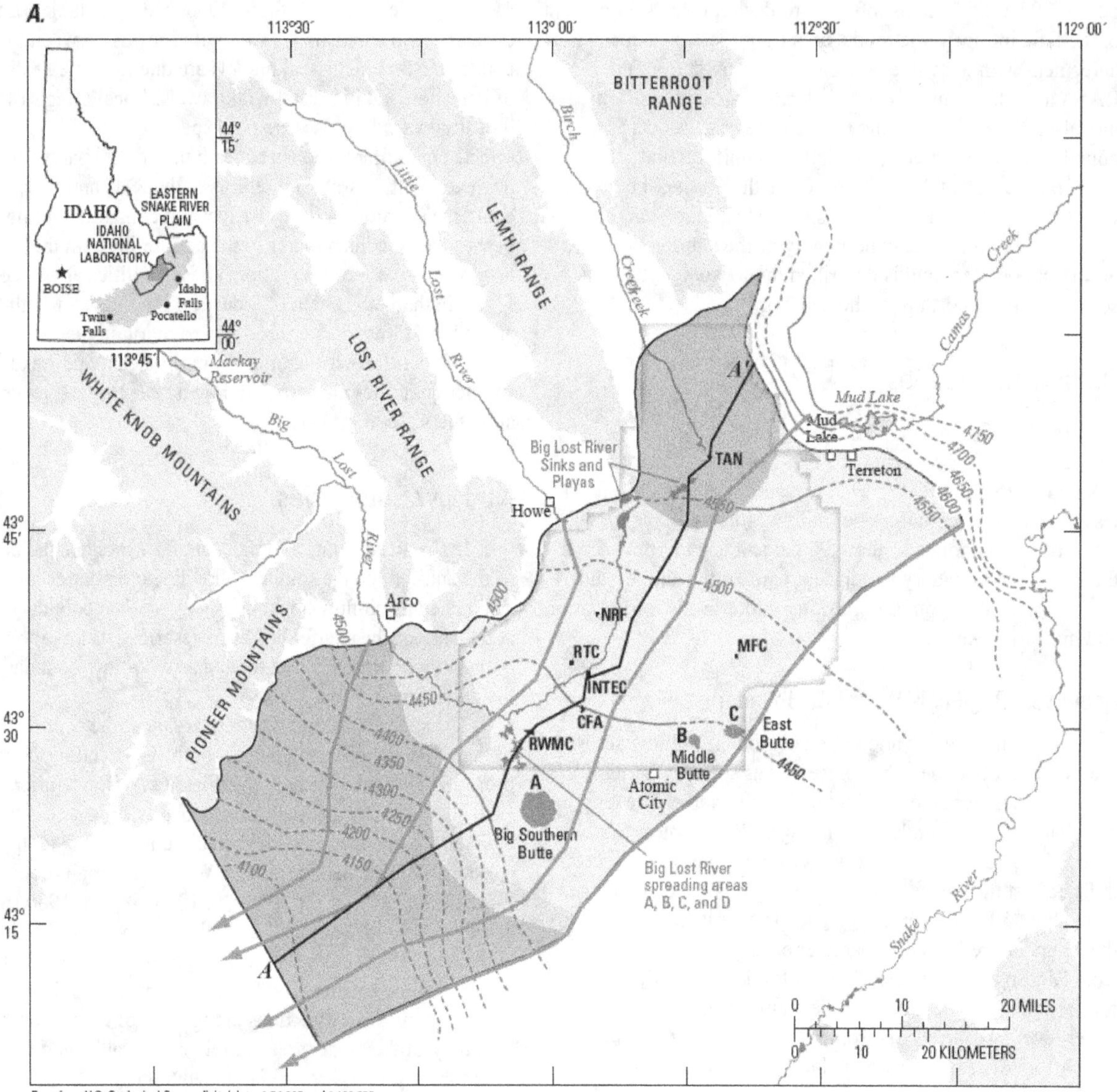

Base from U.S. Geological Survey digital data, 1:24,000 and 1:100,000
Albers Equal-Area Conic projection, standard parallels 42°50′N, 44°10′N;
central meridian 113°00′W; North American Datum of 1927.

EXPLANATION

Hydrogeologic units at the water table

1—Younger rocks consisting of densely fractured basalt and interbedded sediment

2—Younger rocks consisting of massive, less densely fractured basalt and interbedded sediment

3—Intermediate-age rocks consisting of slightly altered fractured basalt and interbedded sediment

Silicic rocks—Includes rhyolite domes Big Southern Butte (A), Middle Butte (B), and East Butte (C). Small unnamed dome and Cedar Butte andesite not shown

Idaho National Laboratory boundary

Model area boundary

A——*A′* **Trace of section**

Site facilities

CFA—Central Facilities Area

INTEC—Idaho Nuclear Technology and Engineering Center

MFC—Materials and Fuels Complex

NRF—Naval Reactors Facility

RTC—Reactor Technology Complex

RWMC—Radioactive Waste Management Complex

TAN—Test Area North

—4200— **Water-table contour**—Shows altitude of water table in March 1980. Interval 50 feet. Dashed where approximately located. Datum is NGVD 29. Modified from Linholm and others, 1988.

Generalized flowpath—Arrow in direction of flow

Figure 13. Distribution of hydrogeologic units and direction of groundwater flow (*A*) at the water table and water-table contours for 1980, and (*B*) along the line of section (*A–A′*), Idaho National Laboratory and vicinity, Idaho. Trace of section shown in figure 13*A*.

B.

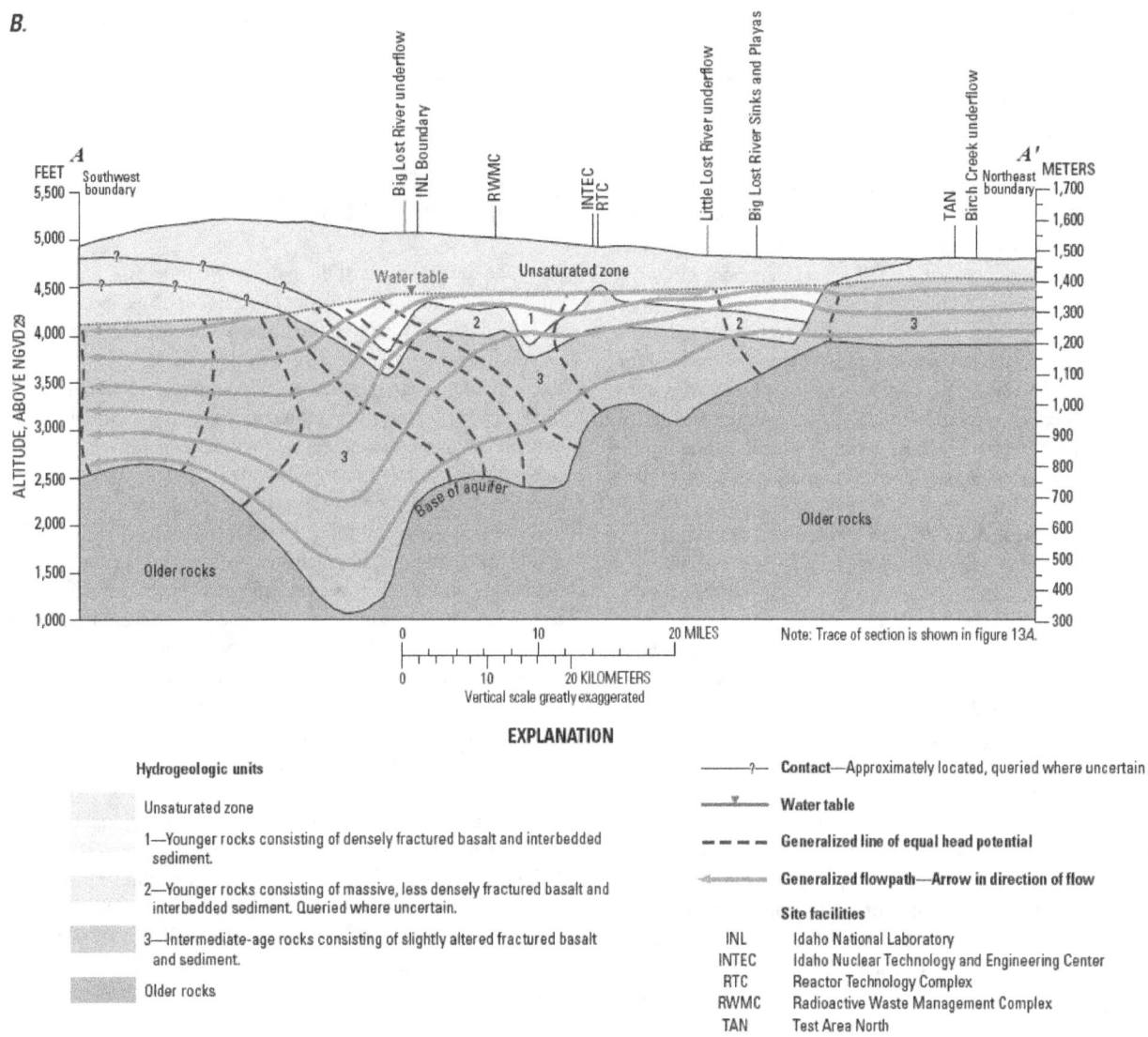

Figure 13.—Continued.

A semi-quantitative distribution of groundwater flow along a northeast to southwest cross section (*A–A'*, fig. 13*B*) through the central part of the model area (Ackerman and others, p. 48), indicates that head decreases and then increases with depth as the aquifer thickens and thins in a direction downgradient of the northeast boundary. Downward flow is indicated in areas where head decreases with depth and upward flow is indicated in areas where head increases with depth. The largest changes in vertical gradient are depicted as occurring southwest of the INL; just upgradient of where the younger, more conductive rocks of hydrogeologic unit 1 pinch out and hydrogeologic unit 2 intersects the water table (fig. 13*B*). This depiction also indicates that flow is dominantly horizontal, but shows slight downward movement of water (note vertical exaggeration in fig. 13*B*) from the younger rocks into the rocks of intermediate age at this location and implies downward movement and deeper circulation of contaminants that migrate offsite.

The conceptual model implies that most contaminant movement beneath the INL takes place in the interflow zones and thin, densely fractured, and highly conductive basalts and interbedded sediments of hydrogeologic unit 1, which compose most of the upper 200 ft of the aquifer beneath most of the INL. This hypothesis is consistent with conclusions reached by earlier investigators who noted that "…waste plumes generally remain as relatively thin lenses in about the upper 250 ft of the aquifer" (Robertson, 1974, p. 6). The hypothesis was based on (1) long-term observation of water levels and contaminant movement in the aquifer, (2) estimates of inflows and outflows across the boundaries of the model area, (3) current interpretations of the 3-D geometry and the distribution of hydrogeologic units within the aquifer, (4) location of contaminant sources in the model area, (5) estimates of the hydraulic properties of the aquifer, (6) history of waste-disposal practices at the INL, and (7) a simplified 2-D model of groundwater flow (Ackerman and others, 2006, p. 48-49) that depicts the effects of the hydrogeologic framework on groundwater flow along a northeast to southeast section through the central part of the model area. The most important implications of the conceptual model for contaminant transport are:

- Preferential flow along basalt interflow zones, which implies a large horizontal to vertical anisotropy.

- Restricted downward movement of water and contaminants from hydrogeologic unit 1 into and across the less conductive massive basalts of hydrogeologic unit 2 beneath and near contaminant source areas (RTC, INTEC, and RWMC).

- Enhanced dispersion and dilution of contaminants in the aquifer resulting from the spatial and temporal variability of streamflow infiltration from the Big Lost River in close proximity to areas of known contamination in the aquifer.

- Inferred downward movement and deeper circulation of water and contaminants beginning near the southern boundary of the INL and downgradient of known locations of contaminants in the aquifer.

Numerical Model

The development of the numerical model was based on the conceptual groundwater flow model of the system and involved (1) selecting the governing equation(s) of groundwater flow constituting the mathematical model and the computer program to solve the mathematical model numerically; (2) translating the conceptual model to a form suitable for numerical modeling including determining the system geometry, discretizing the spatial and temporal domains—designing a spatial grid, selecting time steps and stress periods, and formulating boundary conditions; and (3) selecting measurements of physical properties and hydrologic measurements of aquifer state such as water levels in wells—heads and flow to and from the aquifer. The conceptual model and available data were integrated into the numerical model which was subsequently calibrated, whereby model parameters and boundary conditions were adjusted based on an objective criterion of the match between simulated and observed heads and flows.

Groundwater flow was simulated with MODFLOW-2000, a computer program that simulates three-dimensional groundwater flow through a porous medium using a numerical finite-difference method for solving the governing equations for steady (time invariant) and transient (time variable) flow (Harbaugh and others, 2000, p. 1). MODFLOW-2000 was selected because it (1) simulates 3-D saturated flow, (2) is compatible with other particle tracking and solute transport computer codes that are readily available, (3) is capable of calculating accurate sensitivities (Hill and others, 2000, p. 3), (4) has been widely used by the groundwater scientific community for more than 30 years, (5) is thoroughly documented, and (6) has program developers who were readily accessible for consultation. MODFLOW-2000 is constrained by its governing flow equations to (1) flow of a uniform density and viscosity fluid through saturated porous media and (2) aquifers where the principal axes of the hydraulic conductivity tensor are aligned with the orthogonal model coordinates. The current model is assumed to meet these constraints because (1) water in the aquifer is fresh water within a narrow range of temperature and, therefore, any small fluid-density and viscosity variations are negligible, (2) flow through the fractured parts of the aquifer can be represented as flow through an equivalent porous medium, (3) hydrogeologic units constituting the aquifer are sub-horizontal (not steeply dipping) and, therefore, vertical hydraulic conductivity can be simulated perpendicular to the nearly horizontal hydrogeologic units, and (4) large scale horizontal anisotropy in the basalt aquifer is not indicated.

Model Discretization

The finite-difference method used by MODFLOW-2000 requires that the continuous simulated domain be replaced by a discretized domain. Spatial discretization is defined in terms of layers, rows, and columns that result in discrete rectilinear volumes called cells. This three-dimensional array of cells is known as the model grid. For transient models, time must also be divided into discrete intervals called stress periods and time steps. Stress periods represent intervals over which specified flows in and out of the aquifer are constant. Flows can change from one stress period to the next. Stress periods consist of one or more time steps where simulated heads and flows are calculated for each cell.

The flow models of the INL and vicinity were designed for use in contaminant transport modeling at a subregional scale. The model grid was created in a study-specific Albers equal-area projection and was rotated 31.0605° counterclockwise about the origin of 43° 34′ 35.25″ N. 113° 50′ 32.63″ W., such that it is oriented parallel to the dominant direction of flow (fig. 14); the orientation also minimized the size of the model grid. The projection used the North American Datum of 1927, a central meridian of 113° W., standard parallels of 42° 50′ N. and 44° 10′ N., a false easting of 656,166.67 ft (200,000 meters), and the latitude of the projection's origin was 41° 30′ N. The southwest boundary of the model grid was coincident with a gridline between columns 22 and 23 of the RASA model grid (Garabedian, 1992, p. 38). The coincidence of the study area and the model grid with the RASA model grid was used to facilitate comparison and analysis of the groundwater budget for the conceptual model (Ackerman and others, 2006, p. 36, fig. 17) and analysis of simulated fluxes within sub areas of the model.

Spatial Discretization

Spatial discretization of the model domain was determined by considering suitable representation of (1) hydrogeologic units that extend over hundreds of mi²; (2) the flat, nearly uniform, hydraulic gradient across the INL; and (3) the large areal extent of contaminant plumes (fig. 2) in the southwest part of the INL (for example, in 1998 the plume of tritium for concentrations greater than 500 picocuries/liter (pCi/L) encompassed an area of about 43 mi²). Uniform model grid spacing was used because (1) most groundwater transport models compatible with MODFLOW-2000 require uniform grids over the transport simulation domain, and (2) numerical accuracy is better for regular grids (Anderson and Woessner, 1992, p. 64). Two uniform horizontal model grid spacings

were used in the steady-state and transient flow models: (1) a coarse grid spacing of 0.5 by 0.5 mi (0.25 mi² per model cell) used in early models and (2) a fine grid spacing of 0.25 by 0.25 mi (0.0625 mi² per model cell) (fig. 14) used for the calibrated models described in this report. Even finer model grid spacing might be necessary for a site-scale model of contaminant transport, but was not considered necessary for suitably representing the features described above.

The vertical discretization of the model was referenced to the altitude of the 1980 water table (fig. 13), in that the bottom of model layer 1 (the top of layer 2) was specified 100 ft below the 1980 water-table surface (fig. 15). The vertical model grid spacing consisted of six layers of varying thickness:

- Layers 1 approximately 100 ft thick, varying with the water-table altitude,

- Layer 2, 100 ft thick,

- Layer 3, 0 to 100 ft thick,

- Layer 4, 0 to 200 ft thick,

- Layer 5, 0 to 300 ft thick, and

- Layer 6, 0 to 3,229 ft thick (fig. 15).

Model layers 1, 2, and 3 were kept relatively thin in order to more accurately represent flow in hydrogeologic units 1 and 2, the units that compose most of the upper part of the aquifer beneath the INL (fig. 15) and that contain contaminant plumes. Because the mean thicknesses of basalt flows in hydrogeologic units 1, 2, and 3 are 20, 29, and 23 ft, respectively (table 3), a layer thickness of 100 ft ensured that each of the model layers included several basalt interflow zones. Model layers 3–6 contained the base of the aquifer in some locations and consequently were absent or less than full thickness at those locations.

The model domain is 80 mi northeast to southwest, 44 mi northwest to southeast, and greater than 4,000 ft thick (figs. 14 and 15). The model grid used for the calibrated models described in this report discretizes this volume into 176 rows, 320 columns, and 6 layers to generate 337,920 cells, of which 168,737 were active and 169,183 were inactive (fig. 14). Inactive cells are cells outside the simulated volume but inside the model grid because the irregular model shape does not conform to the rectilinear finite-difference model grid. A coarse model grid, with cells 0.5 by 0.5 mi in the areal dimension also was used in this study, and discretizes this volume into 88 rows, 160 columns, and 6 layers to generate 84,480 cells, of which 42,230 were active and 42,250 were inactive (fig. 14). Vertical discretization was the same for both grids.

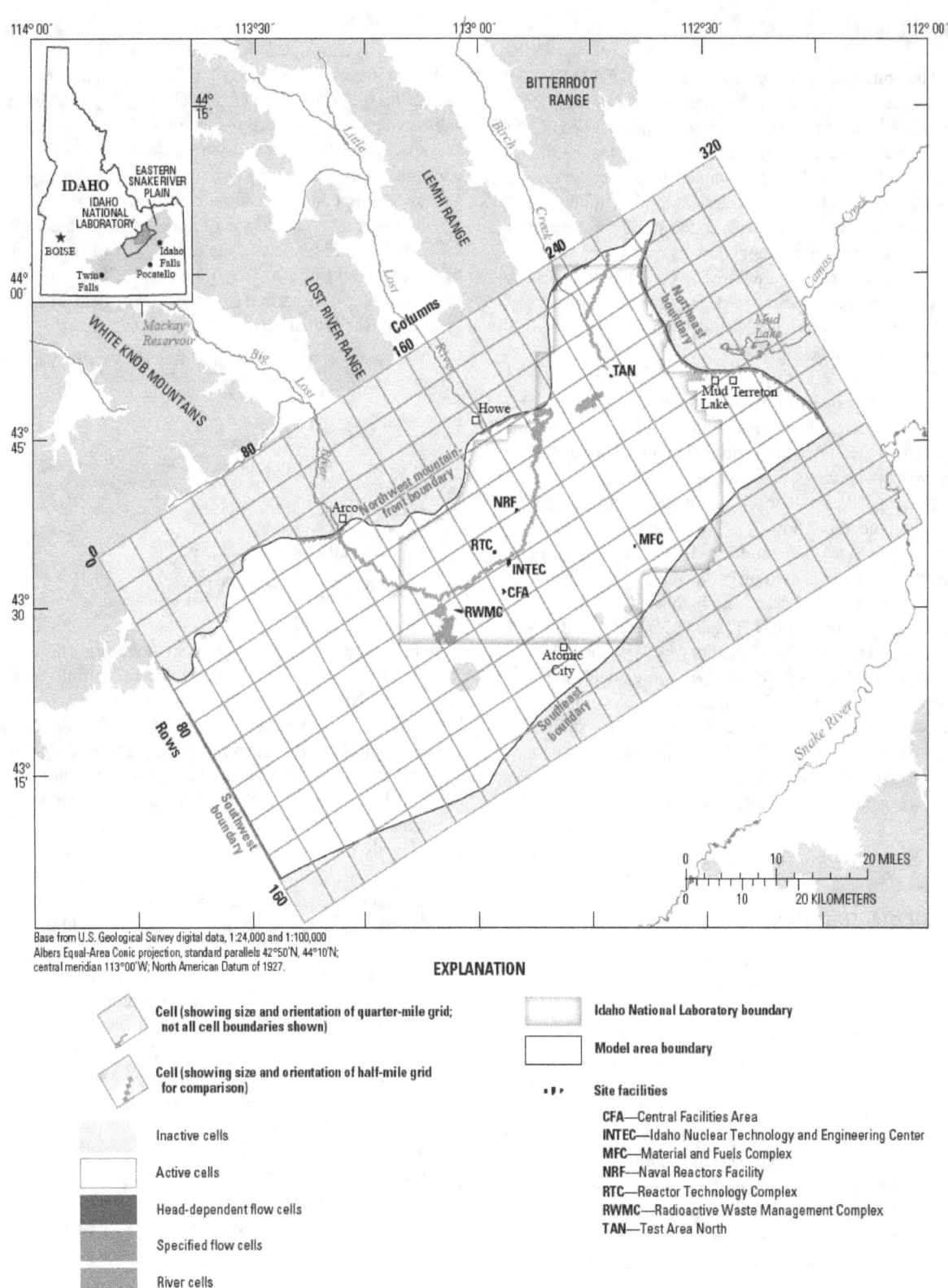

Base from U.S. Geological Survey digital data, 1:24,000 and 1:100,000
Albers Equal-Area Conic projection, standard parallels 42°50'N, 44°10'N;
central meridian 113°00'W; North American Datum of 1927.

EXPLANATION

Cell (showing size and orientation of quarter-mile grid; not all cell boundaries shown)	Idaho National Laboratory boundary
Cell (showing size and orientation of half-mile grid for comparison)	Model area boundary
Inactive cells	Site facilities
Active cells	CFA—Central Facilities Area
Head-dependent flow cells	INTEC—Idaho Nuclear Technology and Engineering Center
Specified flow cells	MFC—Material and Fuels Complex
River cells	NRF—Naval Reactors Facility
	RTC—Reactor Technology Complex
	RWMC—Radioactive Waste Management Complex
	TAN—Test Area North

Figure 14.   Model domain and locations of head-dependent, specified flow, and river boundary conditions for layer 1 used for 0.5- and 0.25-mile model grids for models of groundwater flow, Idaho National Laboratory and vicinity, Idaho.

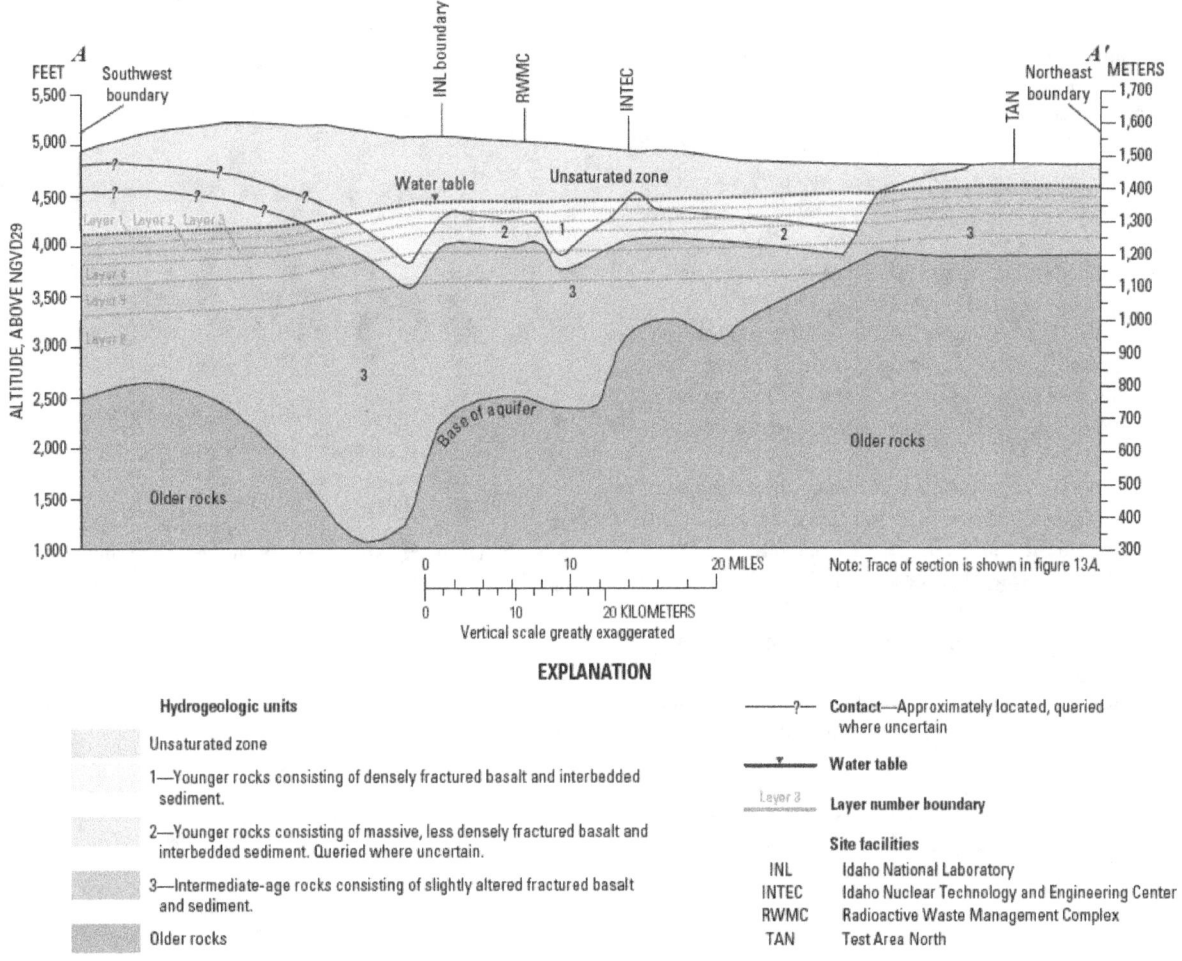

Figure 15. Vertical discretization and distribution of the model layers and hydrogeologic units for models of groundwater flow, Idaho National Laboratory and vicinity, Idaho. Trace of section shown in figure 5A.

Temporal Discretization

Aquifer inflows and outflows in the model area vary due to (1) long-term, decadal changes in climate and water use, (2) short-term (3–8 years) wet or dry climate cycles, and (3) annual or seasonal changes in streamflow and water use. The transient flow model simulates the hydrologic conditions of the aquifer from 1980 through 1995, such that simulations do not capture decadal changes, but do capture a wet (1982–86) and a dry (1987–94) climate cycle and also annual and seasonal hydrologic cycles. The periods 1983–84 and 1987–94 produced the maximum discharge and the longest dry period, respectively, in the Big Lost River during the past 60 years (fig. 10). Unless otherwise stated, all references to years in this report are for calendar years January through December.

Temporal discretization is guided by the timing and duration of simulated stresses. Streamflow infiltration and irrigation water use in the model area generally is between May and August (fig. 16; Spinazola, 1994, fig. 31). In the northeast part of the model area seasonal changes in water levels of 2–3 ft were observed (fig. 17) in wells such as USGS 27 (fig. 1) in response to groundwater withdrawals for irrigation, with water levels rising from September through April and falling from May through August. Changes in

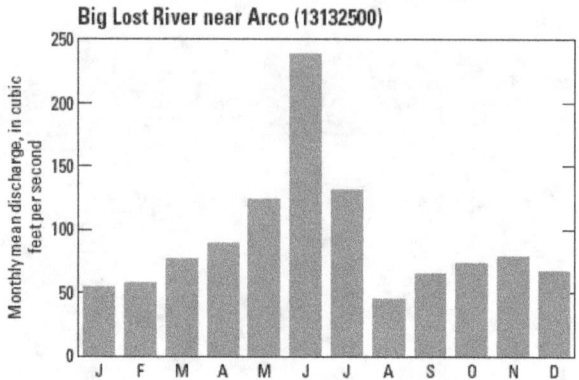

Figure 16.   Monthly mean discharge at streamflow-gaging station Big Lost River near Arco for water years 1946 through 2006.

specified flows into or out of the active part of the model grid were represented with three seasonal (January–April, May–August, September–December) stress periods during each year. The transient model consisted of one annual steady-state stress period representing 1980 followed by 45 seasonal stress periods representing 1981 through 1995.

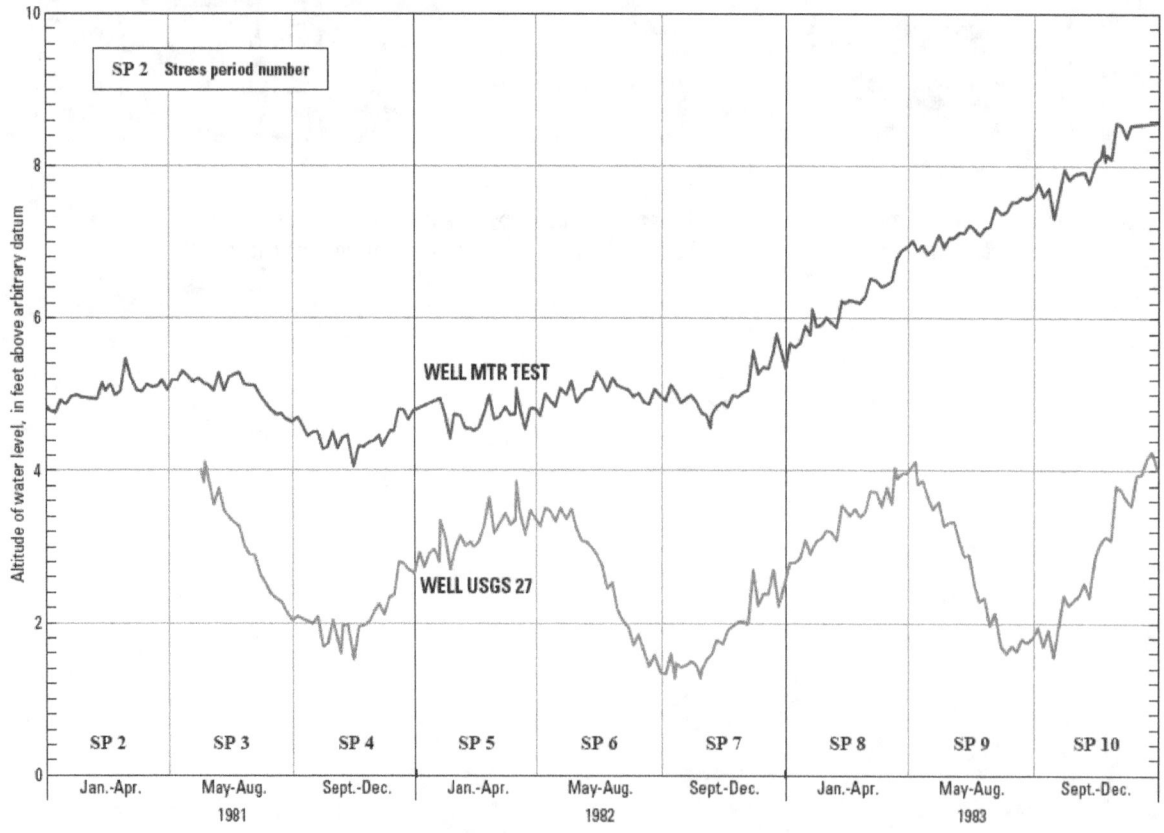

Figure 17.   Stress periods and water levels at wells MTR TEST and USGS 27, 1981 through 1983, Idaho National Laboratory, Idaho.

Boundary Conditions

Boundary conditions in the flow model specify the locations and characteristics of groundwater flow into and out of the active areas of the model grid (fig. 9). Specified flows into the active model grid occur at the northeast boundary, the northwest mountain-front boundary, and the water-table boundary (figs. 14 and 15). Within the model grid, specified inflows occur at injection wells and specified outflows occur at industrial or irrigation wells. These flows represent historical conditions for various boundaries and time intervals between 1946 and 1995. No flow conditions occur at the southeast boundary, a stream surface, and the base of the aquifer. Head-dependent outflow conditions define the southwest boundary (fig. 14). The movement of water in and out of the model grid was simulated with various packages available in MODFLOW-2000 (Harbaugh and others, 2000) selected for their applicability to physical and hydrologic conditions at the boundary and to aid in accounting for water budget terms. McDonald and Harbaugh (1988) document the concepts of most packages used in the model.

Water-Table Boundary

Inflow across the water table is from precipitation, streamflow infiltration, and return flows of irrigation and industrial water. These flows are characterized in the transient models as temporally variable specified flows, except for precipitation, which is characterized as constant specified flow. All specified flows are constant in the steady-state model.

Precipitation Recharge

Recharge from infiltration of precipitation was simulated across the water-table boundary and was applied throughout the model area using the Recharge Package from MODFLOW-2000. Precipitation recharge was allocated to model layer 1 cells at an areally uniform constant rate of 0.04 $(ft^3/s)/mi^2$. Precipitation recharge was simulated as constant because it composes a small proportion of the water budget; most of the changes in water levels are related to precipitation in the tributary basins rather than to precipitation on the plain.

Streamflow Infiltration

Streamflow infiltration from the Big Lost River, Little Lost River, and Birch Creek was simulated as specified flow using the River Package from MODFLOW-2000. Infiltration was allocated to layer 1 cells proportionately to the length of the trace of river channels or to the areas of cells occupied by the Big Lost River Sinks, Playas, and spreading areas (fig. 14). Amounts of infiltration were assigned to reaches of river channels and individual areas. In the steady-state model, differences in discharge between upstream and downstream gaging stations and measured outflows at gaging stations immediately upstream of surface-water impoundments were assumed to represent infiltration. In the transient model representing the period 1980–95, these infiltration estimates were reduced to account for evaporation losses. The periods for averaging streamflow infiltration are different and the level of detail for discussion and implementation of streamflow infiltration are greater for the numerical model than for the conceptual model.

Streamflow estimates for the Big Lost River are based on measured daily mean discharges or when data were not available, on estimates using partial records, miscellaneous discharge measurements, regression analyses, averages from the streamflow loss studies of Bennett (1990), and later more complete discharge records. Gaging stations with useable record, their general location, and their corresponding map numbers (table A2; figs. B1, B2) are: (1) 4 mi southeast of Arco (501), (2) at the inlet to spreading area A (502), (3) at the outlet of spreading area A to spreading area B (503), (4) downstream of the diversion dam used to divert flow into spreading area A (504), (5) at the Lincoln Boulevard bridge near the RTC and INTEC (505), and (6) above the Big Lost River Sinks (506).

Streamflow infiltration from the Big Lost River in the simulated area, which includes sections of the river upstream of the Arco gaging station, was simulated as constant flow (125.3 ft^3/s) for the steady-state model, based on annual mean discharge for a continuous period of record from 1966 to 1980 and average infiltration rates in the Big Lost River channel. This infiltration varies spatially; for example, infiltration from the river channel was estimated to vary from 1 to 4 $(ft^3/s)/mi^2$ between Arco and the Big Lost River Sinks to as much as 28 $(ft^3/s)/mi^2$ at the Big Lost River Sinks (Bennett, 1990, p. 24 and 26). For the transient model, based on average measured discharge and estimated evaporation for 4-month intervals between 1981 and 1995 (table C1), Big Lost River streamflow infiltration ranged from 0 to 641 ft^3/s.

Evaporation losses were calculated and simulated for the Big Lost River Sinks, Playas, channels, and spreading areas. Evaporation rates were calculated using a modified form of equations used by the Idaho Department of Water Resources (Tony Olenichak, Water District #1, written commun., 2006) to estimate water loss from reservoirs and from data collected at the Aberdeen Experiment Station about 40 mi south-southwest of the spreading areas (http://www.usbr.gov/pn/agrimet/webarcread.html).

The modified equations are:

$$EV = PE \times Sc, \tag{1a}$$

or

$$EV = ETr \times Cp \times Sc, \tag{1b}$$

where

EV is daily evaporation rate (ft/d) from large water bodies;

PE is 24-hour pan evaporation (ft/d);

Sc is standard coefficient, 0.7, for converting pan evaporation to large surface area;

ETr is reference evapotranspiration (ft/d) (1982 Kimberly-Penmen equation); and

Cp is correlation factor, 1.18 for converting ETr to Pe.

The average annual evaporation in the model area was calculated using the average evaporation rate for the stress period and the estimated areas for all stream reaches of the Big Lost River during the stress period. Equation 1a and PE from records at the Aberdeen Experiment Station were used to calculate EV for stress periods prior to about 1990, and for 1990 through 1995 EV was calculated with equation 1b and ETr. The average ETr using May through August monthly data for the period 1991–2005 at the Aberdeen Experiment Station was 0.022 ft/d. The calculated value of EV for the same period was 0.018 ft/d. These are similar to a previous estimate of evaporation from the Big Lost River Playas of 0.01–0.02 ft/d (Barraclough, Teasdale, and others, 1967, p. 24). The estimated evaporation ranged from 1 percent of average flow in river channels to more than 7 percent of flow in the spreading areas.

The Little Lost River and Birch Creek have short river reaches on the ESRP (fig. 14) that were estimated to contribute small amounts of infiltration to the ESRP aquifer. Infiltration from the Little Lost River and Birch Creek was simulated as constant flow (3 and 20 ft³/s, respectively) for the steady-state and transient models (table 4). These infiltration estimates were based on uncertain estimates of streamflow onto the ESRP due to discontinuous discharge records at gaging stations upgradient of the ESRP and infrequent observations of flow onto the ESRP. Infiltration to the aquifer from the Little Lost River was simulated along a 0.7 mi reach of river channel near Howe (Map No. 611, fig. 14; fig. B2). The infiltration rate was assumed to be greater than average channel infiltration rates for the Big Lost River because the sediments at the mouth of the Little Lost River are closer to the source area. Infiltration of flow diverted from Birch Creek onto the ESRP was simulated along a 12 mi length of trench and section of an old channel north of TAN (Map No. 612, fig. 14; fig. B2). Streamflow in this reach of the river represents return flows from upstream Birch Creek diversions that are used to operate a hydroelectric power generation facility several miles east of Birch Creek (Ackerman and others, 2006, p. 33). Birch Creek diversions were estimated to be 40 ft³/s using powerplant records from 1988 to 1995 (Ted S. Sorenson, Sorenson Engineering, written commun., March 2001). For 6 months of the year, return flows are diverted to the northeast and are used for irrigation. During the nonirrigation season, return flows are routed to the area north of TAN. Infiltration from the Birch Creek diversion returns was estimated to occur for one-half of the year and was simulated as constant at 20 ft³/s (table 4). Infiltration from the Little Lost River and Birch Creek were simulated as constant flow for the steady-state and transient models. No evaporation losses were estimated or simulated for the Little Lost River or Birch Creek due to the relatively poor estimates of infiltration and short river reaches of these streams in the model area.

Irrigation and Industrial Return Flows

Infiltration of excess applied agricultural irrigation water in the northern part of the study area was simulated at a constant annual rate and applied seasonally in transient stress periods representing the period May through August. Annual infiltration rates of 2.55×10^{-8} and 4.66×10^{-9} ft/s for surface-water and groundwater irrigation areas, respectively, were used and are equivalent to those of Spinazola (1994, p. 32-33, fig. 38) (table 5). Irrigation infiltration rates were applied uniformly to layer 1 cells within individual irrigation areas (fig. 18). Refined estimates of the boundaries and types of infiltration areas used for model simulations resulted in slightly different rates of irrigation infiltration than that given in the conceptual model (tables 4 and 5).

Industrial return flows consisting of infiltration of landscape irrigation at INL facilities, and disposal of industrial waste to ponds and ditches at these facilities were simulated as temporally variable using the Well Package (fig. 18; table D1). The steady-state total industrial return flow was 2.6 ft³/s and transient totals ranged from 1.8 to 6.9 ft³/s per stress period.

Table 5. Steady-state and transient simulated water budgets, Idaho National Laboratory and vicinity, Idaho.

[**Steady-state and Initial condition:** Stress period 1, steady state, 1980 **Maximum inflow:** Stress period 9, May-August 1983 **Minimum inflow:** Stress period 44, January-April 1995 Flow in cubic feet per second, rounded to nearest tenth **Abbreviations**: –, not simulated; NA, not applicable]

Budget component		Boundary condition	Input files	Volumetric budget			
			Steady-state	Initial condition	Maximum inflow	Minimum inflow	
				Transient			
Inflow							
Recharge from precipitation		Specified flow	70.0	70.0	70.0	70.0	70.0
Streamflow infiltration	Big Lost River [1]	Specified flow	125.3				
	Little Lost River		3.0				
	Birch Creek		20.0				
	Streamflow infiltration subtotal		148.3	148.3	143.4	692.1	23.0
Irrigation infiltration		Specified flow	21.6				
Industrial water use returns			5.9				
	Irrigation and industrial subtotal		27.5	27.5	27.5	71.1	4.3
Northwest mountain-front boundary underflow							
	Big Lost River	Specified flow	361.0				
	Little Lost River		223.0				
	Birch Creek		62.0				
Northeast regional-underflow boundary			1,225.0				
Southeast-flowline boundary		No flow	–	–	–	–	–
	Underflow subtotal		1,871.0	1,871.0	1,871.0	1,871.0	1,871.0
Flow across the base of aquifer		No flow	–	–	–	–	–
Decrease in storage			–	–	–	17.2	95.6
	Total inflow		2,116.7	2,116.7	2,111.9	2,721.4	2,064.9
Outflow							
Southwest regional-underflow boundary		Head dependent flow	NA	2,072.0	2,067.2	2,068.6	2,056.0
	Irrigation well discharge	Specified flow	37.2				2,056.7
	Industrial well discharge		7.6				
	Irrigation and industrial subtotal		44.8	44.8	44.8	121.2	5.4
Increase in storage			–	–	–	531.4	2.9
	Total outflow		NA	2,116.8	2,112.0	2,721.2	2,065.0
Budget residual				-0.1	-0.2	0.0	0.1
Volumetric budget difference, percent				.003	.008	.006	.003

[1] Mean annual discharge from 1966 to 1980 for steady-state, and total discharge minus evaporation for all stress periods in transient models

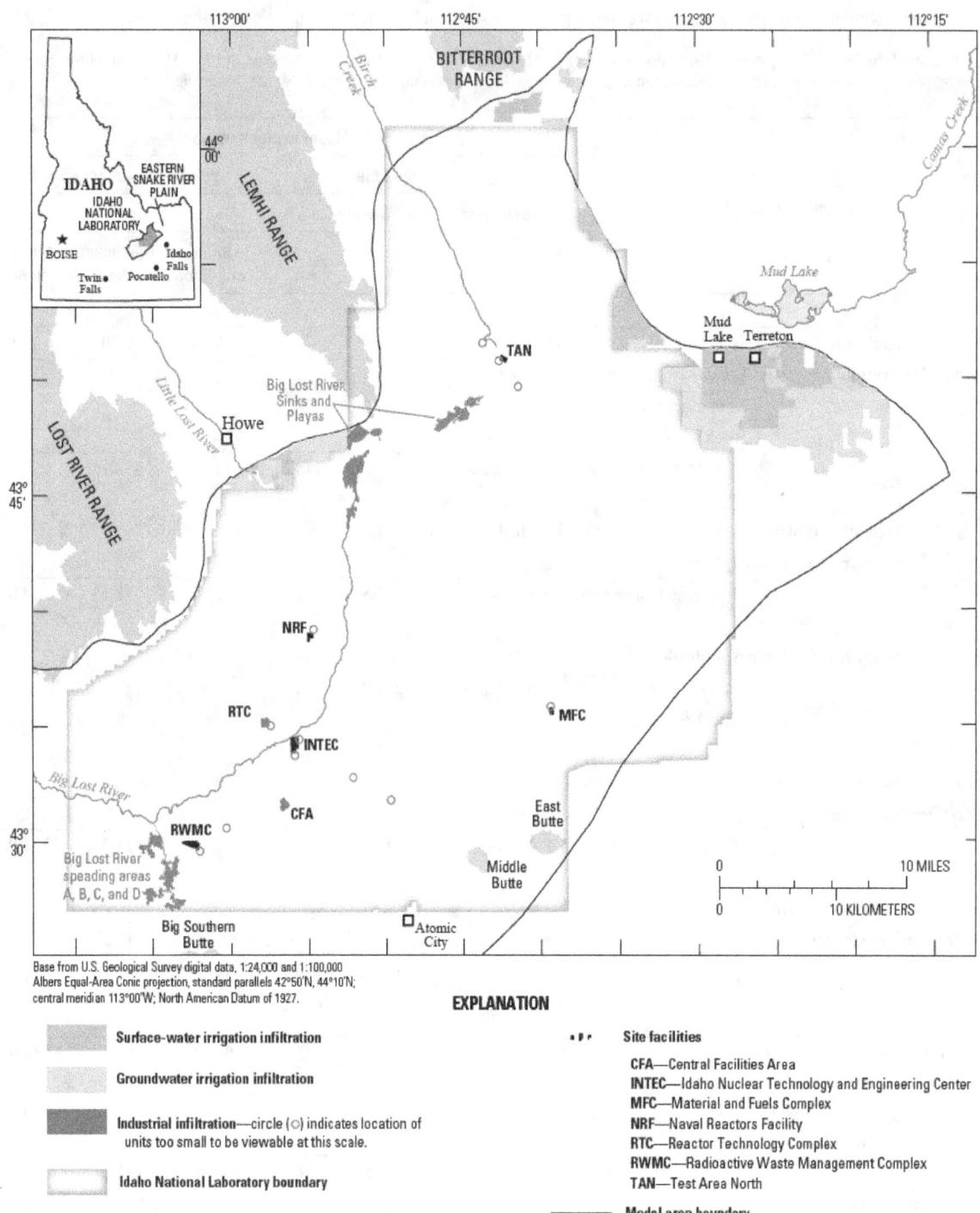

Base from U.S. Geological Survey digital data, 1:24,000 and 1:100,000
Albers Equal-Area Conic projection, standard parallels 42°50'N, 44°10'N;
central meridian 113°00'W; North American Datum of 1927.

EXPLANATION

Surface-water irrigation infiltration

Groundwater irrigation infiltration

Industrial infiltration—circle (○) indicates location of units too small to be viewable at this scale.

Idaho National Laboratory boundary

• ▸ ▸   Site facilities

CFA—Central Facilities Area
INTEC—Idaho Nuclear Technology and Engineering Center
MFC—Material and Fuels Complex
NRF—Naval Reactors Facility
RTC—Reactor Technology Complex
RWMC—Radioactive Waste Management Complex
TAN—Test Area North

——— Model area boundary

Figure 18. Distribution of areas where returned water from irrigation and industrial use infiltrates as used in the calibration of groundwater flow models, Idaho National Laboratory and vicinity, Idaho.

Northwest Mountain-Front Boundary

Inflow to the study area across the northwest mountain-front boundary represents underflow from the alluvial-fill aquifers of the mountain valleys of Birch Creek, Little Lost River, and Big Lost River and underflow from the carbonate and volcanic rocks forming the mountain fronts of the Bitterroot Range, Lemhi Range, Lost River Range, and Pioneer Mountains (fig. 9). Inflow from the mountain fronts was specified as zero. Inflows from the stream valleys were simulated as specified flow at boundary cells corresponding to the locations of the Birch Creek, Little Lost River, and Big Lost River valleys at the model boundary using the Flow and Head Boundary (FHB) Package of MODFLOW-2000 (Leake and Lilly, 1997) (fig. 14). In the transient model, these flows were simulated as constant with time. Underflow from the tributary valleys (table 5) was calculated as basin yield minus simulated streamflow infiltration and distributed across the model boundary surface based on estimates of the widths of the tributary valley aquifers, the thickness and hydraulic conductivity of the ESRP aquifer locally, and knowledge of local conditions. Due to the uncertainty of the vertical location of underflow from the Birch Creek valley, flow entering the model from this valley was distributed uniformly across boundary cells for the entire thickness of the ESRP aquifer, which, at this model boundary, consists of the top four model layers. Inflow entering the model area from the Little Lost River valley was distributed uniformly across boundary cells in the top three model layers. Inflow into the model area was restricted to the top three model layers because these layers constitute the highly conductive hydrogeologic unit 1. Hydrogeologic unit 2, a lower conductivity unit, is in model layer 4 at this location, whereas layers 5 and 6 are not present. Inflow from the Big Lost River valley was distributed only to model layer 1, a simplified representation of the movement of underflow downward in a complex layering of water bearing zones at the mouth of the valley.

Northeast Boundary

Inflow to the model across the northeast boundary represents underflow from the ESRP aquifer upgradient of the boundary (fig. 14) and was simulated as specified flow using the FHB Package. Flow across the boundary was distributed to model layers 1 through 5 (layer 6 is absent along this boundary) in proportion to hydraulic conductivity distribution estimates across the boundary. The presence of highly transmissive hydrogeologic unit 1 along a 20 mi stretch of the southeastern part of the boundary (fig. 13), for example, resulted in the allocation of a large proportion (64 percent) of the flux to this area.

Well Injection and Withdrawal

Industrial water injection through disposal wells and withdrawals from production wells at INL facilities, and withdrawals from irrigation wells were simulated using the Well Package in model layers 1 through 5 at model cells corresponding to well locations and times (tables E1 and F1). Withdrawal and injection rates for industrial wells varied seasonally and were based on the well construction, hydraulic conductivity, and the history of the well. Irrigation water withdrawals in the northern and eastern parts of the model area (fig. 18) were simulated as constant from year to year and were applied during the irrigation season, May to August of each year. Withdrawal rates for irrigation wells were uniform and applied at locations and depths equivalent to those in Spinazola (1994, fig. 29).

Southwest Boundary

Outflow across the southwest boundary was not specified, but was simulated as head-dependent flow (fig. 14) using the Drain Package. This configuration was used in part for reasons of parsimony—to avoid over specifying boundary conditions, and in part because this configuration provides head values at one model boundary, which allows a unique solution for head to be calculated. Outflow occurs across this boundary in all six model layers and is a function of the conductance along, and the head difference across, the boundary. The boundary condition used for the southwest boundary was conceptualized as a downgradient part of the aquifer with identical dimensions as the cells immediately upgradient of this boundary. The conductance for the boundary cells was calculated using the estimated value of hydraulic conductivity for hydrogeologic unit 3, the layer thickness, and cell boundary length. The estimated value for hydraulic conductivity was determined through sensitivity testing of the estimated range of hydraulic conductivity values (850–1,150 ft/d) in this part of the aquifer (Ackerman and others, 2006, p. 35). Results indicate that model budget (mass balance) errors were smallest (less than 0.01) and parameter estimates more stable and efficient (closure with parameter-estimation closure criteria less than 0.015) for a hydraulic conductivity estimate corresponding to the largest (1,150 ft/d) estimated value. The head difference was the calculated decrease in head between individual boundary cells and an assumed head distribution just downgradient of the boundary. The assumed head values were constant, uniform with depth, and equal to the interpolated 1980 altitude of the water table (Lindholm and others, 1988). No flow was simulated across the northwestern most 3.75 mi in layers 1–4 (fig. 14) and 6.25 mi in layers 5 and 6 because the estimated flow direction in this area was approximately parallel to the model boundary (Lindholm and others, 1988).

Model Calibration

Model calibration is an iterative process of adjusting the 3-D distribution or structure of aquifer properties, aquifer property values, or properties of boundary conditions to improve the match between simulation results and observations. The calibration of the numerical models in this study involved (1) selecting and evaluating observed water-levels in wells (hydraulic heads) to be used as *observations* for calibrating the models; (2) defining discrete areas of the model grid, referred to as *zones*, where aquifer properties were assumed to be uniform (a form of parameterization); and (3) adjusting the distributions of zones and the values of aquifer properties in zones, referred to as *parameters*, to provide the best match between observed and simulated hydraulic head.

Model parameters include aquifer properties such as horizontal and vertical hydraulic conductivities, specific yields, and specific storage and may include boundary condition properties such as recharge, conductance, and heads. Observations applicable for calibrating this model include field-measured values of hydraulic heads distributed over space and time.

The models in this study were calibrated by adjusting aquifer properties, either manually by "trial-and-error" or by using an inverse method (*parameter estimation*) to match observed heads or head differences. The trial-and-error method was used initially to calibrate the simplest versions of a steady-state model, primarily to test the overall reliability of the conceptual model and to select important design elements to include in subsequent models.

The Observation, Sensitivity, and Parameter-Estimation (OSP) Processes in MODFLOW-2000 were used to calibrate the more complex versions of the steady-state and transient models (Hill and others, 2000). The OSP Processes use a modified Gauss-Newton nonlinear regression method to adjust values of selected input parameters in an iterative procedure to minimize the value of the weighted least-squares objective function. The objective function used is the sum of squared weighted residuals as described in the section Statistical Methods Used for Model Evaluation. The parameter estimation process used for this report produces statistics that were used to evaluate estimated parameter values and model-calculated values of head or head differences.

Calibration Observations

Steady-state calibration observations were hydraulic heads (water levels) in wells, and transient calibration observations were heads and head differences (hereafter referred to as head observations, head-difference observations, or collectively as observations). Nearly all wells in the model

area (207 well completion depths at 201 well locations) with available water-level data were used in model calibration. Some locations were represented by multiple observations that were attributed to a multi-depth piezometer nest or to modification of open intervals due to collapse, reconstruction, or deepening. Wells that were not used in model calibration (1) were completed below the aquifer, (2) were in close proximity to other wells with more complete records, or (3) had completion or record problems.

Observations for Steady-State Model Calibration

The steady-state model was calibrated using 201 head observations (table G1) from 201 well completion depths at 199 well locations (table A1; appendix B). The last measurement in 1980 was used for hydraulic heads at 108 well locations (110 well completion depths). For 91 other wells, most of which were completed after 1980, head measurements closest to 1980 levels or from 1992 to 1993, a period when water levels were close to levels in 1980, were used as the steady-state observations, and were considered reasonably representative of conditions in 1980.

Observations for Transient Model Calibration

The transient model was calibrated using 328 head observations and 8,171 head-difference observations (table H1) from 206 well completion depths at 200 well locations (table A1; appendix B). All available water-level data were used in the transient calibration except for continuous water-level measurement records. Continuous records were resampled at one observation per month by selecting the measurement closest to the middle of the month. Transient calibration data included (1) 206 initial head observations (110 from 1980, 3 prior to 1980, and 93 after 1980), (2) 8,171 head-difference observations for 151 well completion depths at 146 well locations, and (3) 122 head observations from 6 well locations with few observations or near pumping wells for the period 1980–95. Head-difference observations were calculated internally in the model (Hill and others, 2000, p. 33) as the initial head observation minus the later head observation.

Observation Weights

Not all water-level data were equally suitable for model calibration because some wells (1) were open to multiple layers and hydrogeologic units (table 6), (2) had multiple open intervals, (3) were close to seasonal sources of inflow or pumpage, (4) were in areas of steep local hydraulic gradients, and (5) had different well-datum accuracies. Datum inaccuracies of as much as ±10 ft, caused by using topographic maps with a contour interval of 20 ft, contributed

Table 6. Summary of open intervals for well completions by hydrogeologic unit and model layer for observation wells used in model calibration, Idaho National Laboratory and vicinity, Idaho.

[Zones correspond to the parameter structure for the Big Lost Trough models of steady-state and transient groundwater flow; 207 well completions at 201 well locations]

Hydro-geologic zones	Model layers	Open to one model layer					Open to two model layers					Open to three or more model layers						Subtotal
		1	2	3	4	6	1,2	1,5	1,6	2,3	3,4	1,2,3	1,2,4	2,3,4	3,4,5	1,2,3,4	1,2,3,4,5,6	
							Wells open to one hydrogeologic unit											
1	....	27	---	---	---	---	8	---	---	---	---	2	---	---	---	---	---	37
11	....	36	3	2	---	---	7	---	---	---	---	---	---	---	---	---	---	48
2	....	2	---	2	---	---	---	---	---	---	---	1	1	---	---	---	---	6
22	....	5	---	2	1	---	1	---	---	---	---	7	1	---	---	---	---	17
3	....	7	---	---	---	---	---	---	---	1	---	---	---	---	---	---	---	8
33	....	---	---	---	---	---	---	---	---	---	---	---	---	---	---	---	---	0
4	....	6	---	1	---	---	2	---	---	---	1	5	---	---	---	---	---	15
44	....	22	6	1	1	1	6	1	---	---	---	---	---	1	---	1	---	40
6	....	---	---	---	---	---	---	---	1	---	---	---	---	---	---	---	---	1
Subtotal		105	9	8	2	1	24	1	1	1	1	15	2	1	0	1	0	
							Wells open to multiple hydrogeologic units											
1, 2	....	---	---	---	---	---	5	---	---	---	---	2	---	---	---	---	---	7
11, 22	....	---	---	---	---	---	15	---	---	---	4	5	---	1	---	1	---	26
1, 2, 3	....	---	---	---	---	---	---	---	---	---	---	---	---	---	1	---	1	2
Subtotal				0					24					11				
Total				125					52					30				

the largest source of error for head observations in the steady-state model. To account for these errors and other conditions, all observation data were weighted in proportion to the reciprocal of the variance of their estimated measurement error. Higher-weighted observations have greater influence in the nonlinear regression procedure used to calibrate the model (Hill, 1998, p. 4, 13-14, 45).

For example, using the method given by Hill (1998, p. 46-47), and the assumption that the 90-percent confidence interval for the datum (altitudes determined from topographic maps or surveys) is one-half the contour interval. The variance of the measurement error due to uncertainty in the datum derived from topographic maps with a 20 ft contour interval is 37 ft^2. The variance due to datum errors is added to the variances for other components of measurement error, and the resultant weight is the inverse of the total variance. In this study, application of Hill's method produced a weight of 0.027 ft^{-2} for head observations in the steady-state model at wells with altitudes determined from topographic maps with a 20 ft contour interval, the smallest scaled weight for observations

in this model (table 7). For the steady-state model, one of five observation weights between 0.027 and 2.70 ft^{-2} was applied to each head observation (table 7; table G1). Head observations for years other than 1980 were assigned lower weights.

For the transient model, using head differences as observation data eliminated most of the datum errors from these observations. Other error components, such as equipment accuracy, barometric corrections, and measurement reproducibility dominated the variances for the two measurements used for a head difference. The total variance for the head-difference measurement is the sum of the variance of the two measurements. The observation weights used in the steady-state models were applied to head observations and one weight (60.6 ft^{-2}) was applied to head-difference observations in the transient model (table 7).

Because all head and head-difference observations used in calibrating the models were weighted, all weighted values such as observed heads, simulated heads, and differences between observed and simulated values are dimensionless.

Table 7.   Weights for steady-state and transient model observations, Idaho National Laboratory and vicinity, Idaho.

[**Weighting statistic:** Variance of the measurement error **Abbreviations**: ft, foot; ft², square foot; <, less than; >, greater than]

Weighting level	Features commonly affecting measurement error for hydraulic head	Hydraulic head		Head difference	
		Weighting statistic (ft²)	Weight (1/ft²) Steady-state and transient	Weighting statistic (ft²)	Weight (1/ft²) Transient
1	High-quality datum survey, well completed in one layer; annual water-level fluctuation < 1.5 ft.	0.370	2.70	0.0165	60.6
2	Well completed in two layers; annual water-level fluctuation 1.5–2.5 ft.	1.500	.667		
3	Well completed in three layers; annual water-level fluctuation > 2.5 ft.	3.300	.303		
4	Well completed in four or more layers, well deviation, estimated steady-state water-level for year other than 1980.	9.200	.109		
5	Well datum from topographic map, steep local hydraulic gradient on water-table map.	37.000	.0270		

Statistical Methods Used for Model Evaluation

Several statistical measures and related graphs (Hill, 1994; Hill, 1998; Hill and others, 2000; Hill and Tiedeman, 2007) were used to evaluate the calibrated models and the estimated parameter values. Better models have three attributes: (1) estimated parameter values that are more realistic, (2) better fit of simulated values to observations, and (3) more randomly distributed weighted residuals. Descriptions of the statistical measures and graphical methods are organized in two general groups: (1) diagnostic statistics that quantify the quality of a calibrated model or are useful for selecting between alternative models and (2) inferential statistics that quantify the reliability of parameter estimates. A detailed discussion of these statistical measures and related graphs is available in Hill and Tiedeman (2007).

Evaluation of Model Fit

Evaluation of the model fit, which refers to how well observed and simulated values are matched by nonlinear regression, is one measure of model performance. Model fit was evaluated in the steady-state and transient models by considering the magnitude and distribution (statistically, spatially, and temporally) of the weighted residuals; weighted residuals are the weighted difference between observed and simulated values and represent the fit of the model in the context of the expected accuracy of the observations. Residuals for observations expected to be less accurate are de-emphasized relative to those that are more accurate when weights are considered. (Hill and Tiedeman, 2007, p. 35). Statistical measures and graphs used to evaluate model fit were the (1) sum of squared weighted residuals, (2) average weighted residual, (3) standard error of the regression, (4) distributions of weighted residuals, and (5) normal probability graph and the related correlation coefficient R^2_N.

Sum of Squared Weighted Residuals

The sum of squared weighted residuals is the sum of every squared weighted residual. The relative magnitude of the sum of squared weighted residuals among different models or different calibrations indicates whether a particular set of parameters provides a more or less precise model fit to all the observations. Smaller values indicate a more precise fit, but smaller values also can result when a larger set of parameters is used in the evaluation (Hill, and Tiedeman, 2007, p. 95). The sum of squared weighted residuals was the weighted least-squares objective function that was minimized by the nonlinear regression method used to estimate model parameters.

Average Weighted Residual

The average weighted residual is the simple arithmetic average of the weighted residuals (Hill, 1998, p. 21). An average weighted residual near zero may indicate that the associated set of parameters provides a precise model fit; an average weighted residual that is farther from zero indicates that the associated set of parameters provides a less precise model fit.

Standard Error of the Regression

The standard error of the regression is the square root of the error variance, where the error variance is the sum of squared weighted residuals divided by the number of observations minus the number of estimated parameters. When weights are defined as the inverse of the variance of measurement error, the expected value of the standard error of the regression is 1.0. Successively higher values indicate diminished model fit. Due to model error and unaccounted errors in observations, values typically are greater than 1.0 (Hill, 1998, p. 18-19). The standard error of the regression is not a very intuitive quantification of model fit because it is dimensionless. A more intuitive measure is the fitted standard deviation, defined as the product of the standard error of regression and the standard deviation of measurement error (square root of the variance of measurement error used to define the weights). The fitted standard deviation expresses the average fit of one group of observations in the corresponding units of measurement. This approach (Hill and Tiedeman, 2007, p. 95-96) generally applies only if the fitted standard deviation summarizes the fit to a fairly large number of observations, as is the case for the transient model.

Distribution of Weighted Residuals

Model fit also was evaluated using graphs of the weighted residuals with respect to weighted simulated values and independent variables such as space and time. A valid regression requires that the observation errors be random, have a mean of zero, and that the weighted errors be uncorrelated. Although the actual observation errors are unknown, these error requirements are inferred to be met if the weighted residuals are random, independent, and either are normally distributed or have predictable correlations (Hill, 1998, p. 23).

Graphs of the weighted residuals plotted against weighted simulated values are used to evaluate whether the weighted residuals are independent and normally distributed (Hill, 1994, p. 3-4; Hill, 1998, p.20; Hill and Tiedeman, 2007, p.100-103). Weighted residuals are independent when small and large weighted residuals are not preferentially associated with, respectively, small and large weighted simulated values and when the range of weighted residual values does not increase or decrease with increasing weighted simulated values. Weighted residuals are normally distributed when the uniform distribution of residuals is about zero throughout the entire range of weighted simulated values.

Model fit was evaluated spatially by mapping weighted residuals and inspecting the randomness of the values. For transient models, weighted residuals also are plotted on time-series graphs and maps to evaluate temporal distribution variability. Random distribution of weighted residuals is indicated when the signs and magnitudes of the weighted residuals show no discernable patterns (no clusters of positive or negative residuals or of large absolute values) and seem random.

Normal Probability Graphs and Related Correlation Coefficient

Normal probability graphs and the related correlation coefficient, R^2_N, are used to further evaluate if weighted residuals are independent and normally distributed (Hill, 1998, p. 23). Normal probability graphs are ordered weighted residuals, from smallest to largest, plotted against the standard normal statistic, which is the cumulative probability expected for each value assuming the values are independent and normally distributed. This plot should approximate a straight line—the correlation coefficient, R^2_N, should be close to 1.0—when the weighted residuals are independent and normally distributed (Hill, 1994, p. 5, 19). When R^2_N is significantly less than 1.0, the weighted residuals likely are not independent and not normally distributed (Hill, 1998, p. 23-24, appendix D).

Evaluation of Parameter Estimates

Parameter sensitivity, uniqueness, and uncertainty were measured with composite scaled sensitivities, parameter correlation coefficients, and linear confidence intervals, respectively. To evaluate the validity of linear confidence intervals, the modified Beale's measure was used to test the linearity of the steady-state and transient models. The effect of individual observations on parameter estimates was evaluated with Cook's D and DFBETAS statistics. In addition to the use of statistical measures, parameter values and their confidence intervals were compared to acceptable or expected values gleaned from field and laboratory measurements and estimates, and from published values for similar aquifer materials.

Composite Scaled Sensitivities

Composite scaled sensitivities (CSS) indicate the total amount of information for defining parameters and estimating parameter values provided by all the observations. The CSS are calculated using sensitivities of the simulated equivalent of each observation with respect to one parameter value (Hill, 1998, p. 14). A single sensitivity value indicates the amount that the simulated equivalent would change when the parameter value is changed. The CSS for a parameter that is more than 2 orders of magnitude smaller than the largest CSS, or that is less than a suggested critical value of 1.0, indicates that the parameter is likely to be poorly estimated and will have large confidence intervals (Hill and Tiedeman, 2007, p. 50-51). Larger values of CSS indicate that a parameter likely can be estimated because the information provided by observations dominates the effects of observation error. Illustrations in this report show the normalized CSS, which is the value divided by the maximum CSS for the model.

Parameter Correlation Coefficients

Parameter correlation coefficients indicate whether parameter values can be uniquely estimated by regression. Parameter correlation coefficients are calculated for a pair of parameters as the covariance between the parameters divided by the product of the variance of each parameter. Parameter correlation coefficients range from -1.00 to 1.00, and absolute values less than 0.95 indicate that the parameter pair probably can be estimated uniquely (Hill and Tiedeman, 2007, p. 51-54).

Linear Confidence Intervals

Linear confidence intervals are used to quantify parameter uncertainty. Linear confidence intervals require that the observation errors be independent and normally distributed, which is evaluated by analyzing the distribution of weighted residuals. The width of a confidence interval is a measure of the likely precision of the estimate for a parameter. Greater precision produces narrower confidence intervals (Hill and Tiedeman, 2007, p. 138). The accuracy of linear confidence intervals is compromised if the model is significantly nonlinear, thus a measure of the linearity of the model is necessary to assess the validity of the linear confidence intervals.

Modified Beale's Measure

The modified Beale's measure (N_b) is used to assess model nonlinearity. Calculation of N_b requires generating sets of parameter values near the limits of their confidence regions and using the values to generate simulated equivalents of the calibration observations and linearized estimates of the simulated equivalents. N_b is a measure of the difference between the simulated equivalents of the calibration observations and the linearized estimates of the simulated equivalents. The degree of model linearity is assessed by comparing the calculated N_b with two critical values that are a function of a value from the F probability distribution (F). The model is effectively linear if N_b is less than 0.09/F and nonlinear if N_b is greater than 1.0/F. Values for N_b between these two critical values indicate that the model is moderately nonlinear (Hill and Tiedeman, 2007, p. 142-145).

Cook's D and DFBETAS

Cook's D is a measure of the influence of one observation on a set of parameter estimates (Hill and Tiedeman, 2007, p. 134-136). Weighted residuals and a leverage statistic are used to calculate Cook's D. Cook's D values that are larger than a critical value of 4 divided by the number of observations indicate that the observation has a large influence on the estimation of the set of parameters, relative to other observations with Cook's D values less than the critical value. The likelihood of any observation being influential to a parameter generally decreases as the number of observations increases. The DFBETAS statistic measures the influence of one observation on one parameter. DFBETAS values larger than a critical value of 2 divided by the square root of the number of observations indicate that an observation may be influential in estimating the parameter. The likelihood of any observation being influential to a parameter generally decreases as the number of observations increases (Hill and Tiedeman, 2007, p. 136).

Steady-State Calibration

Steady-state calibration involved (1) selecting a zonation scheme for the model to accommodate the spatial distribution of hydrogeologic units and variations of sediment content within hydrogeologic units; (2) parameterization of hydraulic properties for three different distributions of sediment content within hydrogeologic units; (3) applying parameter estimation techniques to the models representing the three conceptualizations of sediment content (parameter structures); and (4) using statistical measures to select a parameterization with the best model fit and to evaluate model sensitivity to changes in system geometry, boundary conditions, and parameters.

The aquifer in the model consists of hydrogeologic units 1, 2, and 3, which represent groupings of fractured basalts and interbedded sediment, and rhyolite domes that constitute a small remainder of the aquifer volume. Each of the hydrogeologic units are represented as homogeneous, anisotropic porous media. In this representation of the aquifer, the small-scale heterogeneities and anisotropies are not preserved; the hydraulic conductivity of each hydrogeologic unit reflects the aggregate lithology, thickness, and number of basalt flows and sedimentary interbeds in each hydrogeologic unit. Three representations of the hydrogeologic units (parameter structures) within the Big Lost Trough were evaluated. These representations are referred to as the (1) no-sediment model, in which no additional characteristics of the hydrogeologic units are considered other than location in space; (2) Whitehead sediment model (Ackerman and others, 2006, fig. 12*A*), in which the hydraulic properties are estimated within hydrogeologic units according to the relative abundance of sediment as interpreted in the RASA study (Whitehead, 1992); and (3) Big Lost Trough model (Ackerman and others, 2006, fig. 12*B*), in which the hydraulic properties are estimated within hydrogeologic units according to the presence of abundant sediment.

Initial steady-state calibration used trial-and-error methods to determine if a numerical model could reasonably represent the conceptual model and be consistent with estimates of flow across boundaries. Models with increasing complexity were tested starting from a two-dimensional confined model based on a simple distribution of horizontal hydraulic conductivity and progressed to a three-dimensional, three-layer confined model with different distributions of sediment content modifying the hydraulic properties within three hydrogeologic units, and with observations only in layer 1. After examination of the trial-and-error modeling results, the following improvements to the model were determined: (1) represent the southwest boundary as head-dependent flow to improve mass balance; (2) model the influence of sediment content on hydraulic conductivity to improve model quality in the north-central part of the model area (fig. 7); (3) configure head observations to represent the layers represented by the well open intervals so multi-layer flow could be more realistically represented and observations could influence aquifer parameters in lower layers; and (4) subdivide the representation of hydrogeologic unit 3 into north and south zones to improve simulated heads and gradients in the north part of the study area. Trial-and-error experimentation with these simple steady-state models provided confidence that the numerical model reasonably represented the conceptual model.

The steady-state model configurations described in this report are three-dimensional, six-layer unconfined models representing 1980 conditions with average annual values specified for all inflow and all outflow except for flow out of the model area to the southwest. The three different parameter structures—no sediment, Whitehead, and Big Lost Trough— corresponded to separately calibrated steady-state models with individual parameterizations based on hydrogeologic units subdivided into zones according to sediment content.

The implementation of the MODFLOW-2000 groundwater flow model, as discussed throughout this report, used the Layer-Property Flow Package (Harbaugh and others, 2000, p. 22). All calibrated models assumed unconfined conditions, that is, the thickness of the upper layer varied with the altitude of the water table, which ranged between about 4,100 and 4,600 ft. The units of length and time used for model input were feet and seconds. For consistency in presentation with the conceptual model report (Ackerman and others, 2006), the values of hydraulic conductivity are reported in units of feet per day in this report. Values of all parameters were log transformed in the regression procedure used for model calibration. For the final calibration, the criteria for maximum absolute change in head and residual in forward model runs (Hill 1990, p. 14) were set to 0.001 ft and 0.0001 ft³/s for the steady-state model calibration. These criteria were set to 0.001 ft and 0.001 ft³/s for the transient model calibration. The budget error in forward model runs was less than 0.001 percent. For evaluation of alternate parameterizations and for sensitivity analyses used to evaluate budget components and modifications to the geologic

framework, the criteria for maximum absolute change in head and residual (Hill, 1990, p.14) were set to 0.01 ft and 0.001 ft³/s, respectively, unless otherwise noted. In these simulations, the budget error was less than 0.01 percent.

Hydrogeologic Zones

Hydrogeologic zones are groups of model cells with uniform hydraulic properties that compose part or all of a hydrogeologic unit or rhyolite domes (fig. 5). Initial trial-and-error model runs, and examination of CSS for each of the three steady-state parameter structures, indicated that observations of head are sufficient to separate hydrogeologic unit 3 into northern and southern sections. The northern section composes shallower areas of hydrogeologic unit 3 that constitute the entire saturated thickness of the aquifer, and the southern section composes deeper parts of hydrogeologic unit 3. In the northwest part of the model area, hydrogeologic unit 3 was designated unit 3N and in the rest of the model area, hydrogeologic unit 3 was designated as unit 3S. In each parameter structure hydrogeologic units 1, 2, 3S, 3N, and rhyolite domes were represented by zone numbers 1, 2, 3, 4, and 6 respectively. Zone numbers with multiple digits (for example, 11, 333) were used in the Whitehead and Big Lost Trough parameter structures to represent zones containing different amounts of sediments.

The no-sediment parameter structure was conceptualized as sediment having little influence on the hydraulic properties of the aquifer. Five hydrogeologic zones represented the hydrogeologic units and rhyolite domes (table 8). In the Whitehead and Big Lost Trough parameter structures sediment was conceptualized as influencing the hydraulic properties of the aquifer. This conceptualization was summarized by Ackerman and others (2006, p. 22) as:

> …in areas where large amounts of sediment were deposited, such as in the Big Lost Trough, hydraulic conductivity of basalt interflow zones probably is greatly reduced because sediment fills cracks, joints, fissures, and fractures, reducing the original porosity of the basalt and impeding groundwater flow.

In both these parameter structures, hydrogeologic zones represented parts of hydrogeologic units and rhyolite domes based on the thickness of sediment in the stratigraphic section. In the Whitehead parameter structure, 12 zones represented areas with estimated sediment thicknesses of less than 100 ft (zones 1, 2, 3, and 6), 100 to 499 ft (zones 11, 22, 33, and 44), and more than 500 to 999 ft (zones 111, 222, 333, and 444) (table 8). In the Big Lost Trough parameter structure nine zones represented areas with sediment thicknesses of less (1, 2, 3, 4, and 6) or more (11, 22, 33, and 44) than 11 percent of aquifer thickness (fig. 19; table 8).

Table 8.   Hydraulic property parameters and corresponding zones used in parameter structures for steady-state and transient models, Idaho National Laboratory and vicinity, Idaho.

[Parameter values corresponding to zones in **bold** type were optimized, estimated value optimized during model calibration Parameter values corresponding to zones in *italic* indicate parameter is during model calibration **Abbreviations**: HC, hydraulic conductivity; SY, specific yield; SS, specific storage; VANI, ratio of horizontal to vertical hydraulic conductivity; –, not used in parameter structure]

Hydraulic property parameter	Parameter structure				Hydraulic property parameter	Parameter structure			
	Steady-state models			Transient model		Steady-state models			Transient model
	No sediment	Whitehead sediment	Big Lost Trough sediment	Big Lost Trough sediment		No sediment	Whitehead sediment	Big Lost Trough sediment	Big Lost Trough sediment
Hydrogeologic zones where the presence of sediment does not affect hydraulic property estimates					Hydrogeologic zones where the presence of sediment does affect hydraulic property estimates				
HC1	**1**	**1**	**1**	[1]	HC11	–	**11**	**11**	[1]
HC2	**2**	**2**	**2**	[1]	HC111	–	**111**	–	–
HC3	**3**	**3**	**3**	[1]	HC22	–	**22**	**22**	[1]
HC4	**4**	–	**4**	[1]	HC222	–	**222**	–	–
HC6	*6*	*6*	*6*	[1]	HC33	–	*33*	*33*	[1]
SY1	–	–	–	**1**	HC333	–	*333*	–	–
SY2	–	–	–	**2**	HC44	–	*44*	**44**	[1]
SY3	–	–	–	**3**	HC444	–	**444**	–	–
SY4	–	–	–	*4*	SY11	–	–	–	**11**
SY6	–	–	–	*6*	SY22	–	–	–	*22*
					SY44	–	–	–	**44**
					Hydrogeologic properties constant throughout the model domain				
					SS	–	–	–	Fixed
					VANI	Estimated	Fixed	Estimated	[1]

[1] Value from Big Lost Trough steady-state model

A.

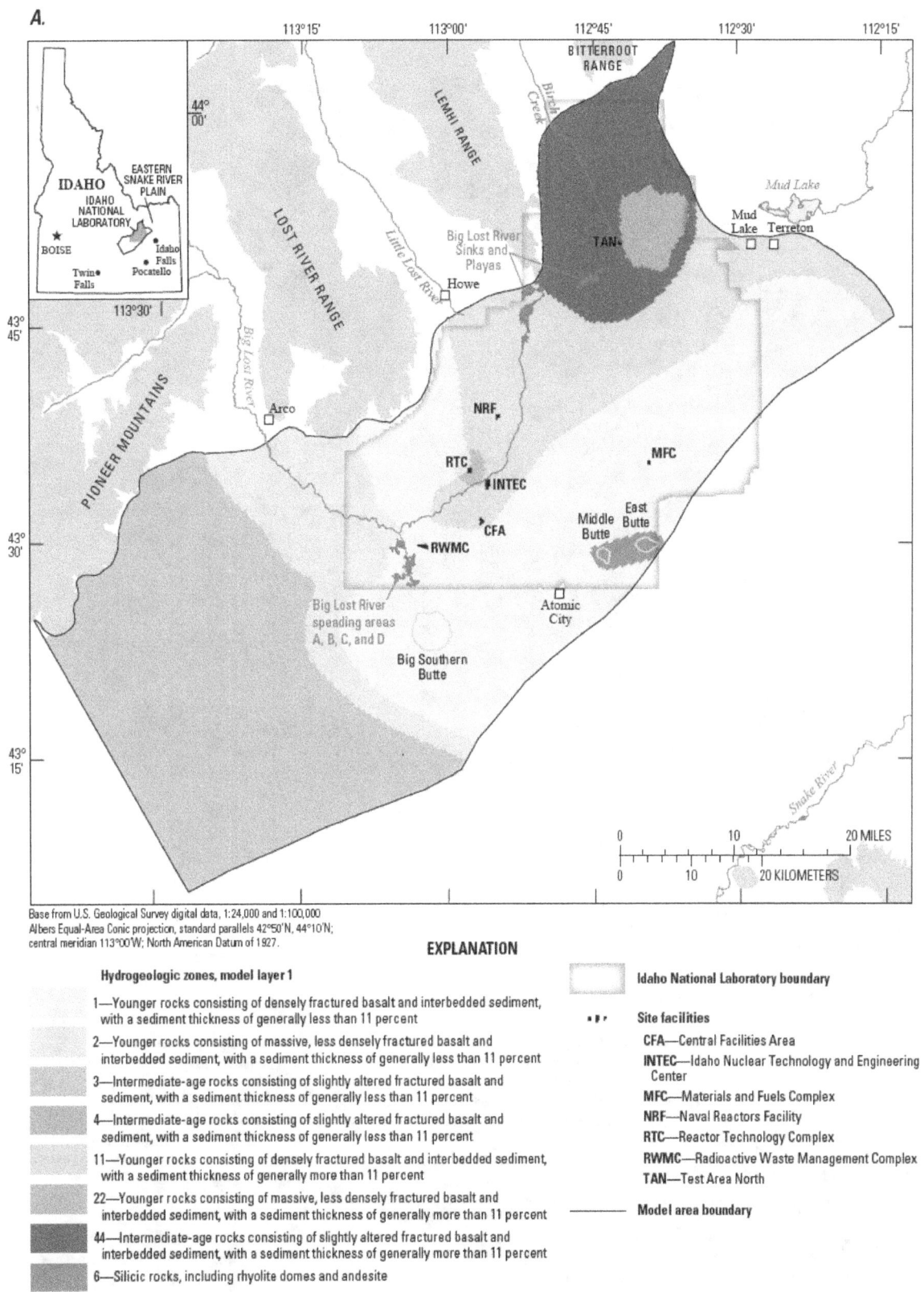

Base from U.S. Geological Survey digital data, 1:24,000 and 1:100,000
Albers Equal-Area Conic projection, standard parallels 42°50'N, 44°10'N;
central meridian 113°00'W; North American Datum of 1927.

EXPLANATION

Hydrogeologic zones, model layer 1

1—Younger rocks consisting of densely fractured basalt and interbedded sediment, with a sediment thickness of generally less than 11 percent

2—Younger rocks consisting of massive, less densely fractured basalt and interbedded sediment, with a sediment thickness of generally less than 11 percent

3—Intermediate-age rocks consisting of slightly altered fractured basalt and sediment, with a sediment thickness of generally less than 11 percent

4—Intermediate-age rocks consisting of slightly altered fractured basalt and sediment, with a sediment thickness of generally less than 11 percent

11—Younger rocks consisting of densely fractured basalt and interbedded sediment, with a sediment thickness of generally more than 11 percent

22—Younger rocks consisting of massive, less densely fractured basalt and interbedded sediment, with a sediment thickness of generally more than 11 percent

44—Intermediate-age rocks consisting of slightly altered fractured basalt and interbedded sediment, with a sediment thickness of generally more than 11 percent

6—Silicic rocks, including rhyolite domes and andesite

Idaho National Laboratory boundary

Site facilities

CFA—Central Facilities Area

INTEC—Idaho Nuclear Technology and Engineering Center

MFC—Materials and Fuels Complex

NRF—Naval Reactors Facility

RTC—Reactor Technology Complex

RWMC—Radioactive Waste Management Complex

TAN—Test Area North

—— Model area boundary

Figure 19. Distribution of hydrogeologic zones for model (*A*) layer 1, (*B*) layer 2, (*C*) layer 3, (*D*) layer 4, (*E*) layer 5, and (*F*) layer 6 for Big Lost Trough parameter structure used to simulate steady-state and transient groundwater flow at the Idaho National Laboratory and vicinity, Idaho.

B.

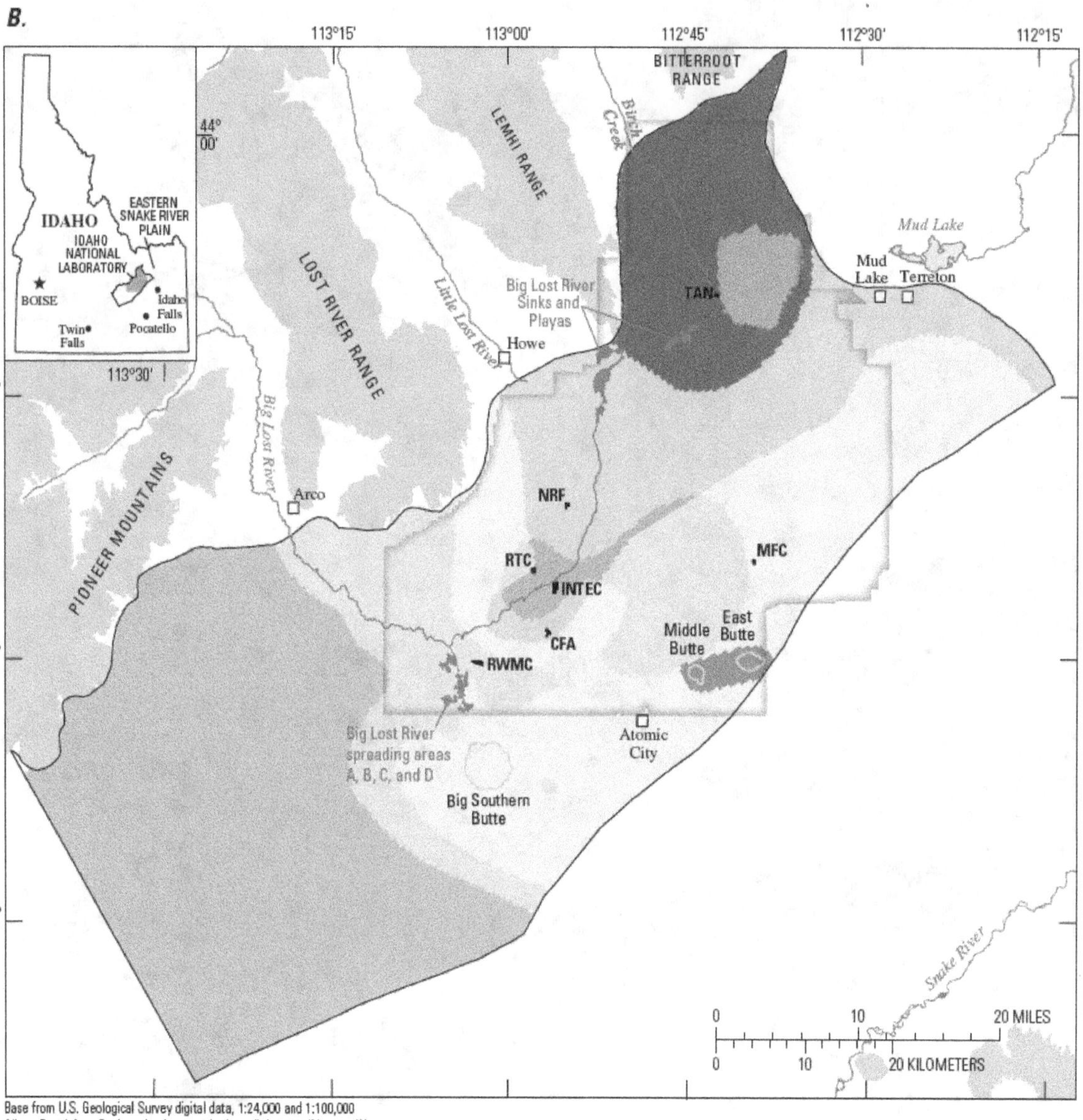

Base from U.S. Geological Survey digital data, 1:24,000 and 1:100,000;
Albers Equal-Area Conic projection, standard parallels 42°50'N, 44°10'N;
central meridian 113°00'W; North American Datum of 1927.

EXPLANATION

Hydrogeologic zones, model layer 2

1—Younger rocks consisting of densely fractured basalt and interbedded sediment, with a sediment thickness of generally less than 11 percent

2—Younger rocks consisting of massive, less densely fractured basalt and interbedded sediment, with a sediment thickness of generally less than 11 percent

3—Intermediate-age rocks consisting of slightly altered fractured basalt and sediment, with a sediment thickness of generally less than 11 percent

4—Intermediate-age rocks consisting of slightly altered fractured basalt and sediment, with a sediment thickness of generally less than 11 percent

11—Younger rocks consisting of densely fractured basalt and interbedded sediment, with a sediment thickness of generally more than 11 percent

22—Younger rocks consisting of massive, less densely fractured basalt and interbedded sediment, with a sediment thickness of generally more than 11 percent

44—Intermediate-age rocks consisting of slightly altered fractured basalt and interbedded sediment, with a sediment thickness of generally more than 11 percent

6—Silicic rocks, including rhyolite domes and andesite

Idaho National Laboratory boundary

Site facilities

CFA—Central Facilities Area

INTEC—Idaho Nuclear Technology and Engineering Center

MFC—Materials and Fuels Complex

NRF—Naval Reactors Facility

RTC—Reactor Technology Complex

RWMC—Radioactive Waste Management Complex

TAN—Test Area North

——— Model area boundary

Figure 19.—Continued.

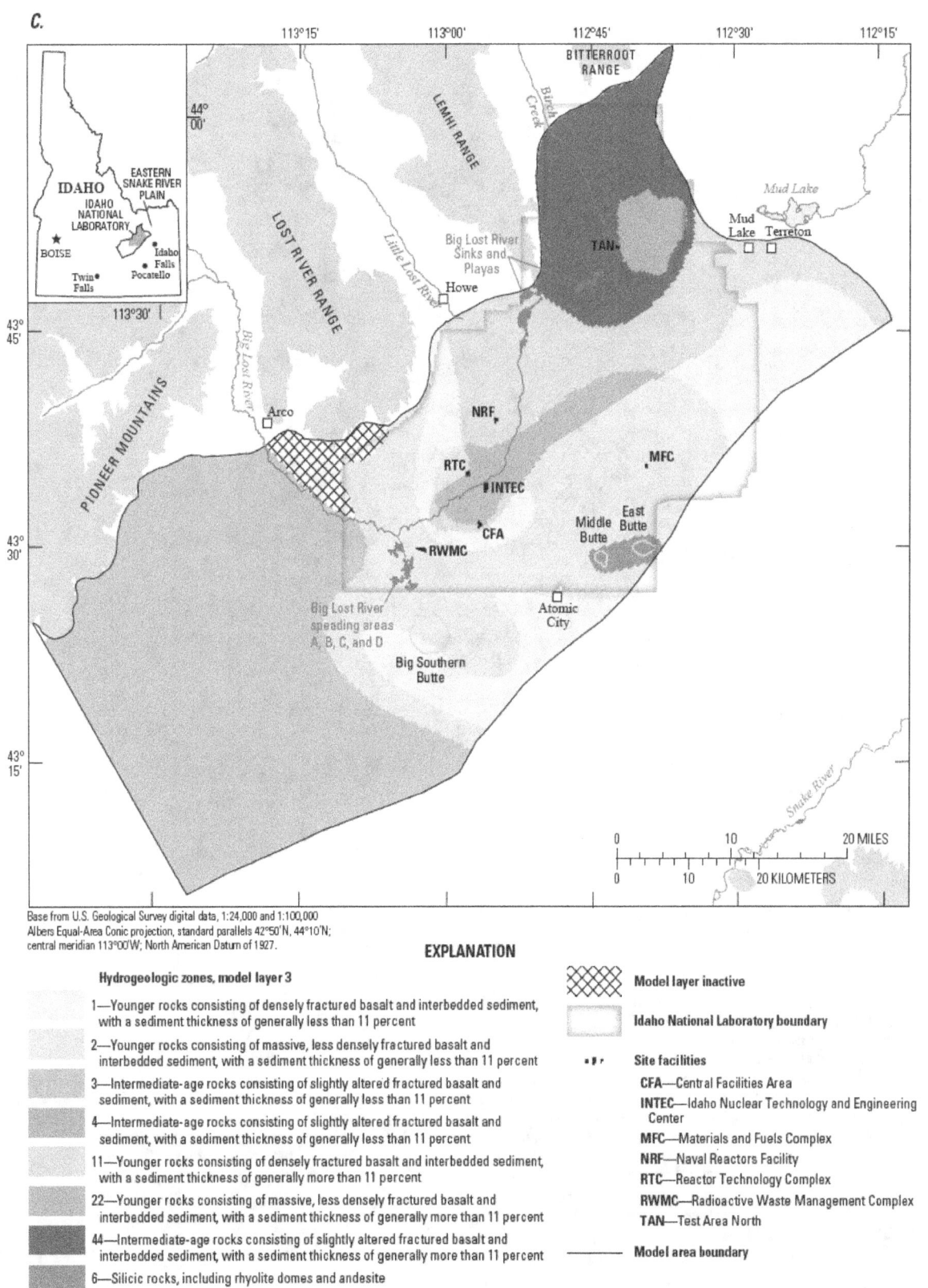

C.

Base from U.S. Geological Survey digital data, 1:24,000 and 1:100,000
Albers Equal-Area Conic projection, standard parallels 42°50'N, 44°10'N;
central meridian 113°00'W; North American Datum of 1927.

EXPLANATION

Hydrogeologic zones, model layer 3

1—Younger rocks consisting of densely fractured basalt and interbedded sediment, with a sediment thickness of generally less than 11 percent

2—Younger rocks consisting of massive, less densely fractured basalt and interbedded sediment, with a sediment thickness of generally less than 11 percent

3—Intermediate-age rocks consisting of slightly altered fractured basalt and sediment, with a sediment thickness of generally less than 11 percent

4—Intermediate-age rocks consisting of slightly altered fractured basalt and sediment, with a sediment thickness of generally less than 11 percent

11—Younger rocks consisting of densely fractured basalt and interbedded sediment, with a sediment thickness of generally more than 11 percent

22—Younger rocks consisting of massive, less densely fractured basalt and interbedded sediment, with a sediment thickness of generally more than 11 percent

44—Intermediate-age rocks consisting of slightly altered fractured basalt and interbedded sediment, with a sediment thickness of generally more than 11 percent

6—Silicic rocks, including rhyolite domes and andesite

Model layer inactive

Idaho National Laboratory boundary

Site facilities

CFA—Central Facilities Area

INTEC—Idaho Nuclear Technology and Engineering Center

MFC—Materials and Fuels Complex

NRF—Naval Reactors Facility

RTC—Reactor Technology Complex

RWMC—Radioactive Waste Management Complex

TAN—Test Area North

Model area boundary

Figure 19.—Continued.

D.

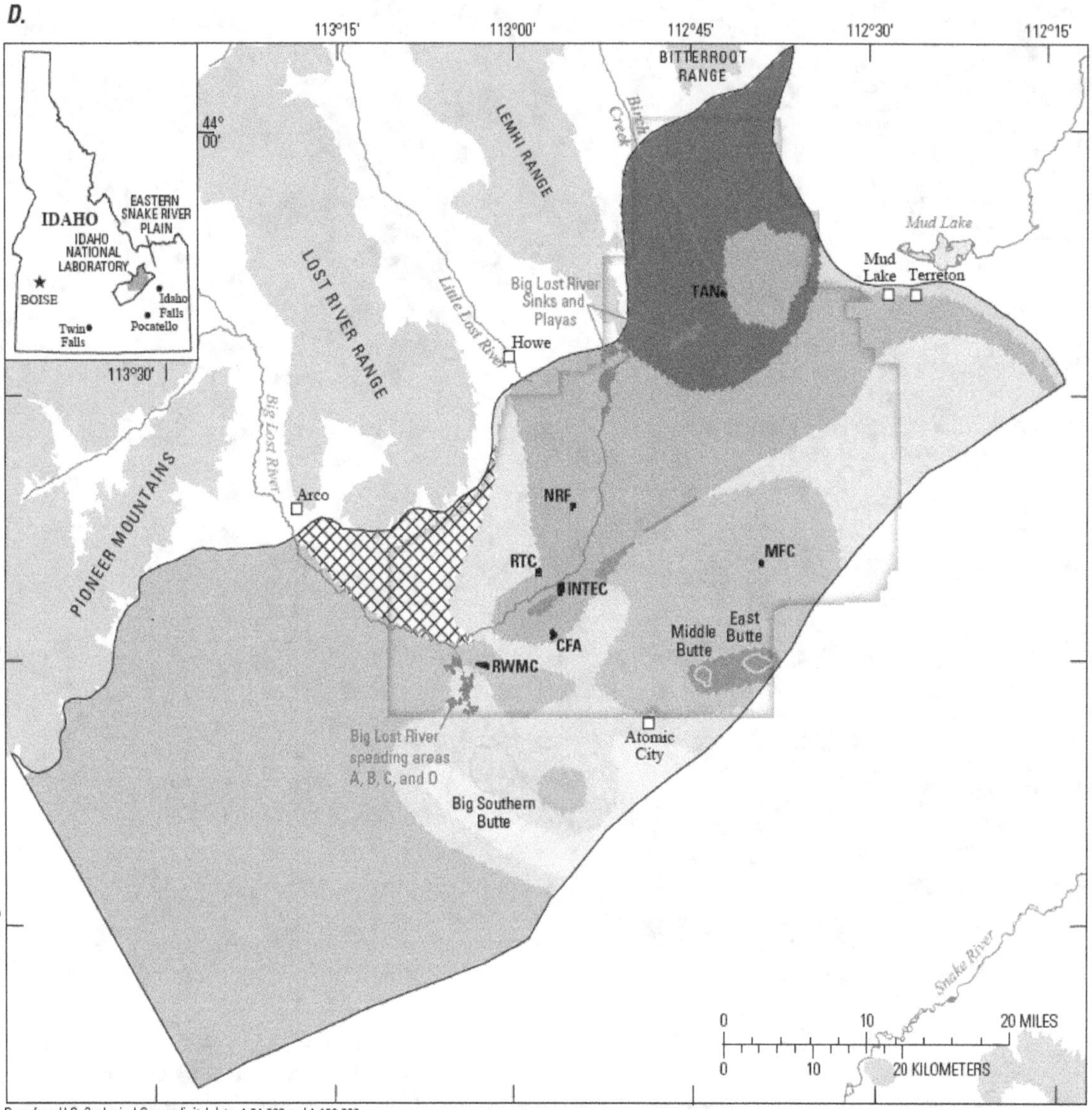

Base from U.S. Geological Survey digital data, 1:24,000 and 1:100,000
Albers Equal-Area Conic projection, standard parallels 42°50'N, 44°10'N;
central meridian 113°00'W; North American Datum of 1927.

EXPLANATION

Hydrogeologic zones, model layer 4

1—Younger rocks consisting of densely fractured basalt and interbedded sediment, with a sediment thickness of generally less than 11 percent

2—Younger rocks consisting of massive, less densely fractured basalt and interbedded sediment, with a sediment thickness of generally less than 11 percent

3—Intermediate-age rocks consisting of slightly altered fractured basalt and sediment, with a sediment thickness of generally less than 11 percent

4—Intermediate-age rocks consisting of slightly altered fractured basalt and sediment, with a sediment thickness of generally less than 11 percent

11—Younger rocks consisting of densely fractured basalt and interbedded sediment, with a sediment thickness of generally more than 11 percent

22—Younger rocks consisting of massive, less densely fractured basalt and interbedded sediment, with a sediment thickness of generally more than 11 percent

33—Intermediate-age rocks consisting of slightly altered fractured basalt and sediment, with a sediment thickness of generally more than 11 percent

44—Intermediate-age rocks consisting of slightly altered fractured basalt and interbedded sediment, with a sediment thickness of generally more than 11 percent

6—Silicic rocks, including rhyolite domes and andesite

Model layer inactive

Idaho National Laboratory boundary

Site facilities

CFA—Central Facilities Area

INTEC—Idaho Nuclear Technology and Engineering Center

MFC—Materials and Fuels Complex

NRF—Naval Reactors Facility

RTC—Reactor Technology Complex

RWMC—Radioactive Waste Management Complex

TAN—Test Area North

Model area boundary

Figure 19.—Continued.

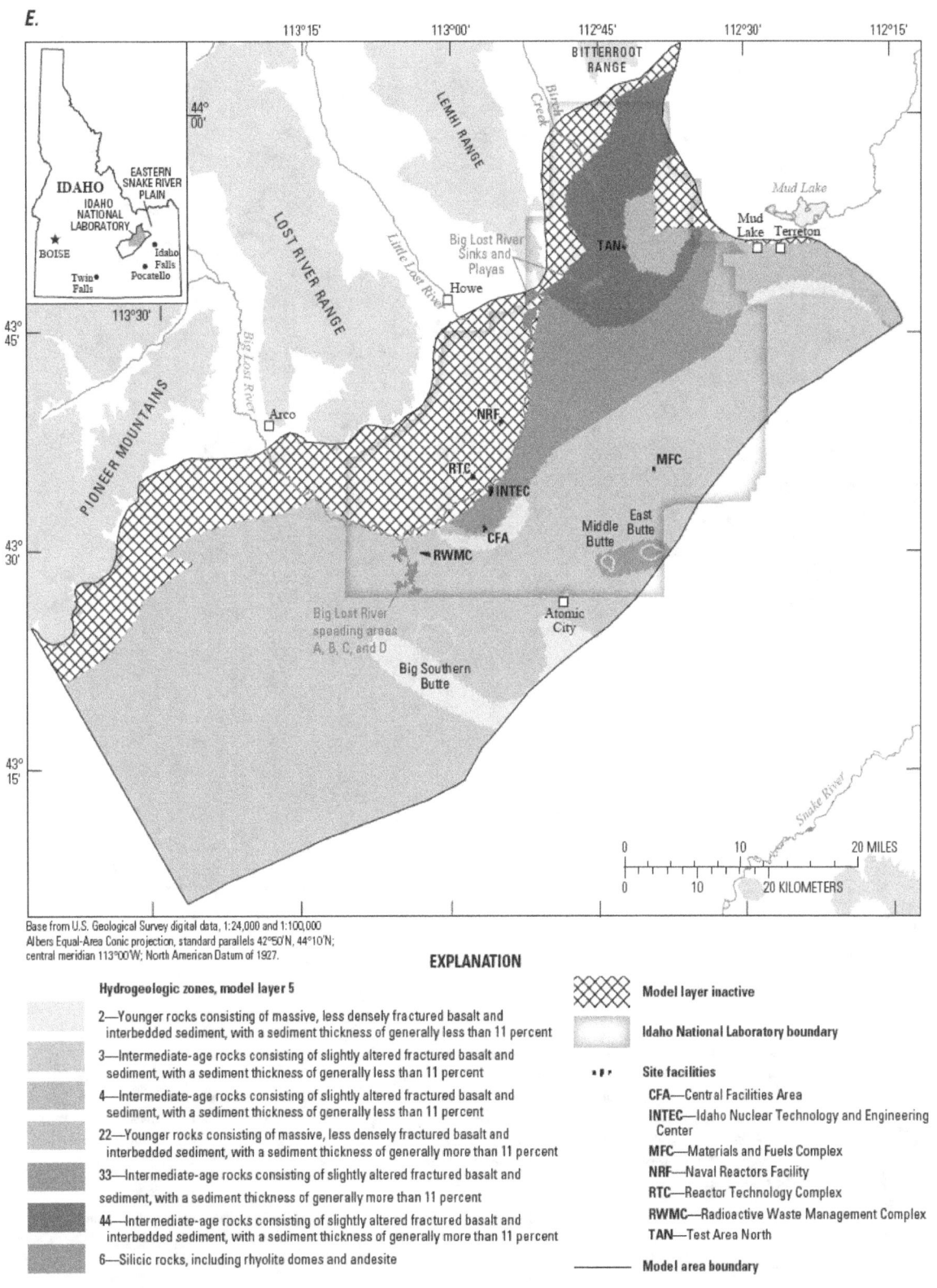

E.

EXPLANATION

Hydrogeologic zones, model layer 5

2—Younger rocks consisting of massive, less densely fractured basalt and interbedded sediment, with a sediment thickness of generally less than 11 percent

3—Intermediate-age rocks consisting of slightly altered fractured basalt and sediment, with a sediment thickness of generally less than 11 percent

4—Intermediate-age rocks consisting of slightly altered fractured basalt and sediment, with a sediment thickness of generally less than 11 percent

22—Younger rocks consisting of massive, less densely fractured basalt and interbedded sediment, with a sediment thickness of generally more than 11 percent

33—Intermediate-age rocks consisting of slightly altered fractured basalt and sediment, with a sediment thickness of generally more than 11 percent

44—Intermediate-age rocks consisting of slightly altered fractured basalt and interbedded sediment, with a sediment thickness of generally more than 11 percent

6—Silicic rocks, including rhyolite domes and andesite

Model layer inactive

Idaho National Laboratory boundary

Site facilities
  CFA—Central Facilities Area
  INTEC—Idaho Nuclear Technology and Engineering Center
  MFC—Materials and Fuels Complex
  NRF—Naval Reactors Facility
  RTC—Reactor Technology Complex
  RWMC—Radioactive Waste Management Complex
  TAN—Test Area North

Model area boundary

Base from U.S. Geological Survey digital data, 1:24,000 and 1:100,000
Albers Equal-Area Conic projection, standard parallels 42°50'N, 44°10'N;
central meridian 113°00'W; North American Datum of 1927.

Figure 19.—Continued.

F.

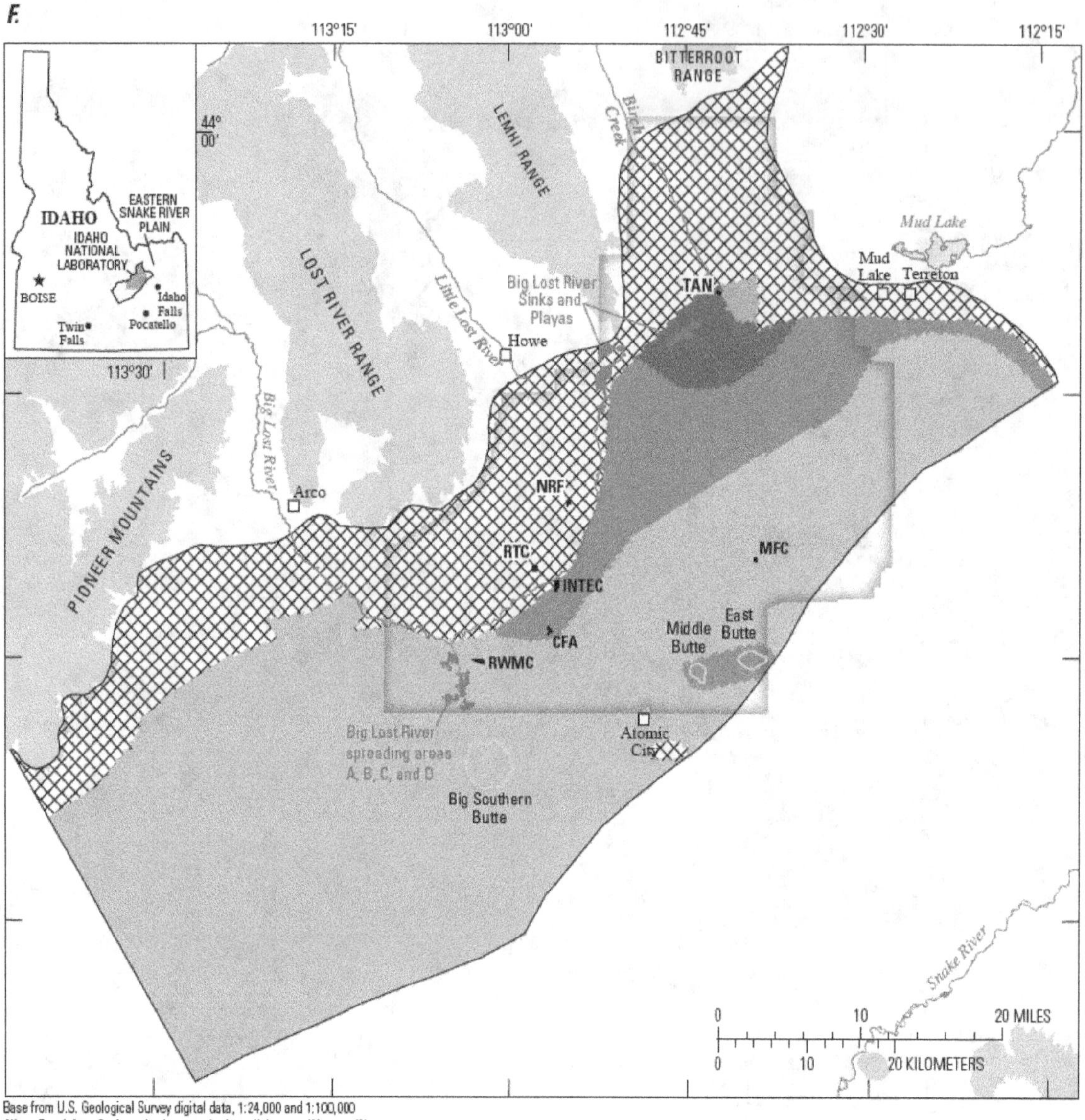

Base from U.S. Geological Survey digital data, 1:24,000 and 1:100,000;
Albers Equal-Area Conic projection, standard parallels 42°50'N, 44°10'N;
central meridian 113°00'W; North American Datum of 1927.

EXPLANATION

Hydrogeologic zones, model layer 6

3—Intermediate-age rocks consisting of slightly altered fractured basalt and sediment, with a sediment thickness of generally less than 11 percent

4—Intermediate-age rocks consisting of slightly altered fractured basalt and sediment, with a sediment thickness of generally less than 11 percent

33—Intermediate-age rocks consisting of slightly altered fractured basalt and sediment, with a sediment thickness of generally more than 11 percent

44—Intermediate-age rocks consisting of slightly altered fractured basalt and interbedded sediment, with a sediment thickness of generally more than 11 percent

6—Silicic rocks, including rhyolite domes and andesite

Model layer inactive

Idaho National Laboratory boundary

Site facilities
  CFA—Central Facilities Area
  INTEC—Idaho Nuclear Technology and Engineering Center
  MFC—Materials and Fuels Complex
  NRF—Naval Reactors Facility
  RTC—Reactor Technology Complex
  RWMC—Radioactive Waste Management Complex
  TAN—Test Area North

Model area boundary

Figure 19.—Continued.

Model Parameters

In this report, a model parameter is defined as a value assigned for a specific hydraulic property for one or more model cells (zones). Model parameters in the steady-state models were horizontal hydraulic conductivity (HC) and vertical anisotropy (VANI), the ratio of horizontal to vertical hydraulic conductivity. When more parameters are defined than can be estimated in a model, some parameter values are not estimated and must be specified (fixed) during model calibration. One parameter value for VANI, applied uniformly throughout the model domain, and multiple parameter values for HC were used in each of the three steady-state models. Single HC parameter values were estimated or fixed for each hydrogeologic zone (table 8). The no-sediment, Whitehead, and Big Lost Trough models had five, eight, and eight estimated parameters and one, five, and two fixed parameters (table 8), respectively. The parameter values not estimated using parameter estimation were set to values within reasonable limits derived from field and laboratory measurements and estimates and published values for similar aquifer materials (table 3). The parameter values not estimated most often were those with lowest relative values of CSS (fig. 20) indicating that less information is provided by observations for defining those parameters and estimating parameter values. Parameters with lower CSS are likely to be poorly estimated and have large confidence intervals. For the Whitehead sediment model, many parameters could not be estimated.

The fixed HC and VANI parameter values (table 8) were determined from initial trial-and-error modeling. Hydraulic conductivity was specified to be 86 ft/d for parameters HC6, HC33, HC333, and HC44 and corresponds to a hydraulic conductivity value near the maximum for basalts that are thicker or cut by dikes (categories 2 and 3 of Anderson and others, 1999, p. 27, figs. 8 and 9). The specified value represents hydrogeologic units composed of slightly altered and thicker basalt flows (HC33, HC333, and HC44), or thicker, less vesicular intermediate composition rocks that have fewer flow contacts (HC6). The value also is equivalent to the geometric midpoint of hydraulic conductivity, 88 ft/d, of the range of values for hydrogeologic unit 3 (table 3). VANI was specified to be 5,600 in the Whitehead model (table 8).

Model Results and Evaluation

The estimated values of horizontal hydraulic conductivity for each hydrogeologic zone, vertical anisotropy of the model area, and simulated hydraulic head and flux for each active cell in the model area constituted the steady-state

model results. Each of the three models (no-sediment, Whitehead, and Big Lost Trough) was calibrated with the same boundary conditions and observations. Results of the estimated parameter values for each of the model calibrations are summarized in table 9. The Big Lost Trough model was determined to be the best of the three models by comparing model fit, parameter correlation coefficients (table 10), CSS (fig. 20), and the visual comparison of simulated heads to observed heads. The results for the Big Lost Trough model were further examined by comparing parameter values to ranges of expected values and by evaluating the sensitivity of the model to changes in system geometry or boundary conditions.

The highest values of the sum of squared weighted residuals and standard error of regression were in the no-sediment model, and the lowest values were in the Big Lost Trough model (table 10). The maximum parameter correlation coefficient for the models was 1.00 (between HC3 and VANI) for the no-sediment model, 0.94 (absolute value, between HC1 and HC2) for the Whitehead model, and 0.92 (between HC3 and VANI) for the Big Lost Trough model. The maximum correlation coefficient of 1.00 between HC3 and VANI in the no-sediment model indicates that these parameters could not be uniquely estimated in this model, whereas the smaller maximum parameter correlation coefficients in the Whitehead and Big Lost Trough models (less than 0.95) indicate that all their parameters could be uniquely estimated (table 10). Examination of water-table (layer 1) contours of simulated head showed that differences in gradient features presumed to relate to the abundance of sediment in the aquifer were reproduced better with the Whitehead and Big Lost Trough models than with the no-sediment model.

Five of six parameters were estimated in the no-sediment model. Eight of 13 and 8 of 10 parameters were estimated in the Whitehead and Big Lost Trough models, respectively, and CSS for all estimated parameters in the Whitehead model and all but HC4 in the Big Lost Trough model were large enough to indicate that these parameters were well estimated (fig. 20).

The Big Lost Trough model was determined to best represent the hydrogeologic zones and model parameters for the steady-state model. The Big Lost Trough model had (1) the smallest sum of residuals, (2) the lowest maximum parameter correlation, (3) large CSS, (4) simulated heads that compared to observed heads as well or better than the other models, and (5) the definitive (interpretations were based on more complete geologic and geophysical logs) and numerous data used to create the distribution of sediment for the parameterization (Ackerman and others, 2006, p. 22). The no-sediment and Whitehead models, therefore, will not be discussed further in this report.

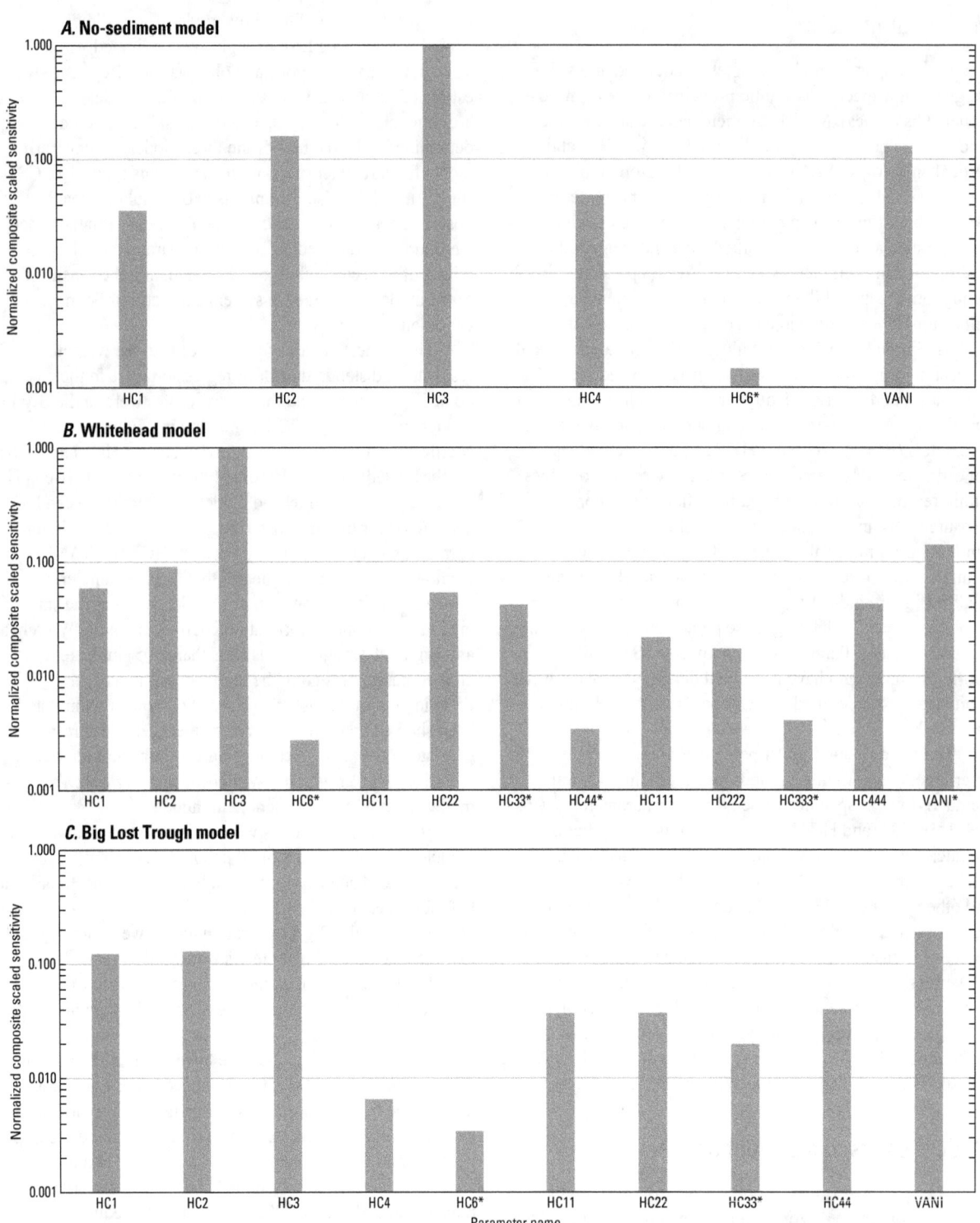

Figure 20. Normalized composite scaled sensitivities for models of steady-state groundwater flow, Idaho National Laboratory and vicinity, Idaho. Asterisks indicate that the parameter value was fixed, rather than estimated.

Table 9. Estimates of hydraulic properties, expected intervals, and 95-percent confidence intervals for calibrated steady-state models of groundwater flow, Idaho National Laboratory and vicinity, Idaho.

[Hydraulic conductivity measured in feet per day Values in *italic* indicate parameter is fixed, not estimated
Abbreviations: HC, hydraulic conductivity; VANI, ratio of horizontal to vertical hydraulic conductivity, dimensionless; –, not used in parameter structure]

Hydraulic property parameter	Expected interval Lower limit	Expected interval Upper limit	No sediment (Estimated value)	Whitehead sediment (Estimated value)	Big Lost Trough sediment (Estimated value)	Big Lost Trough sediment 95-percent confidence interval Lower limit	Big Lost Trough sediment 95-percent confidence interval Upper limit
Parameters corresponding to hydrogeologic zones where the presence of sediment does not affect hydraulic property estimates							
HC1	0.01	24,000	296	1,870	11,700	10,200	13,500
HC2	6.5	1,400	8,240	5,530	384	244	610
HC3	.32	24,000	308	365	435	377	500
HC4	.32	24,000	288	–	9,890	1,730	54,700
Parameters corresponding to hydrogeologic zones where the presence of sediment does affect hydraulic property estimates							
HC11	0.01	24,000	–	231	227	179	296
HC111	.01	24,000	–	512	–	–	–
HC22	6.5	1,400	–	25,422	4,780	3,610	6,140
HC222	6.5	1,400	–	676	–	–	–
HC44	.32	24,000	–	*86*	285	225	365
HC444	.32	24,000	–	459	–	–	–
Parameters corresponding to hydrogeologic property constant throughout the model domain							
VANI	30	1,700	4,720	*5,600*	14,800	7,550	29,100

Table 10. Statistical measures related to parameter uncertainty and overall goodness of model fit for steady-state models of groundwater flow, Idaho National Laboratory and vicinity, Idaho.

[All statistics dimensionless **Abbreviations**: HC, hydraulic conductivity; VANI, ratio of horizontal to vertical hydraulic conductivity; R^2_N, correlation coefficient between ordered weighted residuals and normal order statistics]

Statistic	No sediment	Whitehead sediment	Big Lost Trough sediment
Sum of squared weighted residuals	12,311	7,012	6,082
Average weighted residual [1]	.74	.30	.51
Parameter correlation coefficients [2]	HC3 VANI = 1.00	HC1 HC2 = -0.94	HC3 VANI = 0.92
Standard error of the regression	7.93	6.03	5.61
R^2_N	.945	.960	.954
Modified Beale's measure		6.98	5.85

[1] Positive average weighted residual indicates that, on average, weighted observed values were greater than weighted simulated values

[2] Parameter correlation coefficients greater than 0 90

Estimates of Hydraulic Conductivity and Vertical Anisotropy

For the calibrated Big Lost Trough steady-state model the estimated values of HC parameters ranged from 227 to 11,700 ft/d and the estimated value for VANI was 14,800 (table 9; fig. 21). Confidence in these estimated parameter values was evaluated by (1) comparing the estimated values with expected values, (2) identifying whether parameters were estimated uniquely, (3) examining linear confidence intervals for the parameters, and (4) identifying if parameters were significantly influenced by a single observation. All parameters were log transformed in the calibration process; the distribution and uncertainty of parameters are often best represented by a log-normal probability distribution (Anderson and Woessner, 1992, p. 261; Hill and Tiedeman, 2007, p. 78) so comparisons are best made by viewing the parameters on a logarithmic axis as used on figure 21.

All the estimated HC parameter values were in the high end of the range of expected values (greater than the geometric midpoint) for their corresponding hydrogeologic units (fig. 21) (Ackerman and others, 2006, table 2). All estimates were nearly within 2 orders of magnitude of the maximum expected value in a range that exceeds 6 orders of magnitude. The estimated values are more consistent with the larger expected values of HC parameters derived from large-scale aquifer tests than smaller values of HC parameters derived from small-scale aquifer tests (table 11). The large-scale tests where discharge was greater than 0.52 ft^3/s, were open to longer well intervals, and intersected more interflow zones. In contrast, the small-scale aquifer tests, which were slug tests or straddle-packer tests, with small stresses, were open to smaller intervals, and intersected fewer interflow zones (Welhan and others, 2002, p. 255). Because of the small stress applied and the short interval of the well isolated, the slug and straddle-packer tests characterized only a small volume of rock within a few feet of the well. Consequently, the large-scale tests better represent the magnitude of hydraulic conductivity at the scale of the flow model.

The larger value for HC22 (4,780 ft/d, table 9; fig. 21) relative to HC2 (384 ft/d) and one-half an order of magnitude larger than the upper limit of the expected range of values (1,400 ft/d) is not readily explained. With the exception of hydrogeologic unit 2, estimated values of hydraulic conductivity for zones corresponding to a hydrogeologic unit were larger outside, and smaller inside, the region of abundant sediment. The smaller values for HC parameters inside the region of abundant sediment were consistent with sediment deposition at the land surface between basalt flow events, with the deposited sediment filling void spaces and thereby reducing the hydraulic conductivity of the subsequent interflow zone and the underlying basalt. Because the estimate

of HC22 unexpectedly was larger than the value for HC2, the estimate of HC22 may be more uncertain than most other HC parameter values. The concern about the estimated value for HC22 being larger than the expected range and the value being greater than HC2, however, is tempered by three considerations: (1) the relatively small population of field-estimated values of hydraulic conductivity for hydrogeologic unit 2 (table 3) used to define this range, (2) the relatively small confidence interval for parameter HC2 (fig. 21), and (3) the values for HC22 and HC2 (hydrogeologic unit 2) are less than that of hydrogeologic unit 1 (Ackerman and others, 2006, p. 20).

Estimates of HC22 also may be influenced by the simple discretization of the area of abundant sediment. The influence of sediment content as implemented in the model is very coarse. All zones within the area of abundant sediment contain greater than 11 percent sediment, but may contain as much as 50 percent sediment (Ackerman and others, 2006, p. 22). Recent geostatistical modeling of sediment abundance in the upper 300 ft of the aquifer in the study area (Welhan and others, 2007, p. 25; Stroup and others, 2008) suggest methods for better spatial resolution of relative sediment abundance within the model layers and aquifer parameter zones. This information along with the use of multiplier arrays in MODFLOW-2000 (Harbaugh and others, 2000, p. 13) may provide a parameterization with more spatial resolution of hydraulic conductivity within zones of abundant sediment.

The estimated value of 14,800 ft/d for the VANI parameter exceeded its maximum expected value of 1,700 by nearly an order of magnitude (table 9; fig. 21). The range of expected values of the ratios of horizontal and vertical conductivity are derived from the results of a few aquifer tests that sample only a few interflow zones within a 240 ft open interval (Wood and Norrell, 1996 p. 6-9, table 6-2), straddle-packer intervals of 15 to 21 ft of the aquifer at a distance of 2,600 to 4,200 ft from a production well (Fredrick and Johnson, 1996, p. 16, tables 4–8), or are exclusively in sediment (Spinazola, 1994, table 2). In contrast, estimated VANI represents the large-scale vertical anisotropy of the aquifer that results from the vertical isolation of high permeability interflow rubble zones by intervening layers of less permeable sediment and the dense interior of individual basalt flows. The estimate of VANI is only slightly greater than a range of 2,100 to 13,000 calculated by Welhan and others (2006, p. 26) derived from geostatistical analysis of sediment content and the scaling of hydraulic conductivity with sediment content.

Parameter correlation coefficients were used to evaluate parameter uniqueness. The largest parameter correlation coefficient was 0.92, between HC3 and VANI (table 10). This was smaller than the critical value of 0.95, and indicates that all the parameters were estimated uniquely.

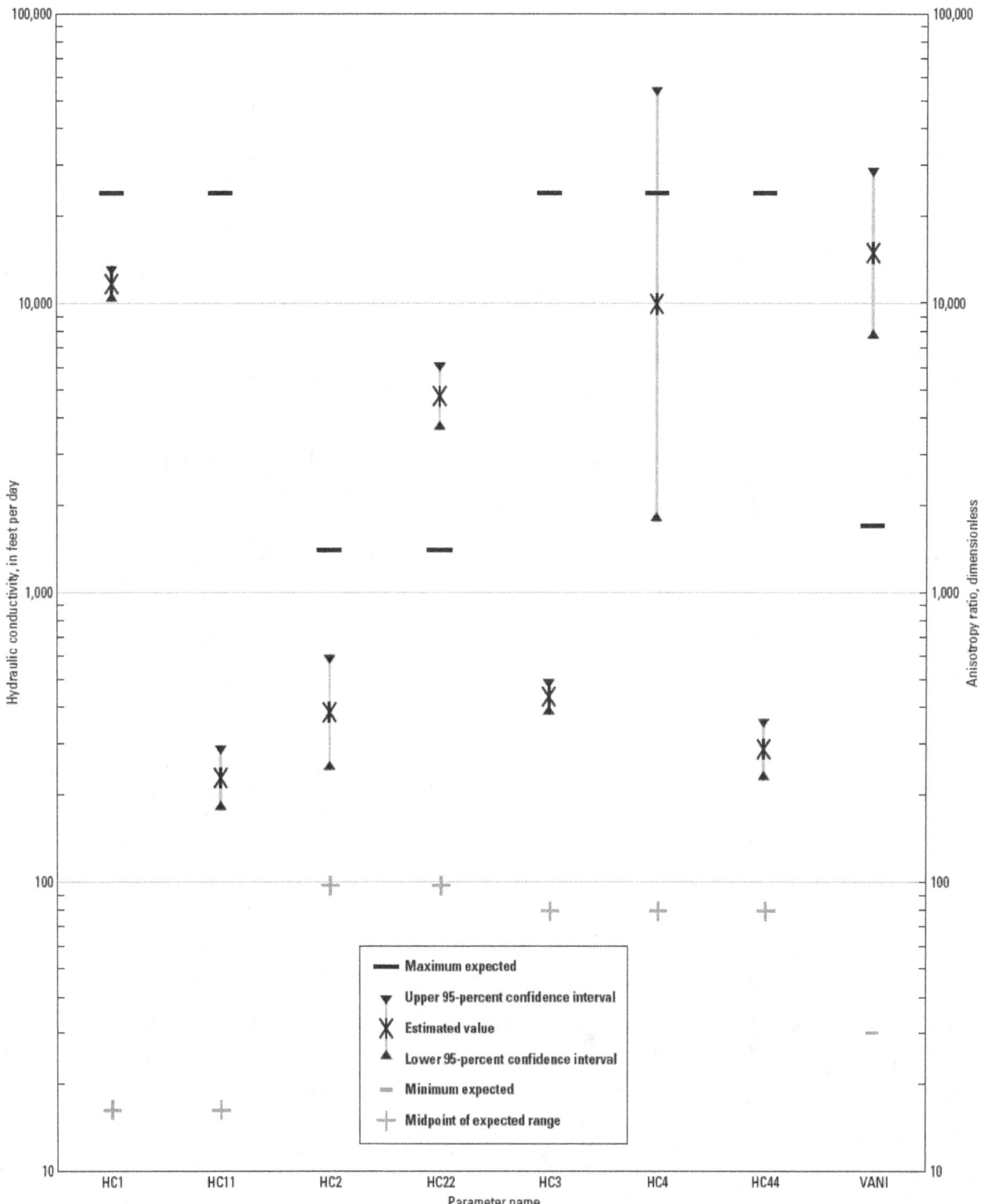

Figure 21. Estimated values of parameters with 95-percent confidence intervals and ranges of expected values for Big Lost Trough model of steady-state groundwater flow, Idaho National Laboratory and vicinity, Idaho. Minimum expected values of hydraulic conductivity parameters are less than 10 feet per day.

Table 11.   Ranges and central values of hydraulic conductivity estimated from aquifer tests, Idaho National Laboratory and vicinity, Idaho.

[Hydraulic conductivity is bulk average hydraulic conductivity in feet per day, averaged over total open, perforated, or test interval length. Values are rounded to two significant figures. **Anisotropy:** Ratio of horizontal to vertical hydraulic conductivity; **Abbreviations** M, multi-well tests; S, single well pumping test; SC, specific capacity pumping test; P, piezometer test; n, number of observations; t, number of tests; w, number of observation wells for multi-well tests; TC, type curve; R, regression analysis; I, iterative solution of equation; NS, numerical simulation; B, basalt; S, sediment; SB, sediment and basalt; INL, Idaho National Laboratory; NE, northeast of INL and East or North of Mud Lake; TAN, Test Area North; ft/d, foot per day; ft³/s, cubic foot per second]

Data set	Count	Hydraulic conductivity (ft/d)						Anisotropy	Test type	Analysis method	Rock type	Location
		Minimum	First quartile	Median	Geometric mean	Third quartile	Maximum					
Medium- to high-capacity irrigation, industrial supply, and test wells (generally 0.7–13 ft³/s)												
1	n = 15, t = 6, w = 1 or 2	600	790	2,900	2,800	13,000	18,000		M	TC		INL
2	n = 10	37	170	520	440	1,100	3,500		S	TC		INL
3	n = 7, t = 3, w = 14, 6, 3	1,600	3,800	6,200	6,100	11,000	22,000		M	TC	SB	NE
4	n = 4, t = 4, w = 4	3,300	3,600	4,300	4,200	5,000	5,200	[1]40 to 1,700	M	TC	B	INL
5	n = 2, t = 1, w = 4	860			2,100		5,300	230	M	NS	B	INL
6	n = 4, t = 1, w = 4	12,000	16,000	18,000	18,000	21,000	24,000	30, 60, 100	M	TC	B	INL
7	n = 2, t = 1, w = 2	140					330	35, 220	M	TC	S	NE
Low- to medium-capacity industrial supply and test wells (generally 0.01–0.1 ft³/s; 30 wells 0.52–9.7 ft³/s)												
8	n = 36	1	200	1,200		3,200	16,000		SC	I	B	NE
9	n = 10	20	600	1,500		4,900	12,000		SC	I	SB	NE
10	n = 6	10	300	780		1,300	3,000		SC	I	S	NE
11	n = 114	0.01	14	360	130	2,000	24,000		S	TC, R		INL
Slug-test data for open intervals or straddle packer intervals for 30 wells												
12	n = 59	0.23	3.6	26	17	120	700		P			TAN
Medium- to high-capacity tests (values less than 0.52 ft³/s excluded)												
[2]13	n = 72	1.1	220	1,100	900	5,300	24,000					INL

[1] Range for 61 estimates corresponding to 22 straddle-packer settings in 4 observation wells.

[2] Data sets 1, 2, 3, 4, 5, 6, 7, and a subset of data set 11 where values of pump discharge were greater than 0.52 ft³/s.

References for data sets 1 through 12

1, 2	Mundorff and others, 1964, table 10.
3	Mundorff and others, 1964, p. 155.
4	Frederick and Johnson, 1996, tables 4–8.
5	Frederick and Johnson, 1996, table 11, p. C3.
6	Wood and Norrell, 1996, table 6-2.
7	Spinazola, 1994, table 2.
8, 9, 10	Spinazola, 1994, fig. 35.
11	Anderson and others, 1999, table 2.
12	Welhan and Wylie, 1997; Welhan and others, 2002, table 1.

With the exceptions of HC22, HC4, and VANI, all parameters had 95 percent linear confidence intervals that were contained within the upper half of the range of expected values (fig. 21). The largest confidence interval (indicating larger uncertainty) was for HC4, and only part of the confidence interval exceeded the maximum expected value. A larger confidence interval also was indicated for parameter VANI, and the entire confidence interval was larger than the maximum expected value for this parameter. The linear confidence intervals were calculated assuming independent and normally distributed errors and model linearity, but some of the statistical measures do not support this assumption, which reduces the accuracy of the confidence intervals. The independence and normality of the weighted residuals were evaluated by plotting the ordered weighted residuals against the standard normal statistic and computing the associated correlation coefficient, R^2_N. On the normal probability graph, the weighted residuals do not form a straight line (fig. 22), which indicates that the residuals deviate from a normal distribution. This is consistent with the R^2_N value of 0.954, which was smaller than the critical value of 0.987 (at the 0.05 significance level), below which residuals are not likely independent or normally distributed. The linearity of the model was evaluated using the modified Beale's measure. The calculated value of 5.85 (table 10) is larger than the critical value for moderate nonlinearity of 0.5, and indicates that the model is highly nonlinear. The non-normal distribution of the weighted residuals and the nonlinearity of the model indicate that the calculated intervals should be considered approximate confidence intervals for the parameters.

The influence of specific steady-state head observations on estimated parameter values was evaluated with Cook's D. Twenty-four head observations had Cook's D values larger than the critical value of 0.02 and thus had the most influence on the set of estimated parameter values. Three observations (TANCH2B_401, USGS7_1280, and USGS13_781) had Cook's D values larger than 0.2 (table 12), and each of these observations were omitted, separately, from the observation data set and the model was recalibrated. In each of the three

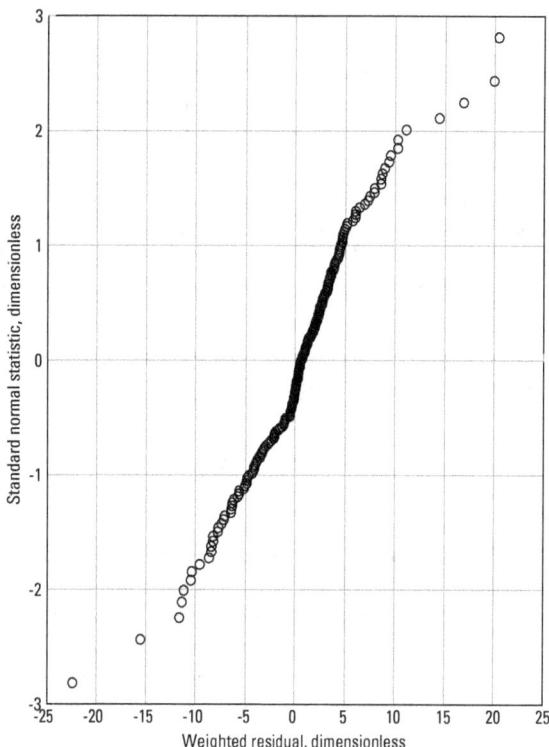

Figure 22. Normal probability plot for Big Lost Trough model of steady-state groundwater flow, Idaho National Laboratory and vicinity, Idaho.

recalibrations no parameter correlations were larger than 0.93, values for the sum of squared residuals were similar, and parameter values did not change substantially except for HC4 (table 12). These large changes in the value of HC4 are consistent with the large uncertainty in this parameter estimate (fig. 21). With the exception of HC4, the large number of observations seems to have reduced the likelihood of a single observation being influential to the set of parameters.

Table 12. Summary of observations and corresponding parameters with highest values of influence measures for steady-state model of groundwater flow, Idaho National Laboratory and vicinity, Idaho.

[Critical values for Cook's D and DFBETAS were 0 020 and 0 141, respectively Parameter associated with DFBETAS statistic Re-calibrated parameter values result from omitting the observation for recalibration; **Abbreviations**: HC, hydraulic conductivity; ft/d, foot per day]

Observation	Cook's D	Parameter	DFBETAS	Hydraulic conductivity	
				Base case model (ft/d)	Recalibrated model (ft/d)
TANCH2B_401	1.01	HC4	2.17	9,890	4,910
		HC44	-1.84	285	329
USGS7_1280	.34	HC4	1.33	9,890	5,260
USGS13_781	.23	HC2	1.34	384	323

Although most parameter values were within acceptable ranges, the unexpectedly large parameter estimates for HC22 and VANI, and the large confidence intervals for HC4 and VANI that were in part or wholly outside the expected range (table 9; fig. 21) allow for a ranking of the confidence in the parameter estimates for the calibrated model. Utilizing the calibrated parameter estimates and the criteria discussed in this section, greater confidence is warranted for parameters HC1, HC11, HC2, HC3, and HC44 than for HC22, HC4, and VANI. The lower confidence in parameter HC4 also is consistent with the low CSS for HC4 (fig. 20).

Comparison of Simulated and Observed Steady-State Heads

Model simulation results also were evaluated based on the agreement between simulated and observed heads. Evaluations included (1) a visual comparison of the altitudes and shapes of the simulated and observed water tables based on 50-ft contour maps of the water table, (2) qualitative comparisons of vertical gradients and vertical flow directions based on simulated heads and piezometer head measurements and intraborehole flowmeter surveys, and (3) quantitative comparisons based on the statistical and spatial distributions of weighted residuals for heads.

Simulated equivalents to observed heads were calculated by MODFLOW-2000 using spatial interpolation in the horizontal plane and calculation of multilayer heads for wells open to multiple model layers (Hill and others, 2000, p. 31-35). Multilayer heads were calculated by specifying a head contribution from each model layer that is in proportion to the transmissivity of the open interval in that layer.

Altitudes and Shapes of Simulated and Observed Water Table

The simulated water-table map was constructed from 30,983 simulated heads distributed uniformly throughout the model area. These heads were in model layer 1 at the center of each grid cell (the equivalent of 16 simulated heads per 1 mi^2). The observed water-table map (Lindholm and others, 1988) was constructed from 66 head observations distributed non-uniformly throughout the model area (the equivalent of 1 head observation per 29 mi^2). Approximately 45 additional head observations, outside of the model area, were used to infer the orientation and altitude of water-table contours near model boundaries. Head information available to construct the simulated water-table map was about 470 times more abundant than that used to construct the observed water-table map, thus the spatial resolution of the simulated water-table map is much finer than that of the observed water-table map. The simulated water-table map should not be considered better than the observed water-table map simply because it has a value for each model cell. Construction of the observed

water-table contour map for the entire model area required considerable interpolation and extrapolation, particularly along the northwest mountain-front boundary and within the southwest one-third of the model area where fewer head observations were available in 1980 (figs. 4D and 8). Water-table contours, contour interval 50 ft, were extrapolated and drawn to reflect the conceptual character of the model boundaries. For example, water-table contours were drawn perpendicular to boundaries or segments of boundaries that were conceptualized as no-flow boundaries and parallel or sub-parallel to boundaries or segments of boundaries that were conceptualized as inflow or outflow boundaries.

The water-table contour maps indicate that the steady-state model simulation was able to reasonably reproduce observed altitudes, orientations, and gradients within the central and northeastern part of the model area where the quality of the head measurements is the most reliable and their spatial density highest. The water-table contour maps also indicate where head observations were weighted in a manner to exert maximum influence on the outcome of model simulations (fig. 23).

Near the northern boundary of the INL, the simulated 4,600 ft altitude of the water table is about 20 ft higher than interpolated water levels from the observed water-table map (fig. 23). Head observations used for calibration in this area are relatively abundant in and around TAN, but are lacking in the immediate vicinity of Birch Creek and the nearby part of the northeast inflow boundary (figs. 4D and 8). Although head observations were more abundant in the TAN area, the distribution was not sufficient to result in improved agreement between the simulated and observed heads in this area because of the lack of head observations near the Birch Creek and northeast inflow boundaries. The orientation of the simulated 4,600 ft contour reflects a complex interplay of inflow from the Birch Creek alluvial aquifer to the northwest, regional aquifer underflow and infiltration from irrigation return flows along the northern end of the northeast boundary, and infiltration from Birch Creek (figs. 14 and 18).

The simulated altitude of the water table near the Little Lost River inflow boundary may be 50 ft higher than indicated by the observed water-table map (fig. 23) and associated observations. Near the mouth of the Little Lost River the simulated water-table contours reflect the combined effects of inflow to the aquifer from the alluvial fill of the Little Lost River; infiltration from the Big Lost River Sinks, Playas, and channel; and infiltration from irrigation return flows east and south of Howe (figs. 14 and 18). Head observations are sparse in this area, many were from wells open to multiple layers, and were insufficient to resolve the local effects of these aquifer inflows on the altitude and shape of the observed water table in this area.

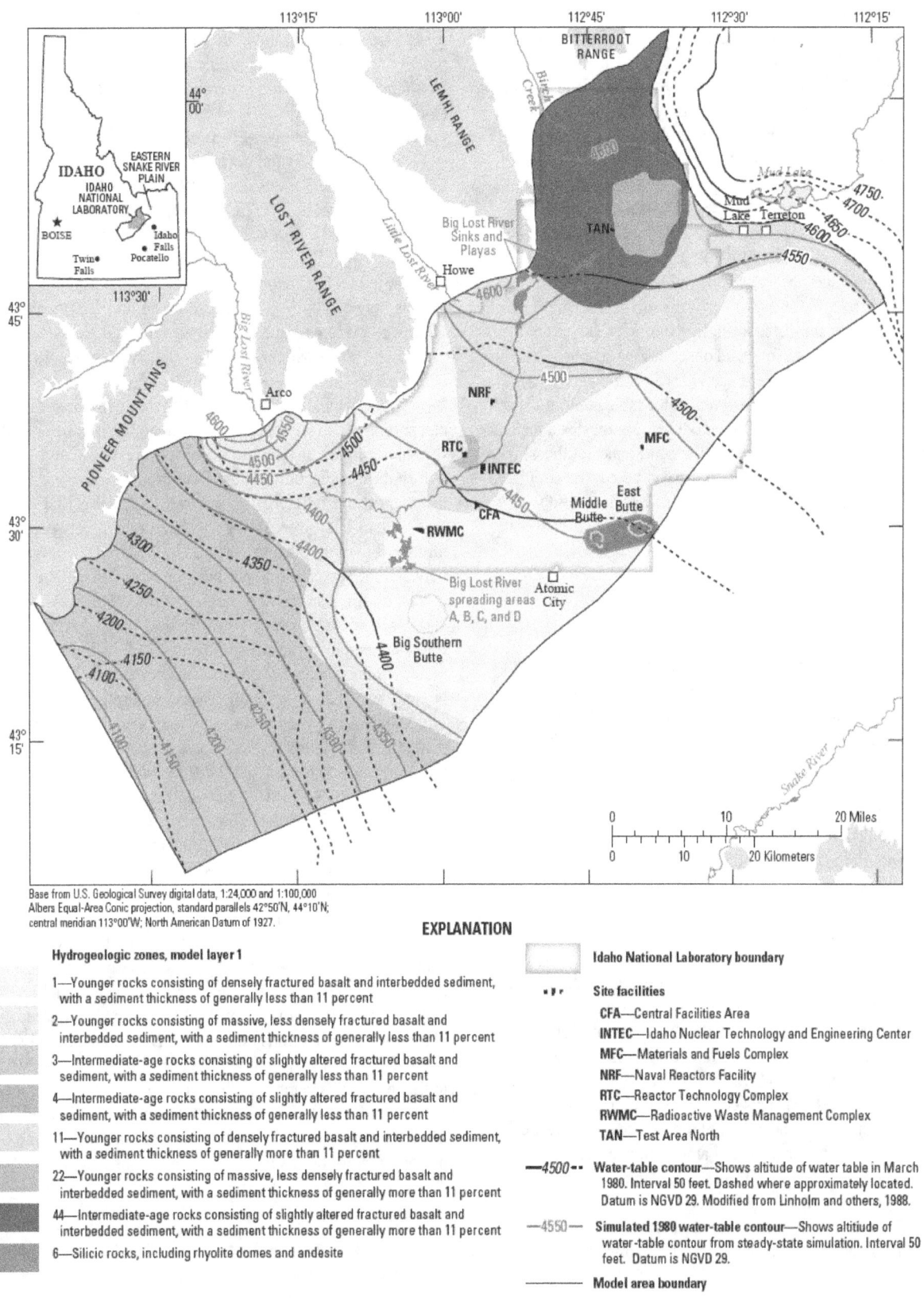

Figure 23. Observed and simulated water table for 1980 showing relation to hydrogeologic zones in layer 1 for Big Lost Trough model of steady-state groundwater flow, Idaho National Laboratory and vicinity, Idaho.

The simulated altitude of the water table near the Big Lost River inflow boundary may be 50 ft higher than indicated by the observed water-table map (fig. 23) and one observation. The altitude of the water table near the mouth of the Big Lost River valley is defined by a single observation well open to layers 2 and 3 (Weaver and Lowe, table A1), 3 mi southwest of where the Big Lost River enters the model area (fig. 4D). This distant single observation with a very low weight (WeavLow_585, table G1) did not restrain the simulated water-table altitudes rising to 4,600 ft and higher near the mouth of the Big Lost River; however, the shapes of the simulated and observed water table are similar.

In the southwest one-third of the model area, simulated and observed water-table altitude differences vary from 0 ft to as much as 115 ft. Ten head observations scattered over an area of about 650 mi^2 were used for calibration (fig. 4D). The shape of the interpreted observed water-table contours within the southwest one-third of the model area indicates a tendency for groundwater to flow toward the center and southeast section of the southwestern boundary, whereas the simulated water-table contours show a more uniform flow field with limited tendency for flow to move preferentially towards the center of the southwest boundary.

Simulated and Observed Vertical Gradients and Vertical Flow Directions

Simulated and observed vertical gradients generally are larger than horizontal gradients; however, flow in the aquifer is predominately horizontal. Observed horizontal gradients at the water table typically range from 0.0002 to 0.002 (Ackerman and others, 2006, p.39) although gradients may be as great as 0.01 at the northeastern boundary or 0.02 near the mountain-front boundary (Barraclough, Teasdale, and Jensen, 1967, p. 22). Measured vertical gradients in the upper 500 ft of the aquifer at two sites in the northeast corner of the INL (table 13) were as large as 0.03 upward and 0.007 downward. Vertical gradients at well Highway 1 (table A1; fig. B2) 10 mi east of the study area are between 0 and 0.0009 downward. Measurement of head gradients was attempted at wells USGS 44, 45, 46, and 59 (table A1; fig. B7) during straddle-packer tests in 1992–94 (Frederick and Johnson, 1996). However, the measurements were below the detection limit of the equipment, which for head differences across the packers, may have been about 0.1 ft (G.S. Johnson, University of Idaho, oral commun., 2004). This corresponds to gradients of ±0.05. These locations are the extent of measurements of vertical gradients in and near the study area. The average simulated vertical gradient between model layers in the upper five layers (800 ft) of the aquifer was 0.04 downward. Seventy-five percent of simulated vertical gradients in the upper five layers were between 0.12 downward and 0.01 upward.

Vertical gradients and flow directions were simulated for the ESRP aquifer, but because there are few wells in the model area where vertical gradients or flow directions have been observed, comparisons are limited between simulated results and observations. The direction of vertical flow in the upper 300 ft (model layers 1–3) of the aquifer was measured at 12 wells (13 intervals) during 1963–65, 1 well (3 intervals) during 1980, and 3 wells during 1991 (table 13). Eleven of these wells are at the INTEC, and the other five wells are southeast of TAN, north of the NRF, and at the RTC (table 13; fig. 1; appendix B). Flow between model layers 1 and 2 in wells at the INTEC and the deepest piezometer pair at USGS 30 was between hydrogeologic zones 11 and 22; flow at the other wells was within a single hydrogeologic zone (table 13).

In the northeast part of the model area, vertical directions and gradients were measured at wells USGS 30 and USGS 4 (fig. 1); model simulation results disagree with most of these measurements. At USGS 30, near the northeast corner of the INL, nested piezometers are in model layers 1, 2, and 4; the two shallower piezometers are in hydrogeologic zone 11, and the deepest piezometer is in hydrogeologic zone 22. Observed gradients in 1980 were 0.019 to 0.022 upward between the middle and upper piezometers and 0.027 to 0.031 upward between the lower and upper piezometers (table 13); the corresponding simulated gradients were 0.0002 downward and 0.044 upward, respectively. At well USGS 4, downward flow between intervals about 50 and 150 ft below the water table was measured using a trace-ejector survey in 1964. In November 1965, an inflatable packer separating these two intervals was installed in the well, and head measurements made through 1966 showed a downward gradient between these intervals ranging from 0.001 to 0.007 (Morris and others, 1965, p. 43-44; Barraclough, Teasdale, and Jensen, 1967, p. 45-47; and Barraclough, Teasdale, and others, 1967, p. 88-90). The 1980 simulated gradient between model layers 1 and 2 at this location, which was entirely within hydrogeologic zone 1, was 0.003 upward.

Simulated and observed flow directions agreed at wells USGS 30, USGS 51, USGS 59, and Site 17, where simulated gradients between layers were larger than 0.01 (table 13). Where simulated gradients between model layers were less than 0.01, only 4 of the 13 wells had simulated and observed flow directions that agreed. Discrepancies between simulated and observed flow directions may be partly due to the temporal separation of observed flows (1963–65 and 1991) and simulated flows (1980), to intermittently operating production and injection wells at site facilities during the period of 1950–80, and to model limitations and error.

Table 13. Observed and simulated steady-state vertical directions and gradients of groundwater flow in open boreholes, piezometer nests, and packer-isolated intervals of boreholes, Idaho National Laboratory, Idaho.

[**Gradient:** Values are dimensionless; simulated gradient was interpolated at well locations within model grid cells **Reference:** A, Morris and others, 1964, p 40-42; B, Morris and others, 1965, p 42-44; C, Barraclough, Teasdale, and Jensen, 1967, p 94-98; D, Barraclough and others, 1967, p 88-90; E Morin and others, 1993, table 1 **Abbreviations**: Pack, temporary packer completion; Piez, piezometer nest; TE, trace ejector; TF, thermal flowmeter; –, no information]

Date	Well name	Direction of flow		Gradient		Corresponding model		Method	Reference
		Observed	Simulated	Observed	Simulated	Layers	Zones		
Northeast Idaho National Laboratory and Southeast of Test Area North									
1964	USGS 4	Down	Up	–	0.003	1 and 2	1	TE	B
1966	USGS 4	Down	Up	0.001 to 0.007	.003	1 and 2	1	Pack	D
1980	USGS 30	Up	Down	[1]0.019 to 0.022	.0002	1 and 2	11	Piez	
1980	USGS 30	Up	Up	[1]0.027 to 0.031	.044	2 and 4	11, 22	Piez	
1963	USGS 31	Up	Down	–	.007	1 and 2	11	TE	A
Idaho Nuclear Technology and Engineering Center									
1965	USGS 42	Up	Up	–	0.001	1 and 2	11, 22	TE	C
1965	USGS 43	Static	Up	–	.001	1 and 2	11, 22	TE	C
1991	USGS 44	Up	Up	–	.001	1 and 2	11, 22	TF	E
1991	USGS 45	Up	Up	–	.002	1 and 2	11, 22	TF	E
1991	USGS 46	Up [2]	Down	–	.001	1 and 2	11, 22	TF	E
1963	USGS 47	Up	Down	–	.003	1 and 2	11, 22	TE	A
1965	USGS 48	Static [3]	Up	–	.005	1 and 2	11, 22	TE	C
1965	USGS 48	Down [3]	Up	–	.008	2 and 3	22	TE	C
1965	USGS 49	Down [3]	Up	–	.006	1 and 2	11, 22	TE	C
1965	USGS 51	Up	Up	–	.011	1 and 2	11, 22	TE	C
1965	USGS 52	Up [4]	Up	–	.004	1 and 2	11, 22	TE	C
1965	USGS 59	Up [4]	Up	–	.012	1 and 2	11, 22	TE	C
North of Naval Reactors Facility									
1964	Site 17	Down	Down	–	0.013	1 and 2	11	TE	B
Reactor Technology Complex									
1964	MTR test	Down [5]	Up	–	0.006	1 and 2	22	TE	B

[1] Observed 1980 range of gradients, water-level data are from http://nwis.waterdata.usgs.gov/id/nwis/gwlevels

[2] Observed gradient is down when nearby production well is pumping

[3] Observed direction may be influenced by nearby injection well

[4] Observed direction may be influenced by nearby injection or production wells

[5] Observed direction may be influenced by nearby infiltration pond

Simulated vertical gradients just downgradient of the INTEC were generally upward from model layers 3 and 2 toward layer 1. This area of upward flow is coincident with the southwest boundary of the Big Lost Trough (figs. 7B and 23). Large downward gradients, greater than 0.05, in the upper 300 ft of the model were coincident with the southwest boundary of hydrogeologic zones 1 and 2. Other areas where larger, greater than 0.2, downward gradients were simulated include (1) near the mouth of the Big and Little Lost River valleys in the upper 300 ft of the aquifer and (2) throughout most of the southwestern part of the INL and south of the INL below an aquifer depth of 400 ft (model layer 4 and below).

Although vertical gradients are larger than horizontal gradients, simulated flow in the aquifer is predominantly horizontal. For example, horizontal fluxes along rows in hydrogeologic zone 1, model layer 1, averaged 1,500 $(ft^3/s)/mi^2$ in a southwest direction, with 75 percent of the values between 880 and 1,900 $(ft^3/s)/mi^2$. Vertical fluxes across the bottom cell faces of the same zone and layer averaged 2.1 $(ft^3/s)/mi^2$ for cells with downward flow and 1.9 $(ft^3/s)/mi^2$ for cells with upward flow. Vertical flow in the aquifer is restricted by relatively impermeable subhorizontal layers of dense basalt and sediment, and the restriction that these stratigraphic layers impose on vertical flow is represented in the calibrated model with a large value for the vertical anisotropy (VANI) of the aquifer (table 9).

Statistical and Spatial Distributions of Weighted Head Residuals

Separate illustrations of the graphical (fig. 24A) and spatial (fig 25A) distributions of weighted residuals corresponding to wells only in model layer 1, which contains the water table, are presented separately from wells not exclusively in layer 1 (figs. 24B and 25B). Separating the two groups enables comparisons of model performance for (1) wells most representative of water-levels at and near the water table and (2) wells in multiple model layers or in layers other than layer 1 that provide information on head differences with depth. All weighted residuals are shown in figures 24C and 25C.

Weighted residuals were plotted against weighted simulated equivalents in order to evaluate if the weighted residuals were randomly distributed. For the 105 wells open only to model layer 1, the distribution of weighted residuals was fairly evenly and uniformly distributed around zero (fig. 24A). There was a slight negative (simulated values greater than observed values) bias for these residuals, however, and the average value of the weighted residuals was -0.30.

For the 96 wells that were open to more than just model layer 1, the distribution of weighted residuals shows a positive (simulated values less than observed values) bias

(fig. 24B). The average value of the weighted residuals was 1.41, including three larger values representing multi-layer completions, which also indicates that, on average, simulated heads were underestimated for model layers below layer 1. The model does not perform as well for estimating heads for multi-layer completions. The model also may be poorly constrained for estimating heads at depth in the aquifer. Uneven spatial distribution of wells and boreholes vertically has been mentioned in the Conceptual Model section as a data limitation for the study. The simulations of heads at depth, vertical gradients, and vertical flow directions could be improved in the study area with the collection of additional head data at multiple levels in the aquifer.

The average value for all 201 weighted residuals (fig. 24C) was 0.52 with nearly equal numbers of larger negative and positive residuals (absolute value of residual greater than 1 standard error of the regression). The weighted residuals generally are evenly distributed about zero with the exception of a slight positive bias for weighted simulated values around 2,200 and a slight negative bias around weighted simulated values of 4,400. For comparison and ease of visualization, weighted residuals also were plotted against (unweighted) simulated values (fig. 24D).

The weighted residuals were plotted on maps of the model area (fig. 25) to evaluate further if they were randomly distributed. The overall distribution of the weighted residuals seems random, although some small areas have clusters of either negative or positive residuals and larger areas have possible patterns of larger residuals, indicating that the model fit to observed data is locally not as good.

Most clusters of negative and positive weighted residuals are near facilities where data density is highest and weighted residuals tend to be small in absolute value at these locations. This clustering likely is caused by representing hydraulic conductivity with a uniform value throughout a hydrogeologic zone. This is a consequence of the parameterization of the model.

In the northern part of the INL, an area at and 2 mi southeast of TAN includes a cluster of small positive weighted residuals for all wells (fig. 25A, B). The model error indicated by this cluster of positive residuals might be small because all but two of these residuals had a value that was less than 1s, the standard error of the regression. This particular cluster, however, is part of a larger trend from negative to positive weighted residuals in the area from 5 mi northwest to 2 mi southeast that results from the model simulating a horizontal gradient of about 0.0001–0.001 in an area with the lowest observed water-table gradient, 0.0002, in the study area (Ackerman and others, 2006, p. 39). Poor model fit in this area is likely related to difficulty in assigning hydrogeologic zones for an area with poor stratigraphic control. Model fit also may not be as good in this area because many of these observations were from periods other than 1980.

A.

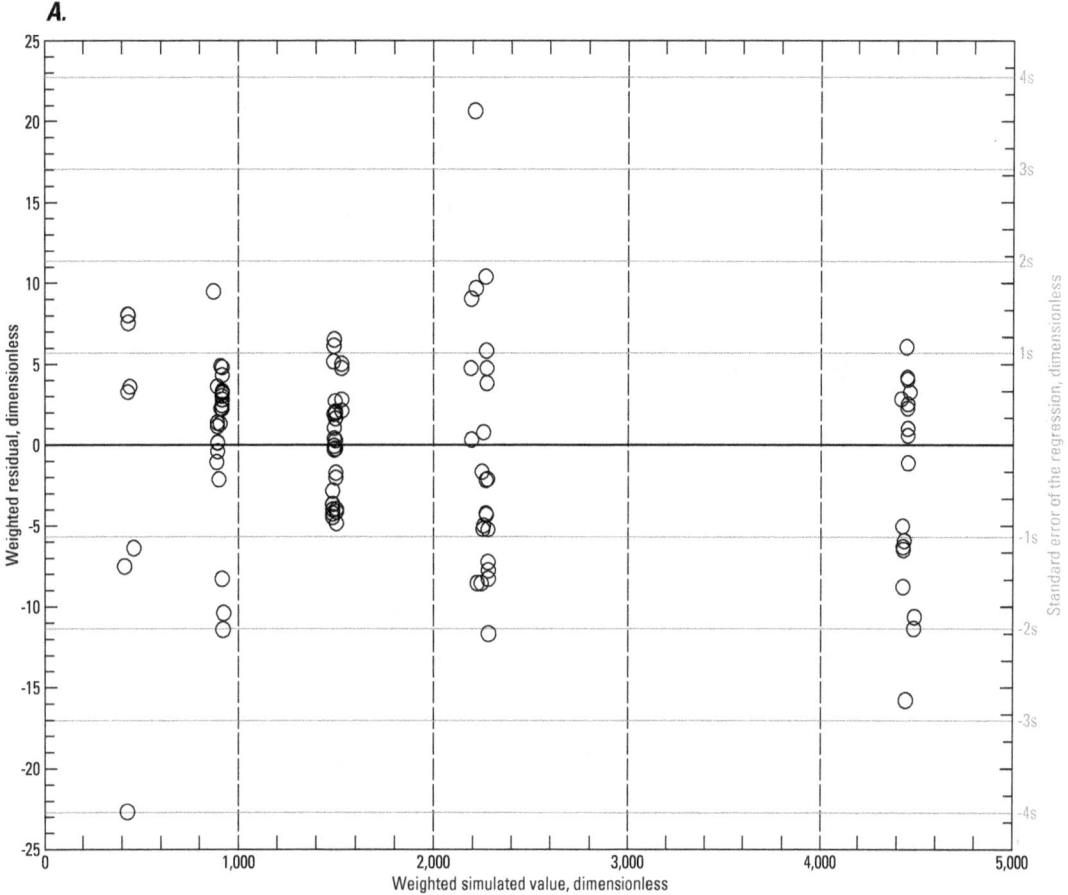

EXPLANATION

○ **Weighted head residual**—in terms of s, standard error of the regression

Note: Positive residual indicates that the weighted observed value is greater than the weighted simulated value.

Figure 24. Weighted residuals as a function of weighted simulated values of head for (*A*) wells open exclusively to model layer 1, (*B*) wells not open exclusively to layer 1, (*C*) all wells, and (*D*) weighted residuals as a function of simulated values for Big Lost Trough model of steady-state groundwater flow, Idaho National Laboratory and vicinity, Idaho.

B.

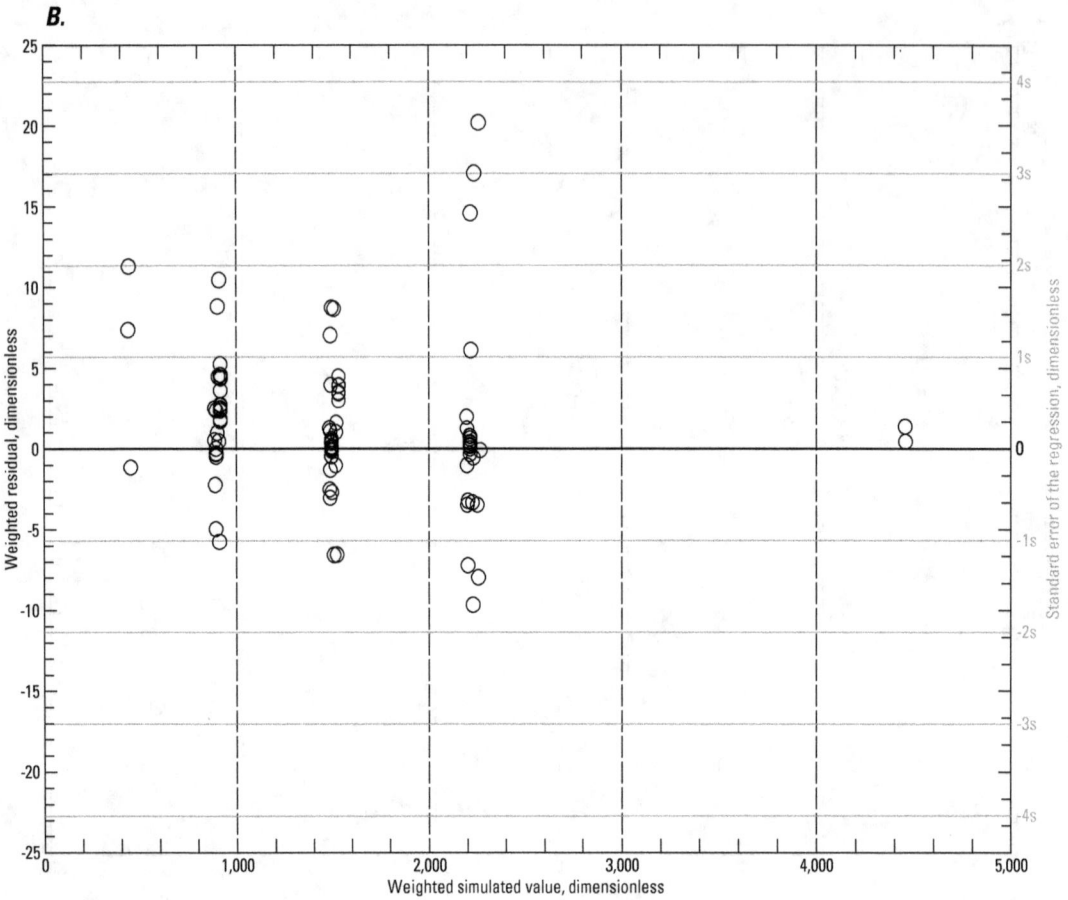

EXPLANATION

○ **Weighted head residual**—in terms of s, standard error of the
regression

Note: Positive residual indicates that the weighted observed
value is greater than the weighted simulated value.

Figure 24.—Continued.

C.

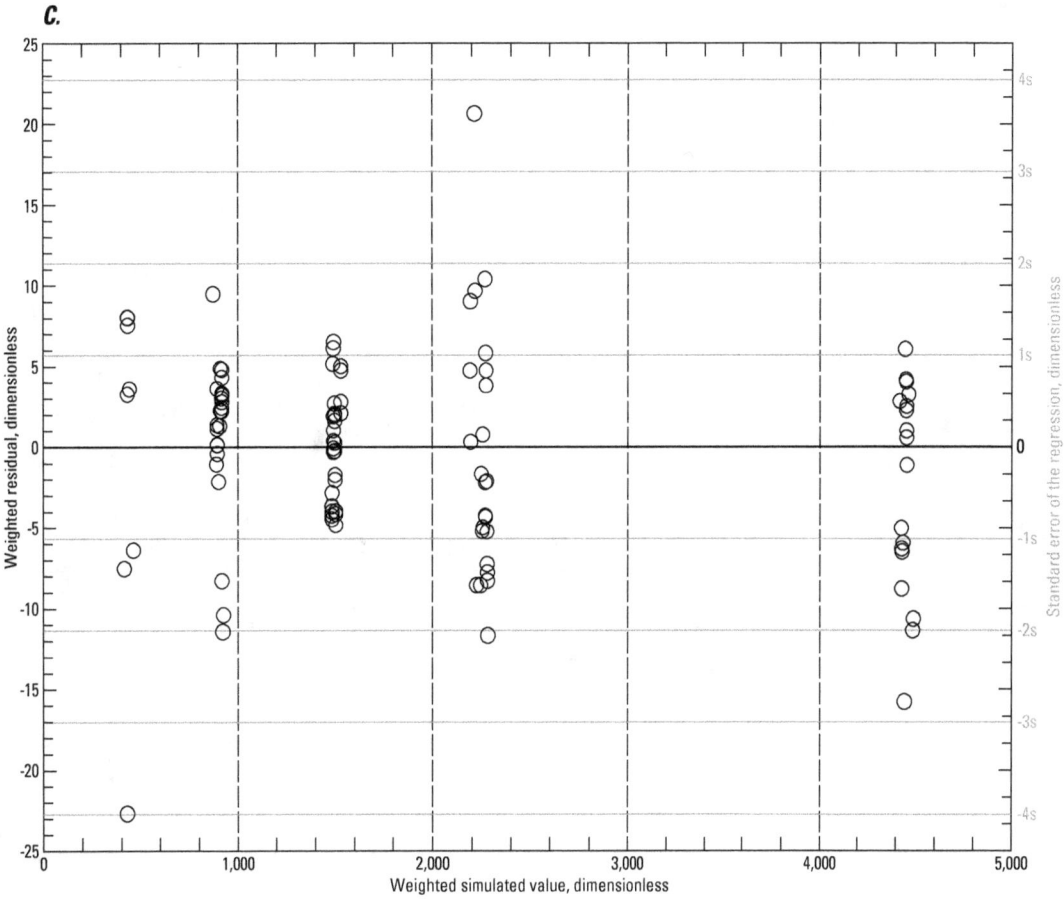

EXPLANATION

○ **Weighted head residual**—in terms of s, standard error of the regression

Note: Positive residual indicates that the weighted observed value is greater than the weighted simulated value.

Figure 24.—Continued.

D.

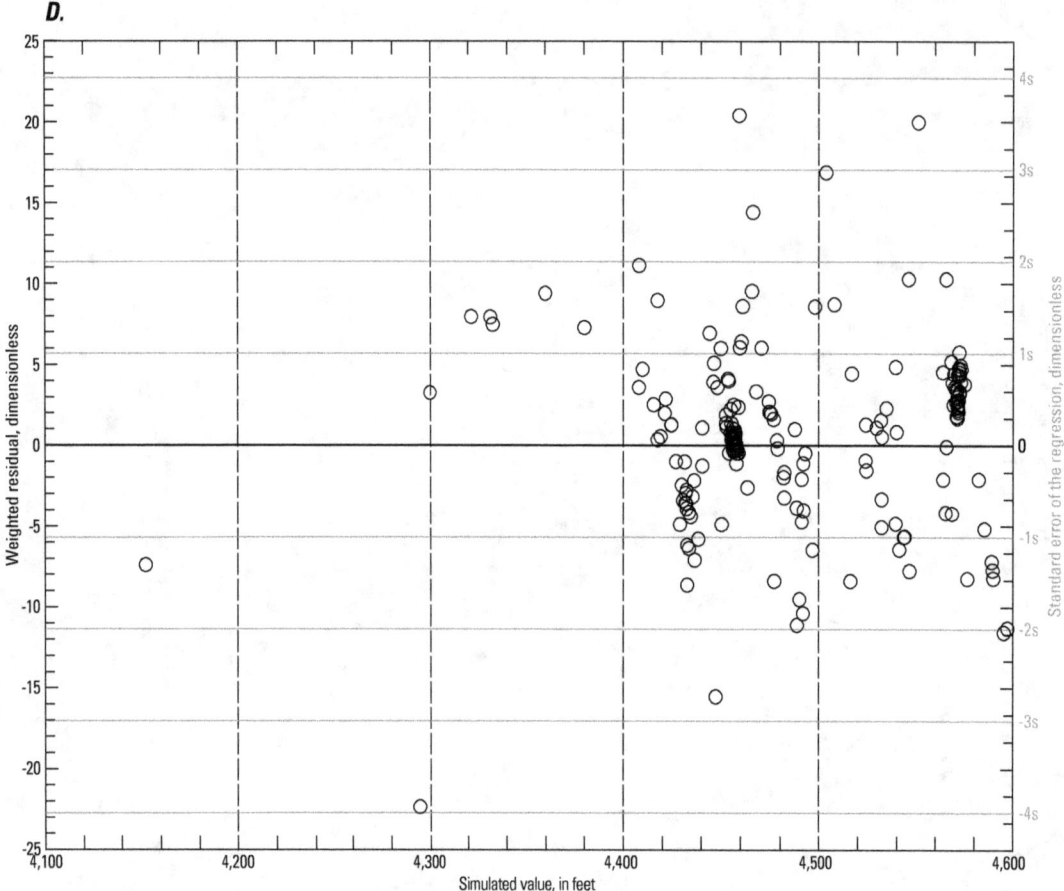

○ **Weighted head residual**—in terms of s, standard error of the regression

Note: Positive residual indicates that the weighted observed value is greater than the weighted simulated value.

Figure 24.—Continued.

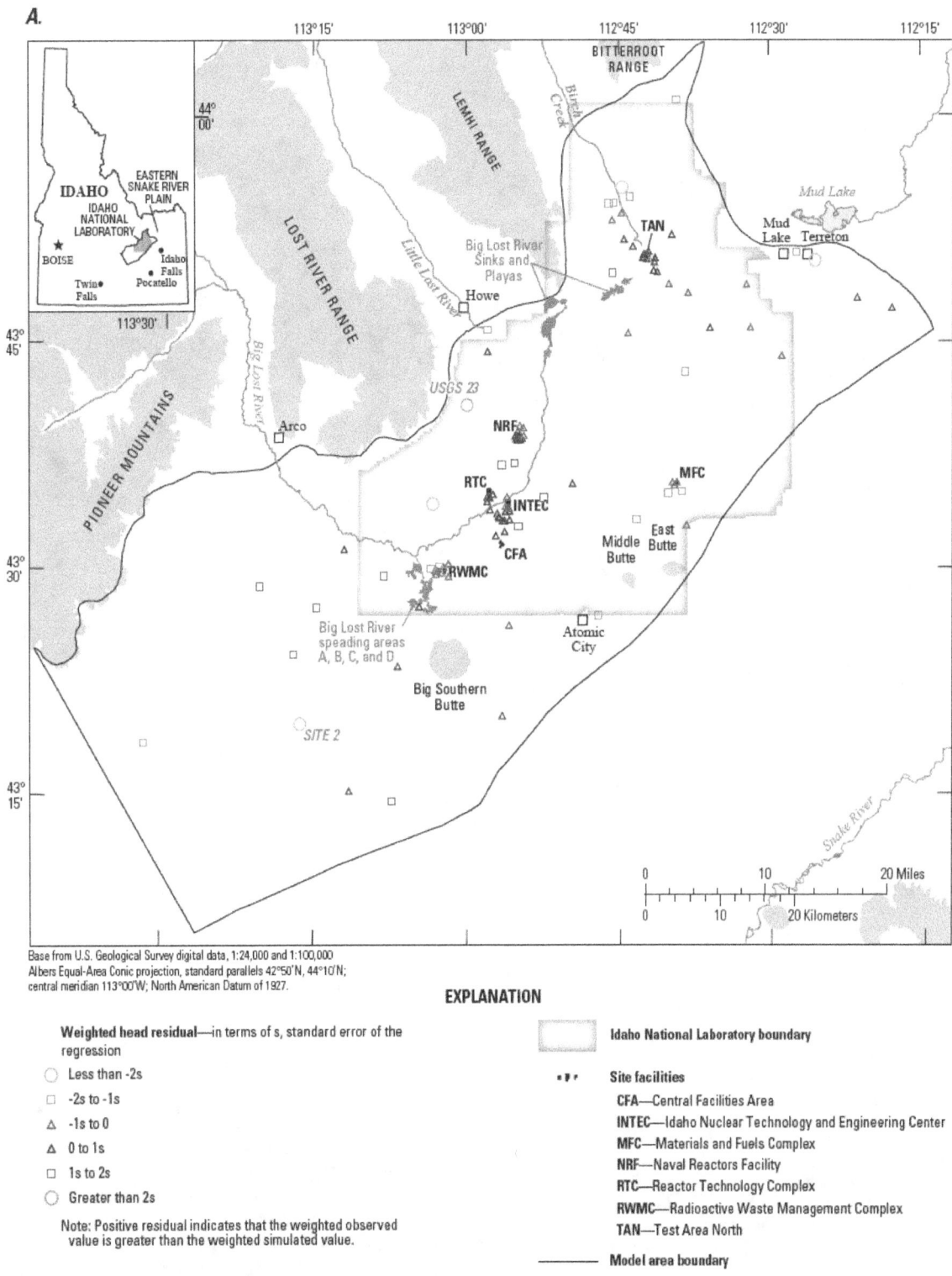

A.

EXPLANATION

Weighted head residual—in terms of s, standard error of the regression

○ Less than -2s

□ -2s to -1s

△ -1s to 0

▲ 0 to 1s

□ 1s to 2s

◯ Greater than 2s

Note: Positive residual indicates that the weighted observed value is greater than the weighted simulated value.

▭ Idaho National Laboratory boundary

•▪▶ Site facilities

CFA—Central Facilities Area

INTEC—Idaho Nuclear Technology and Engineering Center

MFC—Materials and Fuels Complex

NRF—Naval Reactors Facility

RTC—Reactor Technology Complex

RWMC—Radioactive Waste Management Complex

TAN—Test Area North

—— Model area boundary

Base from U.S. Geological Survey digital data, 1:24,000 and 1:100,000
Albers Equal-Area Conic projection, standard parallels 42°50'N, 44°10'N;
central meridian 113°00'W; North American Datum of 1927.

Figure 25.   Distribution of weighted residuals of head for (*A*) wells open exclusively to model layer 1, (*B*) wells not open exclusively to model layer 1, and (*C*) all wells for Big Lost Trough model of steady-state groundwater flow, Idaho National Laboratory and vicinity, Idaho.

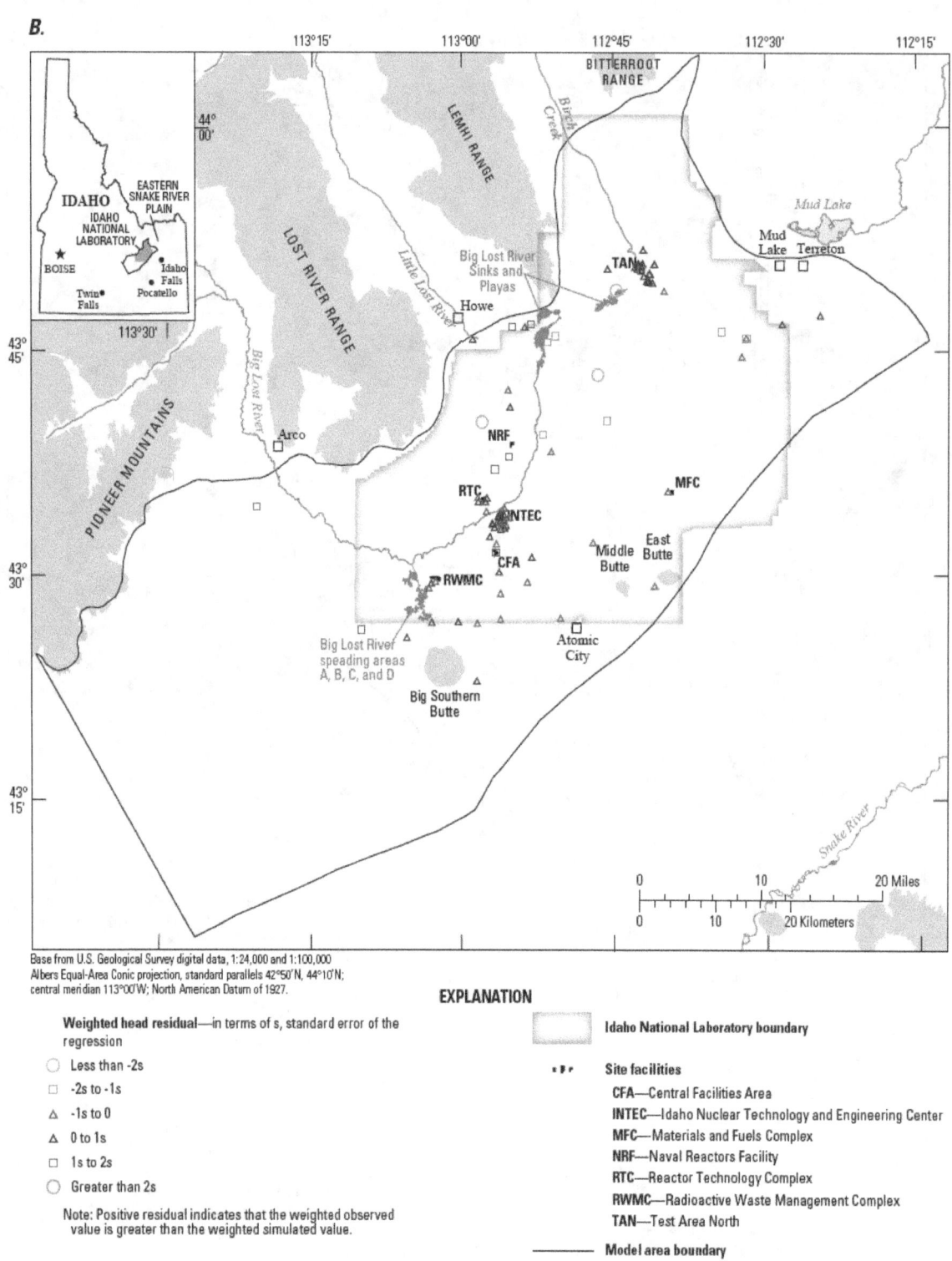

B.

Base from U.S. Geological Survey digital data, 1:24,000 and 1:100,000
Albers Equal-Area Conic projection, standard parallels 42°50'N, 44°10'N;
central meridian 113°00'W; North American Datum of 1927.

EXPLANATION

Weighted head residual—in terms of s, standard error of the regression

○ Less than -2s

□ -2s to -1s

△ -1s to 0

▲ 0 to 1s

▣ 1s to 2s

◯ Greater than 2s

Note: Positive residual indicates that the weighted observed value is greater than the weighted simulated value.

▢ Idaho National Laboratory boundary

• • • **Site facilities**

CFA—Central Facilities Area
INTEC—Idaho Nuclear Technology and Engineering Center
MFC—Materials and Fuels Complex
NRF—Naval Reactors Facility
RTC—Reactor Technology Complex
RWMC—Radioactive Waste Management Complex
TAN—Test Area North

——— Model area boundary

Figure 25.—Continued.

C.

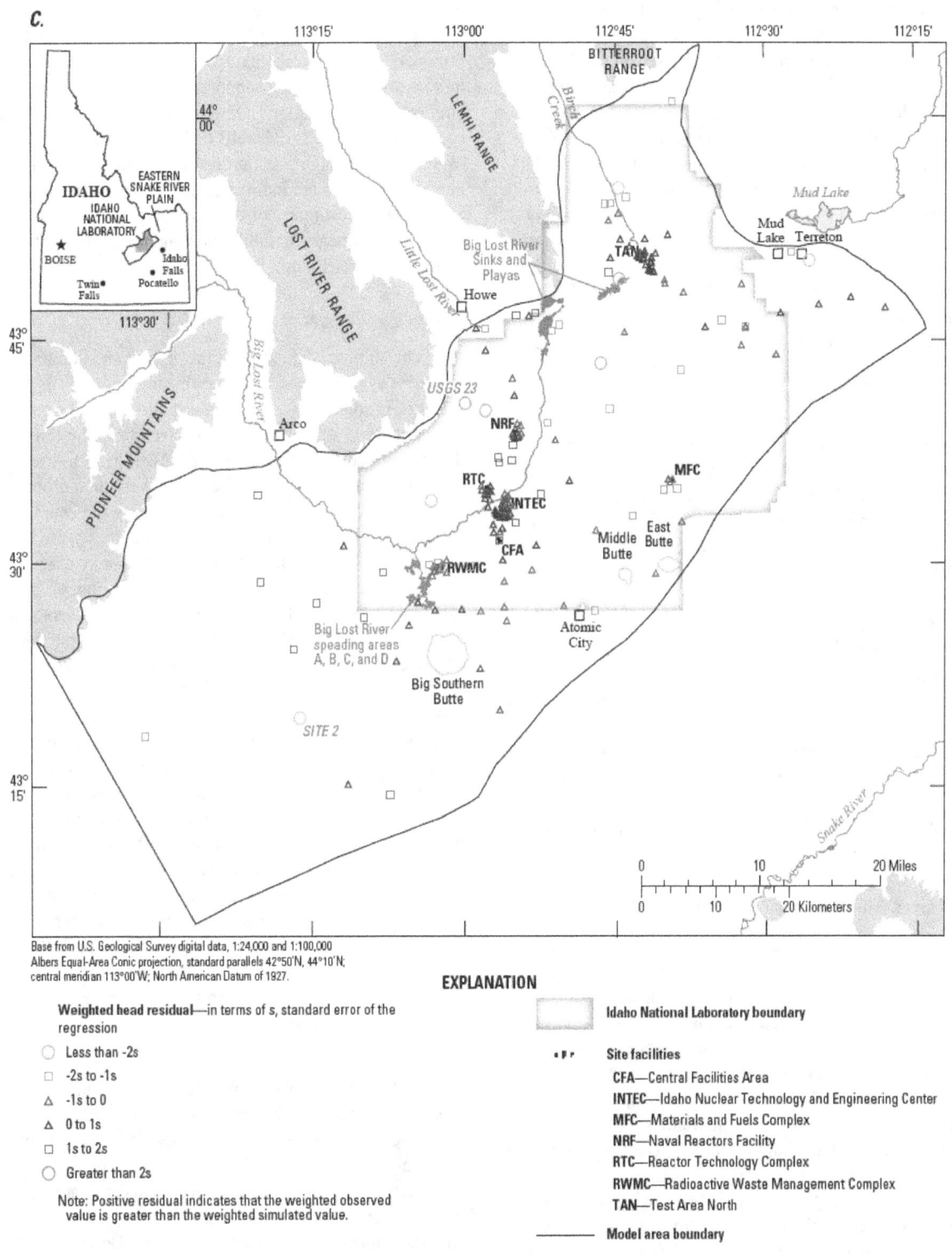

Base from U.S. Geological Survey digital data, 1:24,000 and 1:100,000
Albers Equal-Area Conic projection, standard parallels 42°50'N, 44°10'N;
central meridian 113°00'W; North American Datum of 1927.

EXPLANATION

Weighted head residual—in terms of s, standard error of the
regression

○ Less than -2s

□ -2s to -1s

△ -1s to 0

△ 0 to 1s

□ 1s to 2s

○ Greater than 2s

Note: Positive residual indicates that the weighted observed
value is greater than the weighted simulated value.

☐ Idaho National Laboratory boundary

◦ ⌐ ꞁ **Site facilities**

CFA—Central Facilities Area

INTEC—Idaho Nuclear Technology and Engineering Center

MFC—Materials and Fuels Complex

NRF—Naval Reactors Facility

RTC—Reactor Technology Complex

RWMC—Radioactive Waste Management Complex

TAN—Test Area North

——— **Model area boundary**

Figure 25.—Continued.

The largest weighted residuals generally are in areas mentioned in the section Altitude and Shapes of Simulated and Observed Water Table. The largest positive weighted residual in the model, 20.4 for well USGS 23, was near the mouth of the Little Lost River valley (fig. 25A). Poorer model fit at this location was a result of few water-level data available to constrain the simulated water table and few geologic data available to constrain the conceptualization of the complex geologic transition from shallow mountain-valley aquifers to the deeper ESRP aquifer. Water levels from just one well near the mouth of the Big Lost River valley and from only a few wells near the mouth of the Little Lost River valley were available for contouring water levels and changes in gradient in these areas. Large weighted residuals resulted from the inability to represent the complex hydrogeology accurately near the mouths of these tributary valleys.

The four largest negative residuals (fig. 25A) are scattered throughout the aquifer, but the largest negative residual, -22.4 at well Site 2, is southwest of the INL half way to the southwest border. In the southwest one-third of the model area all maps show a broad area of positive weighted residuals. The model error reflected by this area of positive residuals is a result of the limited geologic and water-level data for this area. The uniform estimated value for HC3 in this area strongly influenced the simulated water table, which had a fairly uniform gradient to the southwest of the INL (fig. 23). This is in contrast to the water-table gradients estimated from field observations, which ranged from steep to moderately flat (fig. 23).

In the southeast corner of the INL near the southeast boundary of the model area (fig. 25) scattered values of dominantly negative weighted residuals may indicate a possible bias toward higher simulated values in this area, but most values are relatively small, between +1s and -1s.

Analysis of Alternative Model Conceptualizations

A series of alternative model conceptualizations were evaluated to examine the effect of certain aspects of the conceptual model and numerical model characteristics and assumptions on the model simulation results. The calibrated Big Lost Trough steady-state model, or base case steady-state model, was evaluated by examining the response of the model to alternate implementations of the parameter structure, hydrogeologic framework, or boundary conditions. The initial comparison of different representations of the sediment content (parameter structures) was the first such analysis (tables 9 and 10; fig. 21). For the base case model, conceptualizations of two hydrogeologic framework alternatives and eight changes to boundary condition quantities and areas were evaluated. Eight of these alternative model analyses are compared quantitatively (table 14). The effects of the alternative representations primarily are quantified

in relation to the base case values of hydraulic conductivity (HC) parameters. The implementation of the hydrogeologic framework in the conceptual model was examined using two modifications, which are supported by geologic evidence and alternative interpretations. Other than measured withdrawals or injection for industrial wells and infiltration of streamflow estimated from gaging station records, estimates of flux at model boundaries are indirect estimates from various sources. Some of these estimates of flux may be of greater importance than other estimates for calibration and the analysis of contaminant transport.

Thickness of Aquifer

The altitude of the base of the aquifer and aquifer thickness as interpreted from geophysical surveys (fig. 6) exceeds, in some areas, the values of altitude of the base of the aquifer and aquifer thickness derived from the few wells that penetrated the full aquifer thickness (fig. 4A). Alternate conceptualizations of the aquifer framework arise from using the geophysical surveys or from interpreting the data from the available data from deep wells. To test the sensitivity of the model to the aquifer thickness (alternative model A), the saturated thickness was set to a maximum of values (1,200 ft) interpreted from well information, (Anderson and Liszewski, 1997, p. 11). This adjustment limited the thickness of model layer 6 to 400 ft. The alternative model was similar in most calibration criteria and parameter values (table 14). Most estimated values for parameters were slightly smaller or larger. Parameter HC3 was greater than the base case value (27 percent different, 0.12 orders of magnitude) and outside the confidence interval for the base case model. This difference did not alter the dominance of horizontal flow in the uppermost part of the aquifer at the INL and downward flow downgradient of the INL. The model was not sensitive to a change in aquifer thickness where the depth to the base as interpreted from geophysical surveys was deeper than any thickness derived from borehole logs. Expensive drilling to gain additional data about the thickness of the aquifer would therefore not seem to be necessary for improving the model.

Domes and Buttes

Early in the calibration process using parameter estimation, the sensitivity of the model was evaluated for including the presence of a less permeable hydrogeologic unit near East and Middle Buttes corresponding to the intermediate composition rocks represented by parameter HC6. The presence in the aquifer of these rocks associated with the buttes is documented only in well Corehole 1 (fig. 1). Because of a slight improvement in calibration criteria, and because this change simulated an observed (Ackerman and others, 2006, p.22), albeit slight, deflection in the

Table 14. Regression statistics and estimated values of parameters for base case and alternative steady-state models of groundwater flow, Idaho National Laboratory and vicinity, Idaho.

[Parameter values in foot per day except for VANI Estimated parameter values for alternative models in **bold** are outside the confidence intervals for the calibrated model **Alternative model:** A, thickness of the aquifer limited to largest value from well data; B, flow across NE bound × 1 2; C, Flow across NE bound × 0 8; D, flow across LLR bound × 1 2; E, flow across LLR bound × 0 8; F, flow from BLR × 0 85; G, SW bound head increases with depth; H, SW bound head decreases with depth **Abbreviations:** LLR, Little Lost River; BLR, Big Lost River; VANI, ratio of horizontal to vertical hydraulic conductivity]

Regression statistics	Base case model	Alternative model								
		A	B	C	D	E	F	G	H
Sum of squared weighted residuals	6,082	5,742	6,144	6,050	6,017	6,238	6,068	6,111	6,052
Average weighted residual	0.51	0.53	0.55	0.50	0.52	0.56	0.51	0.52	0.53
Standard error of the regression	5.61	5.45	5.64	5.60	5.58	5.68	5.61	5.63	5.60

Estimated parameter values

Hydraulic property parameter	Calibrated value	95-percent confidence interval		Estimated value							
		Lower limit	Upper limit								
HC1	11,700	10,100	13,400	11,900	13,300	10,100	12,100	11,300	11,800	11,600	11,700
HC11	227	176	294	264	253	211	253	210	228	229	228
HC2	384	243	606	471	450	317	401	364	386	386	376
HC22	4,780	3,670	6,240	5,270	**6,510**	**2,940**	4,420	4,980	4,610	4,740	4,740
HC3	435	377	500	**571**	**509**	**365**	440	437	431	459	414
HC4	9,890	1,740	56,200	9,860	16,200	5,870	8,840	11,400	9,980	9,740	9,960
HC44	285	224	364	329	318	257	279	295	285	286	287
VANI	14,800	7,510	29,200	26,600	18,300	11,700	13,400	17,000	14,800	14,900	14,700

water-table gradients in the area, the inclusion of the hydraulic conductivity zone representing these rocks was adopted without further analysis in the calibrated model and in all alternative models.

Northeast Boundary Flux

The largest component of flow into the aquifer is across the northeast boundary. To assess the effect of uncertainty in the quantity of flow for this boundary condition, flow was increased (alternative model B) and decreased (alternative model C) by 20 percent and the alternative model was recalibrated. Multiplying the boundary flow by a factor of 1.2 produced a recalibration that did not improve regression statistics compared to the base case model, and simulated higher hydraulic conductivities and VANI (table 14). Alternative model B values were well within the confidence intervals of the base case model except for parameters HC22 and HC3. Confidence intervals for HC22 and HC3 in

alternative model B overlapped those of the base case model and alternative values were larger by 32 and 16 percent (0.14 and 0.07 orders of magnitude), respectively.

Multiplying the boundary flow by a factor of 0.8 (alternative model C) resulted in a slight improvement in the sum of squared weighted residuals and lower simulated hydraulic conductivities and VANI compared to the base case scenario. Confidence intervals for parameters HC22 and HC3 overlapped those of the calibrated model and alternative values were smaller by 46 and 17 percent (0.20 and 0.08 orders of magnitude), respectively.

The model calibration is apparently sensitive to a decrease in the inflow along the northeast boundary. Of the estimates of flow entering the system in the model area, the underflow from the aquifer upgradient of the model area (northeast) is most important because the underflow has some measurable influence on parameter (hydraulic property) estimates. This change did not alter the dominance of horizontal flow in the uppermost part of the aquifer at the INL and downward flow down gradient of the INL.

Little Lost River Tributary-Valley Underflow

The Little Lost River tributary-valley underflow is positioned upgradient of and across the Big Lost River from the central area of concern for contaminant transport modeling. The confidence in the quantity of inflow across the boundary, estimated from basin budget analysis, is poor. In comparison, the estimates for the Big Lost River and Birch Creek tributary valleys are fair (Kjelstrom, 1986). To evaluate the potential sensitivity of the Little Lost River tributary-valley underflow boundary condition, inflow was increased (alternative model D) and decreased (alternative model E) 20 percent in a manner similar to that for the northeast boundary. Recalibration of the model with increased underflow from the Little Lost River valley improved the sum of squared errors and standard error of the regression. Alternative model D and E parameter values were smaller and larger than base case values (table 14). No alternative parameter values were outside of the confidence intervals for the base case parameters and all were within 0.05 order of magnitude of the calibrated values. Recalibration of the model with decreased underflow from the Little Lost River valley did not improve calibration criteria. No alternative parameter values were outside of the confidence intervals for the calibrated parameters, and all were within 14 percent (0.07 order of magnitude) of the base case values. Early in the calibration process using both trial-and-error analysis and parameter estimation, the distribution of underflow from the Little Lost River tributary valley aquifer was tested using different rates and proportions of flow among the model layers at the boundary. The simulations of underflow only to model layer 1 resulted in much higher heads in layer 1. Simulations with underflow in model layer 4 also were unsatisfactory due to poor model convergence. Limited water-table head data from Little Lost River Valley and sparse information on variations in head with depth hinder improvement of the model in this area.

Big Lost River Streamflow Infiltration Area

Local-scale experiments documenting horizontal movement of infiltration from the Big Lost River in the unsaturated zone of as much as 0.8 mi (Nimmo and others, 2002) indicate that the discretization of infiltration may be better represented with an increased area of influence. Early in the calibration process using parameter estimation, the model was recalibrated with infiltration applied to areas increased by 1 and 2 mi from the areal traces of rivers or boundaries of spreading areas, sinks, and playas. No substantial change in parameter values was observed. No further evaluation was done for the calibration and sensitivity analysis for the model. Due to the cell size used for discretization, the implementation of the conceptual model inherently treats the area of influence for river channels with a minimum of 1,320 ft. River channel width in the model area averages 38 ft and ranges from 23 to 48 ft (Nace and Barraclough, 1952, p. 15, table 5). Calibration of the steady-state model is insensitive to extending the footprint of the Big Lost River within the range of 0.25 to 2 mi.

Big Lost River Infiltration and Evaporation

The steady-state infiltration of streamflow is a locally significant proportion of inflow to the aquifer (Ackerman and others, 2006, p. 33). The sensitivity of the model to reductions in estimates of infiltration from Big Lost River due to evaporation or inaccurate measurements (alternative model F) was estimated by recalibration with a 15 percent decrease in infiltration. The reduction of streamflow infiltration by evaporation losses was not considered in the conceptual model. Consideration of evaporation from the Big Lost River is included in the refinement of the conceptual model in the transient calibration.

A system-wide decrease of 15 percent in the total amount of infiltration from the Big Lost River, whether representing evaporation loss or error in estimating streamflow infiltration, does not have much influence on the calibrated values of hydraulic conductivity parameters. Alternative model F was nearly identical in calibration criteria and parameter values (table 14). No alternative model F parameter values were outside of the confidence intervals for the calibrated parameters and all were within 3 percent (0.01 order of magnitude) of the calibrated values. This supports the assumption of steady-state conditions for streamflow infiltration being represented by a long-term average and that large departures from the estimated average would not affect steady-state results for the estimation of hydraulic conductivity.

Head Distribution on Southwest Boundary

Little data are available to describe the hydrology of the aquifer near the downgradient boundary (figs. 4D and 8). Data on head differences with time are sparse, and data on head differences with depth are non-existent. The sensitivity of the model to head differences with depth along the southwest model boundary was tested by recalibrating the model with an increase (alternative model G) and a decrease (alternative model H) in head of 2.5 ft per 100 ft of depth, which is a vertical gradient of 0.025, similar to the vertical gradient (0.021–0.031) measured at the USGS 30 piezometer nest (tables 13 and 18). The estimated parameter values in the alternative models were nearly identical to the values in the calibrated model (table 14) with a small, 6 percent (0.02 order of magnitude), increase or decrease in the value of parameter HC3. All other parameters were within 1 percent (0.01 order of magnitude) of the calibrated values. No alternative model parameter values were outside of the confidence intervals for the base case parameters. The model calibration is only slightly sensitive to changes in the head with depth near the downgradient (southwest) boundary.

Transient Calibration

Understanding the aquifer response to transient stresses is important because groundwater and contaminant movement through the aquifer may be substantially affected by these stresses. To investigate the aquifer response, the transient flow model was calibrated over a 16-year period of record (1980–95) that included a 5-year wet cycle (1982–86) followed by an 8-year dry cycle (1987–94) (fig. 10). Modeling groundwater flow during this period provided an opportunity to evaluate the relation between transient stresses and temporal changes in the altitude of the water table and groundwater flow directions. Modeling temporal changes in the altitude of the water table also allowed the specific yield of the hydrogeologic units present at the water table to be estimated. These values of specific yield can be used along with field measurements of porosity and other information to approximate the effective porosities of the hydrogeologic units, which is a necessary hydraulic parameter for simulating the velocities and travel times of groundwater flow and groundwater contaminants.

The three-dimensional, six-layer unconfined transient model was calibrated with nonlinear regression methods. The steady-state model provided the initial conditions for the transient model, and the hydrogeologic zones and model parameters are the same as for the steady-state model. The calibration observations included heads during the steady-state modeling period and head differences (Hill 1998, p. 33) during the transient modeling period. Qualitative and quantitative evaluation of the model results was performed through (1) statistical analyses of the estimated parameters, (2) visual comparison of the distribution of observed and simulated head differences and flow directions throughout the model area, and (3) statistical analyses of model fit and the distribution of weighted residuals for head differences.

Initial Conditions

Initial head conditions for the transient model were the 1980 conditions simulated by the steady-state model. The presence of approximate steady-state aquifer conditions in 1980 was supported by observed fluxes in 1980 that were nearly equal to average fluxes for the period 1950–80, a period that followed initial development of the aquifer for agriculture (Garabedian, 1992; Ackerman, 1995). The period after 1980 was characterized by some of the wettest and driest years on record (Pittman and others, 1988; Bartholomay and others, 1995, 1997, 2000) and provided a clear contrast with the period prior to 1980, in terms of hydraulic stresses and responses. The top of model layer 1 was increased 25 ft to accommodate water-table rise during transient simulations.

Model Parameters

Transient model parameters included specific yield (SY), specific storage (SS), HC, and VANI (table 8). The parameters SS and VANI had constant fixed values for the entire model domain. The HC parameter definition was the same as in the steady-state model. An SY parameter was defined for each hydrogeologic zone at the water table in model layer 1. Parameter-estimation model runs were attempted with (1) HC parameters specified and SY parameters estimated and (2) HC and SY parameters estimated.

For the model runs in which HC parameter values were specified, five SY parameters (SY1, SY2, SY3, SY11, and SY44) were estimated; all other model parameters (all HC parameters, VANI, SS, SY4, SY6, and SY22) were fixed (table 8). SY33 has no parameter because hydrogeologic zone 33 is absent in model layer 1. The HC and VANI parameter values were fixed at the values estimated or specified for the steady-state model. The fixed value for parameter SS was set at 0.000015 ft^{-1}. This value is the average of the geometric means of the minimum and maximum estimated SS values for fissured and jointed rock and dense sand (Domenico, 1972, table 5.1), materials that compose the bulk of the aquifer. Simulations with a range of values for the parameter SS indicated that the model was less sensitive (little change in model fit) for values corresponding to the range of fissured and jointed rock and very sensitive (decrease in model fit and parameter estimation would not converge) with larger values of SS in the range of dense sand (fig. 26). Fixed parameter values for SY were set at 0.05 for SY4 and SY6 and 0.15 for SY22. These values reflect the assumption that the SY of basalts in hydrogeologic zones 4 and 6 is small and that the SY of hydrogeologic zone 22 is large because of the presence of abundant sediment in this zone.

Regression modeling with SY and HC parameters estimated (joint parameterization) was attempted but was unsuccessful. One five-parameter simulation (three HC and two SY parameters) with a lowered closure tolerance and reduced maximum parameter change slightly improved the simulation of differences, but with one parameter correlation of 0.98 and strongly degraded the simulation of head. In the joint parameterization regression models, the observations of head difference modified hydraulic conductivity values smoothing the water-table gradients and degrading the effect of sedimentary features on the head distribution. The observations were only sufficient to estimate one specific yield parameter corresponding to areas of sediment influence. The combination of head and head-difference observations supported too few model parameters in joint parameterization to match the overall spatial quality of simulated head from the steady-state HC parameters. Consequently, all the following calibrated transient model results were derived from the regression run in which only SY parameters were estimated.

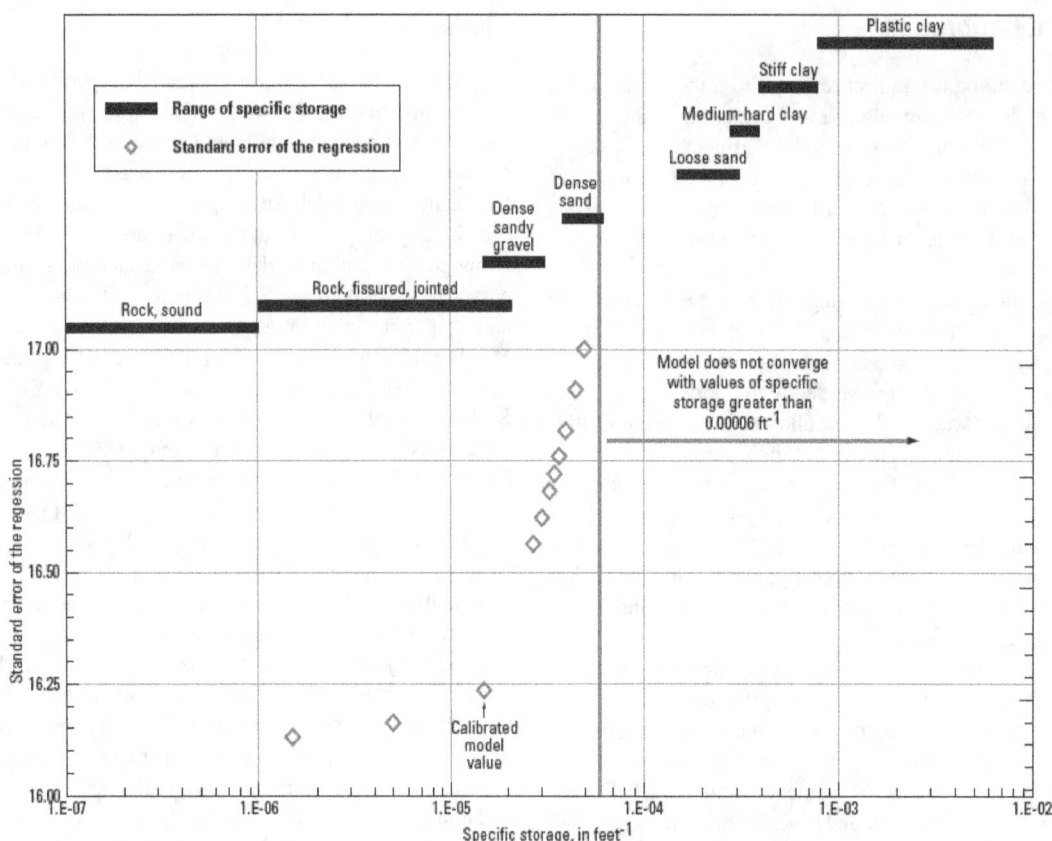

Figure 26. Ranges of values and sensitivity of the objective function to values of specific storage for model of transient groundwater flow, Idaho National Laboratory and vicinity, Idaho. Specific storage data from Domenico (1972, table 5.1).

Model Results and Evaluation

Transient model results included estimated values for specific yield and the simulated hydraulic head and flux for each active cell in the model area. Methods for evaluating model results and the conceptualization and implementation of the model were similar to those described in the section Steady-State Calibration.

Estimates of Specific Yield

Estimated values of SY parameters ranged from 0.028 to 0.115 (table 15). In a manner similar to the steady-state model, confidence in the estimated parameter values was evaluated by (1) comparing field or laboratory values, (2) checking parameter uniqueness with parameter correlation coefficients, (3) comparing linear confidence intervals to expected ranges, and (4) identifying whether parameters were significantly influenced by a single observation using Cook's D and DFBETAS influence statistics. Using these criteria, greater confidence is warranted for the estimates of parameters SY1, SY11, and SY44 than for the estimates of SY2 and SY3.

Table 15. Estimates of specific yield, expected intervals, and 95-percent confidence intervals for calibrated model of transient groundwater flow, Idaho National Laboratory and vicinity, Idaho.

[Specific yield is dimensionless **Abbreviation**: SY, specific yield]

Hydraulic property parameter	Expected interval		Estimated value	95-percent confidence interval	
	Lower limit	Upper limit		Lower limit	Upper limit
Parameters corresponding to hydrogeologic zones in areas where the presence of sediment does not affect hydraulic property estimates					
SY1	0.01	0.30	0.072	0.068	0.077
SY2	.01	.30	.115	.099	.133
SY3	.01	.30	.055	.039	.078
Parameters corresponding to hydrogeologic zones where the presence of sediment does affect hydraulic property estimates					
SY11	0.01	0.30	0.072	0.066	0.077
SY44	.01	.30	.028	.023	.035

All the estimated SY parameter values were within the range of expected values (fig. 27) for aquifers, 0.01 to 0.30 (Freeze and Cherry, 1979, p. 61), and the range of values for porosity of basalt and other estimates of specific yield for basalt (table 3; Ackerman and others, 2006, p. 20-23). The values also were consistent with ranges of values of porosity for dense, vesicular, and fractured basalts in the Columbia River Plateau (Freeze and Cherry, 1979, table 4.1). The values for SY1 and SY2, 0.072 and 0.115 (table 15), respectively, are the reverse of expected relative values for these parameters based on an expectation that porosity and specific yield are related. Conceptually, hydrogeologic unit 2 consists of thicker, denser, basalt flows, fewer fractures, and fewer interflow zones than hydrogeologic unit 1. Based on these hydrogeologic features and on published values of porosity for dense and fractured basalt in the Columbia River Plateau

(Freeze and Cherry, 1979, p. 162, table 4.1) the porosity, and consequently the SY, of the fractured basalt of hydrogeologic unit 1 may be expected to be larger than the porosity of the dense basalt of hydrogeologic unit 2. No data are available to support this expectation. The amount of the total porosity that drains will be related to interconnectedness and the capillary characteristics of water in the pore and fracture structure. Aquifer tests, often used to estimate specific yield, have not yielded much information on the storage properties of the aquifer. The availability and problems associated with large-scale aquifer tests in the vicinity of the INL are described in Ackerman (1991).

Parameter correlation coefficients for the transient model parameters were all less than 0.58. These small correlation coefficients are well below the critical value of 0.95, and indicate that all the parameters were estimated uniquely.

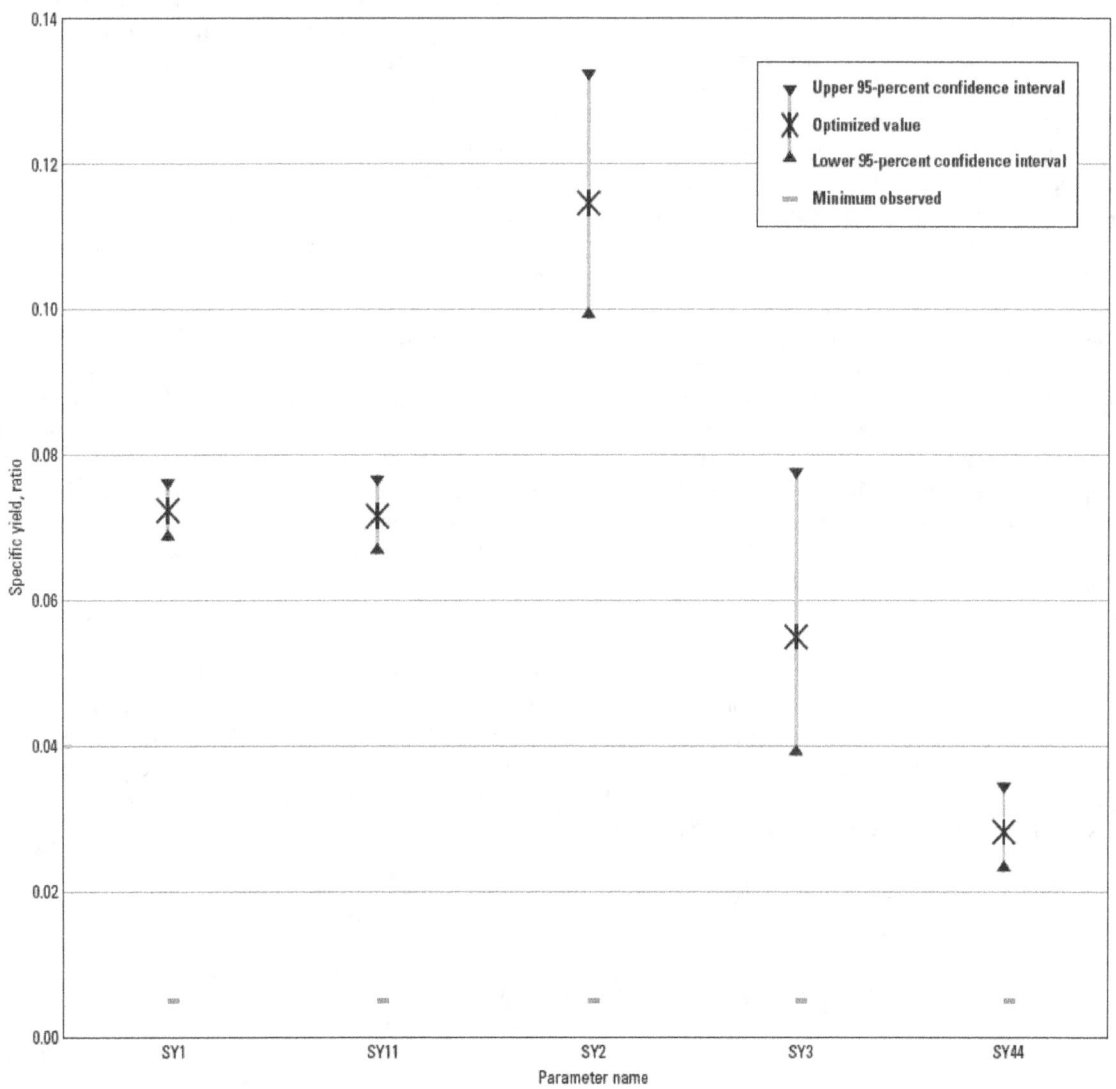

Figure 27.   Estimated values of specific yield with 95-percent confidence intervals for model of transient groundwater flow, Idaho National Laboratory and vicinity, Idaho. Maximum expected values of specific yield are greater than 0.14.

All the 95 percent linear confidence intervals of the SY parameters were within the ranges of expected values for these parameters (fig. 27). Confidence intervals for SY2 and SY3, however, were two to three times larger than confidence intervals for SY1, SY11, and SY44, but still cover relatively narrow ranges of values. These large confidence intervals generally are associated with SY parameters of hydrogeologic zones that contain the fewest observations (table 6) and have the smallest CSS (SY2, SY3, and SY44; fig. 28). The SY parameter values not estimated were most often those with lowest relative values of CSS (fig. 28) indicating that less information is provided by observations for defining those parameters and estimating parameter values. The smaller confidence intervals for SY1 and SY11 also may reflect their proximity to the Big Lost River (fig. 19A). Large changes in water-table altitudes occur near the Big Lost River in response to highly variable amounts of streamflow infiltration. These large changes in water-table altitude increase the sensitivity of simulated head differences in these areas.

As with the steady-state model, the linear confidence intervals were calculated assuming independent and normally distributed errors and model linearity, and the accuracy of the confidence intervals decreases if these assumptions are not met. The independence and normality of the weighted residuals were evaluated using a normal probability plot of the ordered weighted residuals and the associated correlation coefficient, R^2_N. On the normal probability graph, the weighted residuals nearly form a straight line (fig. 29), which indicates that the residuals deviate from being normally distributed. R^2_N was 0.959, which was smaller than the critical value of 0.987 (at the 5 percent significance level), below which residuals are not likely independent or normally distributed. The linearity of the model was evaluated using the modified Beale's measure. The calculated value of 0.078 is less than the criteria of 0.44 for non-linearity, but greater than the criteria of 0.040 for effective linearity (table 16). The nearly normal distribution of the weighted residuals and the moderate nonlinearity of the model indicate that the calculated linear confidence intervals for the parameters may be fairly accurate.

The influence of specific transient head observations on the set of estimated parameter values and on individual parameter estimates was evaluated with Cook's D and the DFBETAS statistic, respectively. Cook's D values greater than the critical value of 0.00047 were computed for 684 transient head observations and DFBETAS values greater than the critical value of 0.022 were computed for 1,454 observations. All these observations had the potential to significantly influence parameter values.

The 10 largest Cook's D values and 10 of the 11 largest DFBETAS values corresponded to the 5 observations at well USGS 8 between May and September of 1984, a period that includes the peak of the water-level rise in 1984 at this well. The Cook's D and DFBETAS values also correspond to the five observations at well Site 2 between April 1993 and April 1995, a period that includes the lowest water levels at this well during the simulation period. Well USGS 8 is west of the INL, about 1 mi southwest of the Big Lost River where it enters the INL (fig. 1). The aquifer near the water table, model layer 1, in this location consists of hydrogeologic zone 2 (figs. 13 and 19A). Well Site 2 (fig. 1; fig. B1) is approximately 10 mi southwest of the southwest corner of the INL, and the aquifer near the water table in this location consists of hydrogeologic zone 3 (figs. 13 and 19A).

Two model recalibrations were done, the first without the five observations from well USGS 8 and the second without the observations from well Site 2. The first recalibration resulted in an increase in the 95 percent confidence intervals of all the estimated parameters and an increase in parameter value SY2 from 0.115 to 0.102 (table 17). The new estimated value for SY2 was inside the original confidence interval for this parameter (fig. 27). The second recalibration resulted in a small increase in the 95 percent confidence interval for SY3 and an increase in SY3 from 0.055 to 0.081 (table 17). The new estimated value for SY3 was outside the original confidence interval for this parameter (fig. 27), but within the range of expected values. All other new estimated parameter values were within the original 95 percent confidence intervals for the parameters. Parameters SY2 and SY3 are influenced by single observations more than are other estimated parameters, likely because hydrogeologic zones 2 and 3 have only 2 and 7 wells, respectively, open only to model layer 1, whereas zones 1, 11, and 44 have 27, 36, and 22 wells, respectively, open only to model layer 1 (table 6). Although the 95 percent confidence intervals and influence statistics show that the estimate of SY2 is more uncertain than the estimates of parameters SY1, SY11, and SY44, the estimate of SY2 still is considered reasonably well constrained because its confidence interval spans a relatively small range of values.

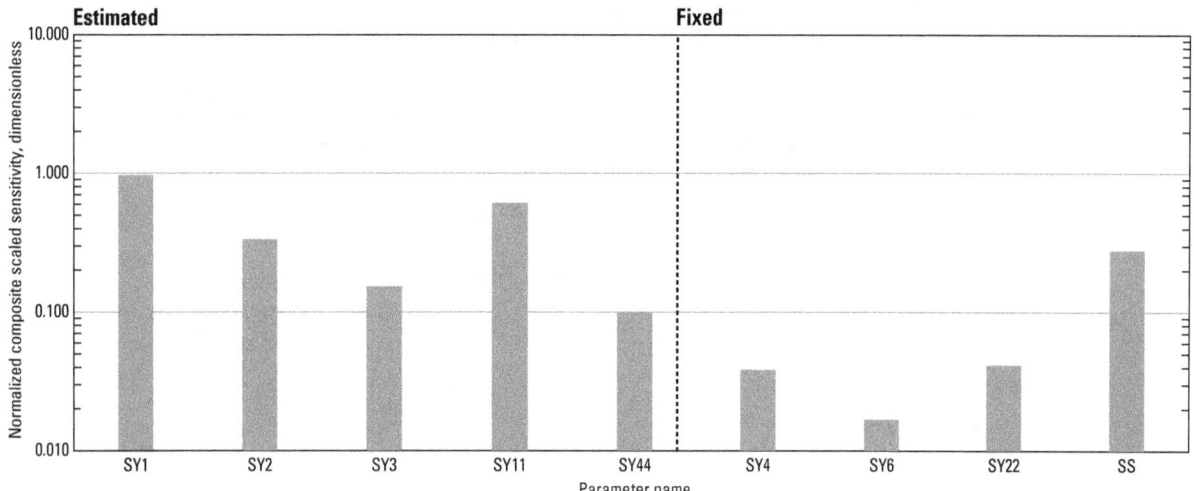

Figure 28. Composite scaled sensitivities for model of transient groundwater flow, Idaho National Laboratory and vicinity, Idaho.

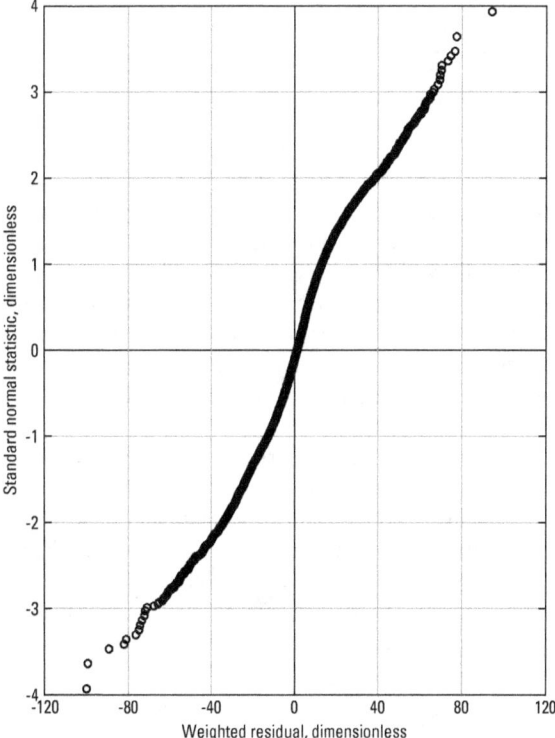

Figure 29. Normal probability plot for model of transient groundwater flow, Idaho National Laboratory and vicinity, Idaho.

Table 16. Statistical measures related to parameter uncertainty and overall goodness of model fit for model of transient groundwater flow, Idaho National Laboratory and vicinity, Idaho.

[All statistics dimensionless R^2_N, correlation coefficient between ordered weighted residuals and normal order statistics]

Statistic	Parameter structure
	Big Lost Trough sediment
Sum of squared weighted residuals	2,226,390
Average weighted residual [1]	.80
Parameter correlation coefficients [2]	None
Standard error of the regression	16.32
R^2_N	.959
Modified Beale's measure	.078

[1] Positive average weighted residual indicates that, on average, weighted observed values were greater than weighted simulated values

[2] Parameter correlation coefficients greater than 0 90

Table 17. Summary of observations and corresponding parameters with highest values of influence measures for model of transient groundwater flow, Idaho National Laboratory and vicinity, Idaho.

[Critical values for Cook's D and DFBETAS were 0 00047 and 0 022, respectively Recalibrated parameter values result from omitting five observations for recalibration **Parameter:** Associated with DFBETAS statistic **Abbreviations**: SY, specific yield]

Observations		Cook's D	Parameter	DFBETAS	Specific yield	
Well	Date				Base case model	Recalibrated model
USGS 8	July 1984	0.0374	SY2	-0.431	0.115	0.102
USGS 8	June 1984	.0290		-.380		
USGS 8	August 1984	.0277		-.372		
USGS 8	September 1984	.0209		-.322		
USGS 8	May 1984	.0205		-.319		
Site 2	April 1995	.0290	SY3	-.372	.055	.081
Site 2	September 1994	.0270		-.359		
Site 2	September 1995	.0253		-.347		
Site 2	September 1993	.0217		-.322		
Site 2	March 1994	.0207		-.314		

Distribution of Head and Inferred Flow Directions

Comparison of simulated and observed heads and flow directions were used to evaluate how well the transient model represented recharge and flow in the ESRP aquifer. A qualitative evaluation of the model was made through comparison of simulated and observed (1) changes in water levels and horizontal flow directions and (2) changes in vertical gradients and flow directions. Quantitative evaluation of the model was done by analyzing the model fit and the distribution of weighted residuals for head-difference observations.

As in the steady-state model, simulated equivalents to observed heads were calculated by MODFLOW-2000 and used spatial interpolation in the horizontal plane and calculation of multilayer heads for wells open to multiple model layers. In the transient model, simulated equivalents to observed heads also were calculated by temporal interpolation. Simulated head differences were calculated from these values.

The quality of the transient model can be demonstrated by its ability to simulate the head difference with time and the differing magnitude and timing of water-level changes in response to changes in locations of infiltration. The quality of the model also is demonstrated by comparing the simulated changes in vertical gradients and the distribution of residuals in time and space. Simulation of transient conditions gave an accurate representation of observed changes in the flow system resulting from episodic infiltration from the Big Lost River. The simulation facilitated understanding and visualization of the relative effects on heads and flows caused by (1) extended periods of flow and drought related to climate variability; (2) differences in infiltration between channels, playas, and spreading areas; and (3) changes in distribution of infiltration resulting from regulating diversion to the spreading areas.

Simulation of transient conditions did not reproduce observed annual fluctuations of water levels over the northeast one-third of the model area and produced limited annual fluctuations of water levels in the northeast corner of the model area. These regular observed annual fluctuations, believed to originate in response to irrigation in the northeast corner of the model area and irrigation to the north of the model area, are superimposed on the other responses, which were demonstrated by the model. Overall, the quality of the simulation of water-level change is good, especially in areas with abundant head data that coincides with the area of greatest interest near observed contamination of the aquifer in the southwestern part of the INL.

Changes in Water Levels and Flow Directions

As discussed in the previous section on steady-state calibration, observed water-level altitudes were not matched exactly. A comparison of observed and simulated heads (fig. 30) shows that general geometry and altitude of the water table is well simulated in the model. At any particular well location, however, simulated and observed heads will differ. Rather than compare just simulated and observed water-levels for transient model analysis, hydrographs (fig. 31) also show residuals of head differences (differences between simulated head difference and observed head difference).

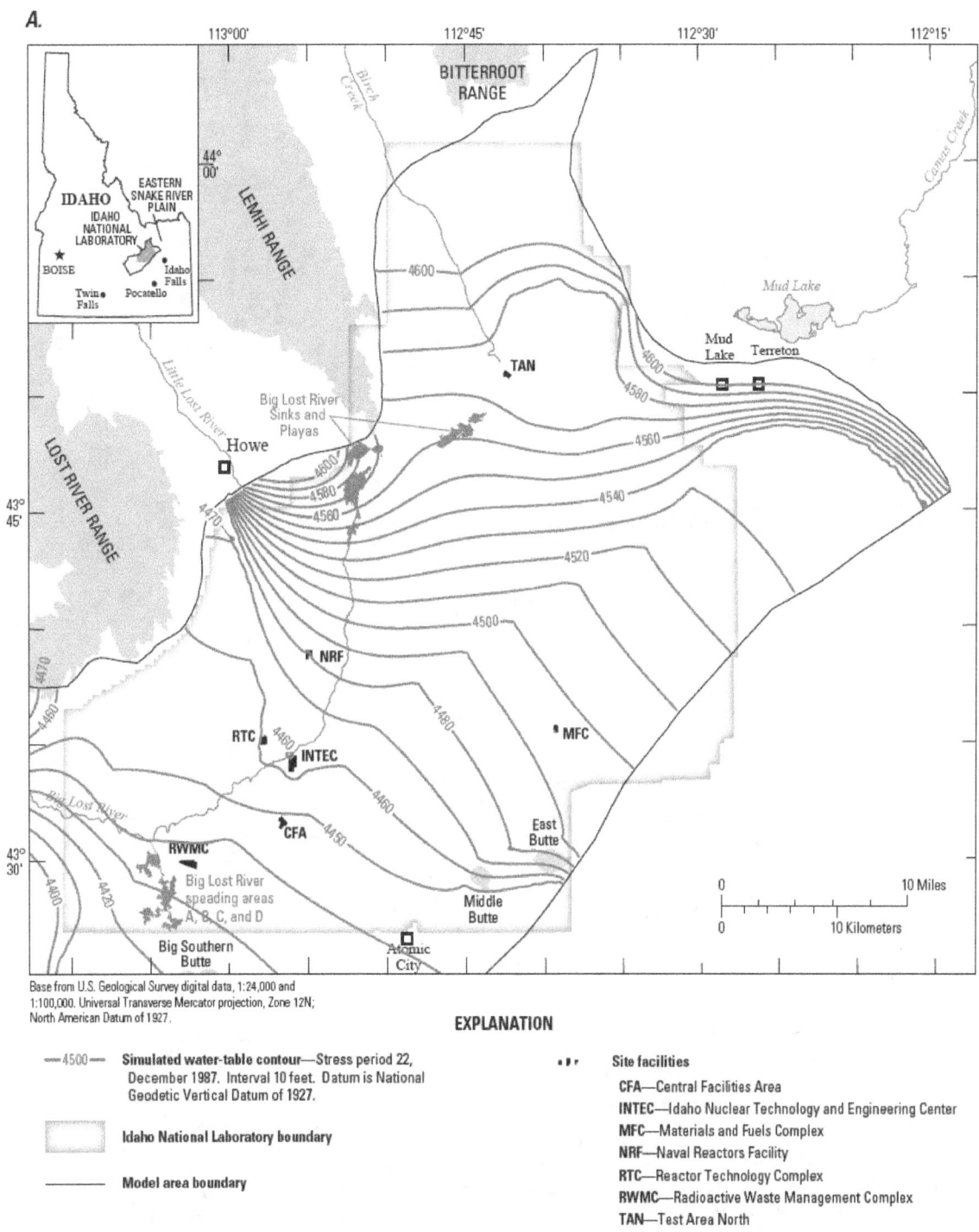

A.

Base from U.S. Geological Survey digital data, 1:24,000 and
1:100,000. Universal Transverse Mercator projection, Zone 12N;
North American Datum of 1927.

EXPLANATION

—4500— Simulated water-table contour—Stress period 22,
December 1987. Interval 10 feet. Datum is National
Geodetic Vertical Datum of 1927.

Idaho National Laboratory boundary

Model area boundary

◦ ▸ ▸ Site facilities

CFA—Central Facilities Area
INTEC—Idaho Nuclear Technology and Engineering Center
MFC—Materials and Fuels Complex
NRF—Naval Reactors Facility
RTC—Reactor Technology Complex
RWMC—Radioactive Waste Management Complex
TAN—Test Area North

Figure 30.   Comparison of (*A*) simulated water table, December 1987 and (*B*) observed and simulated water table, April 1995, Idaho National Laboratory and vicinity, Idaho.

B.

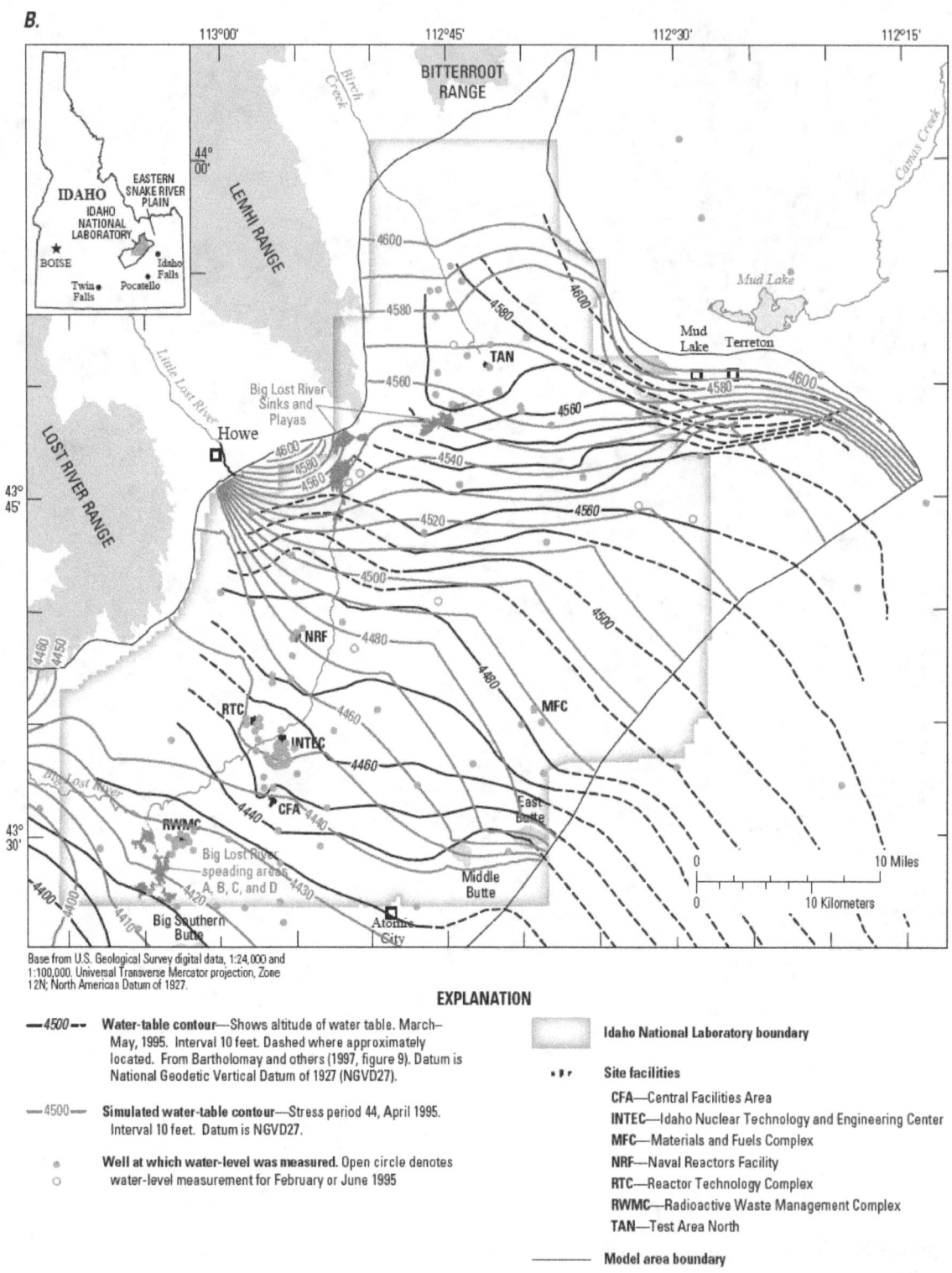

Base from U.S. Geological Survey digital data, 1:24,000 and
1:100,000. Universal Transverse Mercator projection, Zone
12N; North American Datum of 1927.

EXPLANATION

—*4500*--   **Water-table contour**—Shows altitude of water table. March–
May, 1995. Interval 10 feet. Dashed where approximately
located. From Bartholomay and others (1997, figure 9). Datum is
National Geodetic Vertical Datum of 1927 (NGVD27).

—*4500*—   **Simulated water-table contour**—Stress period 44, April 1995.
Interval 10 feet. Datum is NGVD27.

    ●   **Well at which water-level was measured.** Open circle denotes
    ○   water-level measurement for February or June 1995

    Idaho National Laboratory boundary

● ▪ ▼   **Site facilities**

    CFA—Central Facilities Area
    INTEC—Idaho Nuclear Technology and Engineering Center
    MFC—Materials and Fuels Complex
    NRF—Naval Reactors Facility
    RTC—Reactor Technology Complex
    RWMC—Radioactive Waste Management Complex
    TAN—Test Area North

———   **Model area boundary**

Figure 30.—Continued.

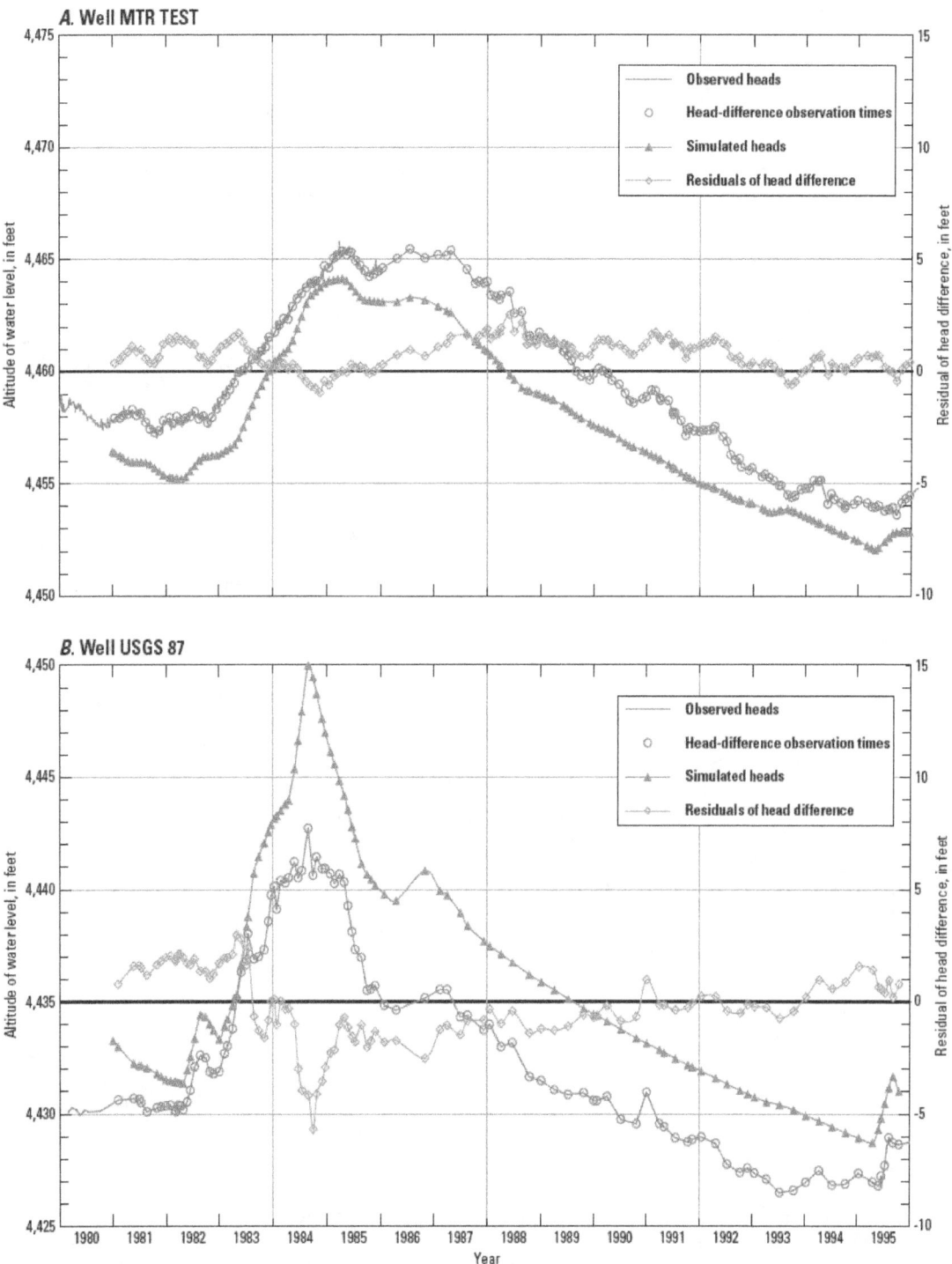

Figure 31. Observed and simulated hydraulic head, head-difference observation times, and residuals of head difference for wells (*A*) MTR TEST, (*B*) USGS 87, (*C*) USGS 25, and (*D*) USGS 18 at various locations in the transient model of groundwater flow, Idaho National Laboratory, Idaho. Residuals of head difference are differences between simulated head difference and observed head difference.

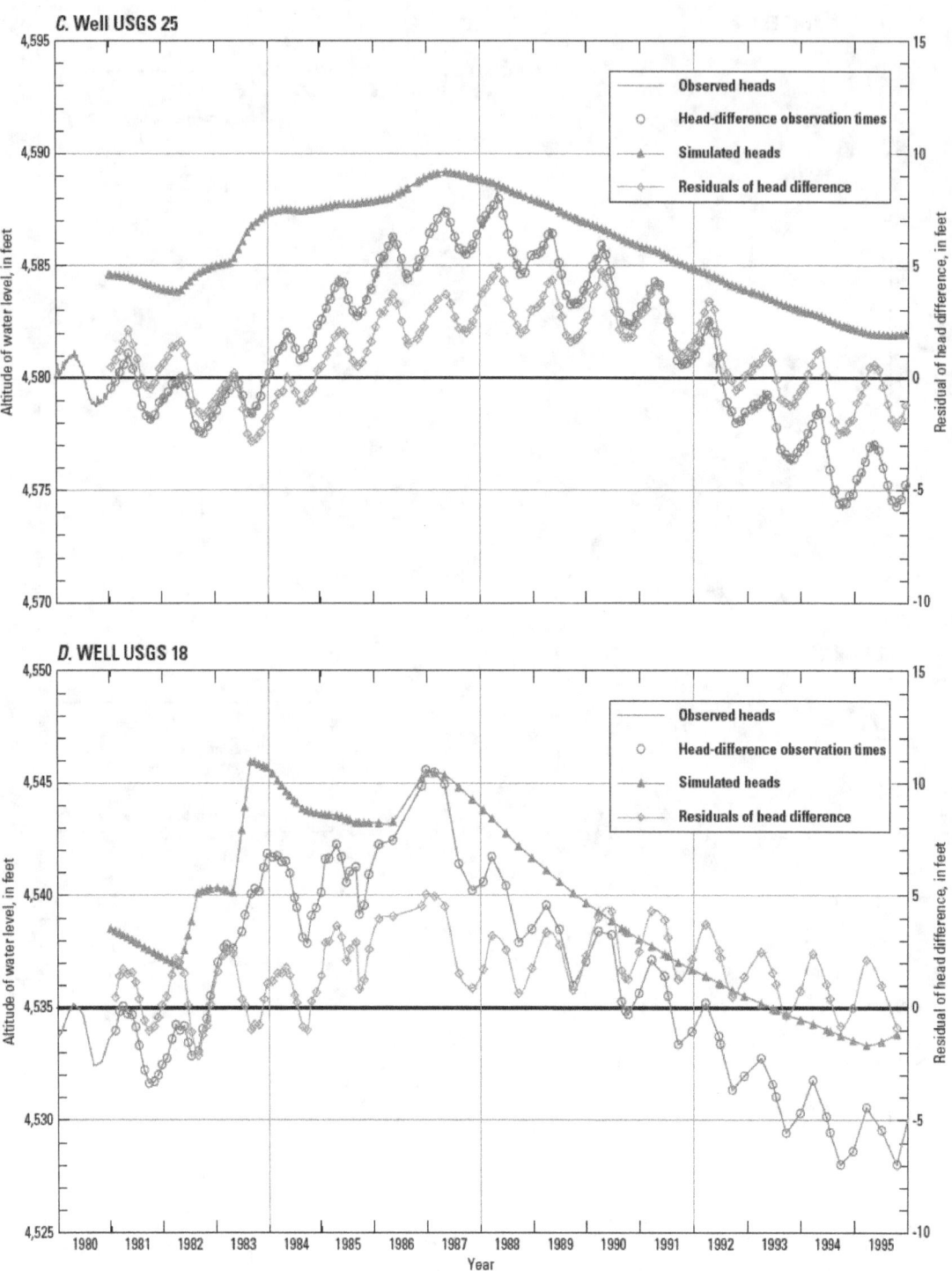

Figure 31.—Continued.

The hydrograph of well MTR TEST (figs. 1 and 31A) is the most complete record representative of water levels near a major area of aqueous waste disposal. The hydrograph shows two peaks, one peak in 1985, and a broad peak from 1986 to 1987. The response in this area corresponds to the pattern of initially higher infiltration and the accumulating effects of infiltration from the channel of Big Lost River between the INL diversion and the Big Lost River Sinks and in the sinks and playas from the beginning of the simulation to 1986 (table C1). Figure 31A demonstrates model fit, the ability of the model to simulate the head difference with time, and the differing magnitude and timing of water-level changes in response to changes in locations of infiltration.

The hydrograph for well USGS 87 (figs. 1 and 31B), just north and east of the spreading areas (fig. 1) shows that simulated heads peaking in the late summer 1984 match the timing of major diversion of streamflow to the spreading areas in 1984 (fig. 10; table C1). Overall, the simulated head differences provide a good match to the observed changes. Overestimates and underestimates of the magnitude of water-level change in this area of the model probably are related to uncertainty in the distribution of infiltration of ungaged flow below the outlet of spreading area A to spreading area B (stream gage 503, fig. B1).

The hydrograph of well USGS 18 (figs. 1 and 31D), representative of water-table conditions at a location upgradient of the major area of aqueous waste disposal and just east of the terminal playas of the Big Lost River, shows the simulation of two peaks as a result of infiltration from the Big Lost River Sinks and Playas. Differences in the magnitude of some of the observed and simulated peaks may be caused by errors in the estimation of the relative amounts of infiltration in the sinks and the separate playas (fig. 10; table C1) in an area with few discharge records. The amount of data from wells representing only model layer 1 (water-table conditions) and the coarse definition of the influence of sediment on hydrologic parameters in the model area also limits the ability of the model to estimate the local variations in specific yield, which in turn affect the simulation of changes in water level in individual wells.

In well USGS 18 (figs. 1 and 31D) and others to the north and east such as USGS 25 (figs. 1 and 31C) water levels reflect a slightly more subdued cycle of changes in water-levels corresponding to the wet period and following dry period simulated in the model compared to changes measured in wells near the spreading areas (fig. 31B). The effects of the episodic infiltration, likely focused in the sinks and playas (fig. 10; table C1) of the Big Lost River, diminish with distance from the Big Lost River and become indistinguishable from a basin-wide change in streamflow infiltration that originates from precipitation in the mountainous tributary basins (Lindholm, 1996, p. 41). Although the simulated response of these wells demonstrates the ability of the model to simulate multi-year processes, they

do not reproduce regular annual fluctuations. In these wells and others to the north and east, the water levels reflect an annual cycle likely due to the seasonal effects of irrigation in the northeast corner of the model area and areas north and east of the model. The influence of irrigation is likely complex due to withdrawals from wells and infiltration of excess applied irrigation from surface water and groundwater sources.

Wells in and adjacent to the northeast corner of the model area have annual water-level fluctuations of approximately 4–7 ft. Wells in the northeast area of the INL, such as wells USGS 27, USGS 25, and USGS 18 (figs. 1, 17, 31C, 31D), have annual fluctuations of between 2 and 4 ft. With increasing distance from the irrigated area in the model, annual fluctuations of water levels diminish. Annual fluctuations are about 2.5 ft in the east-central area of the INL (well USGS 21), 2 ft (well USGS 6) in the central INL, and less than 2 ft (well USGS 2) southwest of MFC (fig. 1). Regular annual fluctuations of water levels are less than 0.5 ft to indistinguishable south and west of a line from NRF to the southeast corner of the INL. For example, the wells MTR TEST and USGS 87 (fig. 31A, B) are west of that line in the area where most waste was disposed and nearly all medium to long half-life contaminants were detected. A fine detail comparison of the responses for wells USGS 27 and MTR TEST is shown in figure 17.

Changes in Vertical Gradients and Flow Directions

Throughout most of the model area, the simulated directions of vertical gradients between model layers rarely changed with time. Exceptions are near INTEC and along the Big Lost River from the INL diversion to the sinks, where simulated directions of vertical gradient between model layers 1 and 2 changed to downward during the extended wet period 1983–87, and then reverted to upward during the following dry period. These changes in directions were caused by changes in pumpage and disposal at INTEC and the influence of infiltration from the Big Lost River channel (table 18).

Changes in simulated vertical gradients in the immediate vicinity of INTEC are influenced by (1) the simulated injection of waste into model layers 1 and 2 from the beginning of the simulation through 1984, (2) recharge from infiltration of industrial wastewater, (3) withdrawals from production wells, and (4) the relatively constant infiltration of streamflow in the channel from the INL diversion to the Big Lost River Sinks from the beginning of simulation to early 1987. The level of discretization used in the calibrated model is sufficient to allow simulated injection of waste, infiltration of wastewater, and industrial withdrawals in separate cells and to avoid summing of simulated inflows and outflows that result in a smaller net stress within a single cell. A larger cell size may not appreciably affect the calibration of the flow model as shown in a sensitivity analysis, use of the model for advective transport analysis or advective/dispersive transport model calibration may benefit from the current cell size.

Table 18. Observed and simulated transient directions and gradients of groundwater flow in open boreholes, piezometer nests, and packer-isolated intervals of wells, Idaho National Laboratory, Idaho.

[Methods, qualifications, and references for observed data in table 13 **Gradient:** Values are dimensionless; data for flow between layers 1 and 2 except as noted **Simulated values:** December 1980, steady-state, stress period 1; August 1982, stress period 9; April 1995, stress period 44 **Abbreviation**: –, no information]

Date	Well name	Observed		Simulated values					
				December 1980		August 1982		December 1995	
		Direction of flow	Gradient	Direction of flow	Gradient	Direction of flow	Gradient	Direction of flow	Gradient
Northeast Idaho National Laboratory and southeast of Test Area North									
1966	USGS 4	Down	0.001 to 0.007	Up	0.003	Up	0.002	Up	0.002
1980–95	USGS 30	Up	[1]0.018 to 0.032	Down	.0002	Down	.022	Down	.023
1980–95	USGS 30	Up	[1]0.021 to 0.032	Up [2]	.044	Up [2]	.044	Up [2]	.043
1963	USGS 31	Up	–	Down	.007	Down	.004	Down	.011
Idaho Nuclear Technology and Engineering Center									
1965	USGS 42	Up	–	Up	0.001	Down	0.006	Up	0.006
1965	USGS 43	Static	–	Up	.001	Down	.010	Up	.010
1991	USGS 44	Up	–	Up	.001	Down	.008	Up	.008
1991	USGS 45	Up	–	Up	.002	Down	.007	Up	.007
1991	USGS 46	Up [3]	–	Down	.001	Down	.007	Up	.007
1963	USGS 47	Up	–	Down	.003	Down	.007	Up	.007
1965	USGS 48	Static [4]	–	Up	.005	Down	.002	Up	.003
1965	USGS 48	Down [4]	–	Up [5]	.008	Up [5]	.008	Up [5]	.011
1965	USGS 49	Down [4]	–	Up	.006	Up	.000	Up	.004
1965	USGS 51	Up	–	Up	.011	Up	.009	Up	.002
1965	USGS 52	Up [6]	–	Up	.004	Down	.000	Up	.010
1965	USGS 59	Up [6]	–	Up	.012	Up	.009	Up	.003
North of Naval Reactors Facility									
1964	Site 17	Down	–	Down	0.013	Down	0.006	Down	0.010
Reactor Technology Complex									
1964	MTR test	Down [7]	–	Up	0.006	Up	0.006	Up	0.012

[1] Range of gradients observed from 1980 to 1995 Water-level data are from http://nwis.waterdata.usgs.gov/id/nwis/gwlevels

[2] Simulated flow between model layers 2 and 4

[3] Observed gradient is down when nearby production well is pumping

[4] Observed direction may be influenced by nearby injection well

[5] Simulated flow between model layers 2 and 3

[6] Observed direction may be influenced by nearby injection or production wells

[7] Observed direction may be influenced by nearby infiltration pond

Analysis of Residuals

As discussed in the previous section on steady-state calibration, weighted residuals that are not randomly distributed can be an indication of model error. Although most discussion in this section describes weighted residuals, which are by definition dimensionless, a more intuitive measure of model fit is available that can be expressed in the original dimensions of measurement (feet). Because all observations of difference in head have the same weight, the fitted standard deviation can be calculated for the transient model. The fitted standard deviation for differences in heads (2.10 ft) is small compared to the range of differences in heads (36 ft). This indicates that errors are only a small part of the overall model response.

Weighted residuals of differences in head were plotted against weighted simulated equivalents (fig. 32) to evaluate if the weighted residuals were randomly distributed. The mean weighted residual of differences in head was near zero (0.82 for 8,171 observations). The weighted residuals within ±2s were randomly distributed about zero for all weighted simulated values. For weighted residuals greater than the absolute value of 2s, weighted simulated values between -50 and 0 have a negative bias, and values between about 10 and 60 have a positive bias.

Residuals of head differences were plotted against time (fig. 33) to evaluate bias with time. The residuals were generally close to zero, but for some wells, residuals varied according to annual cycles or reflected apparent systematic model error due to the resolution of inflow from infiltration of streamflow in playas, sinks, and spreading areas. The residuals with absolute values greater than 2s, as mentioned previously, often correspond to high or low points in the year on the hydrographs with strong annual fluctuations. The trend in residuals with time, if any, is difficult to determine using the scatter plot of residuals in figure 33. The trends in observed head difference and residuals are better summarized by annual summaries (fig. 34). The slight variation in average residuals (fig. 34B) from year to year does not follow the large head differences from year to year (fig. 34A) in magnitude or timing indicating that model error was generally independent of the large simulated changes in streamflow that caused large simulated head differences. The number of observations per year averaged 558 and ranged from 356 in 1986 to 715 in 1984.

The areal distribution of weighted residuals (fig. 35A) shows the location where observed head difference for seven wells is underestimated (positive weighted residual) in the west-central area of the INL (near the NRF) for a time corresponding to the end of the extended wet period. This area generally is where the distribution of infiltration from streamflow is not well documented, the observed head difference is largest, and few wells are completed in model layer 1 only. The highest weighted residuals, for well USGS 98 (77.4) and nearby wells (44.4), may be caused by local sediment interbeds acting as confining beds within the first model layer at and near the water level. The presence of confining beds at or just below the water table has been noted in well completion documents in this area. A temporary change in the configuration of well USGS 98 near NRF in January 1994 resulted in a several foot drop in water level and a downhole video log showed strong downward flow between perforations above and below a 29-ft thick sediment bed. Near the end of the simulated dry cycle, weighted residuals were uniformly smaller and showed little areal bias (fig. 35B).

Hydrographs for several wells (fig. 31) further illustrate the nature and location of higher residuals and probable causes for the errors. Residuals for well MTR TEST (fig. 31A) are only slightly positive for most of the simulation indicating possible minor errors in storage or distribution of infiltration near this well in the area of greatest interest. This simulated response is typical in areas where cyclical annual variations of water levels are absent and infiltration of streamflow is relatively well known.

Residuals for well USGS 87 (fig. 31B) are positive early in the simulation and then negative later, perhaps in response to small errors in the steady-state and initial transient values for infiltration of Big Lost River flow to the spreading areas. Weighted residuals for well USGS 18 (fig. 31D) are generally positive indicating head differences are underestimated, possibly caused by local variations in the distribution of hydraulic conductivity or specific yield that is not represented in the model. The area near well USGS 18 is more abundant with sediment. The weighted residuals show an annual fluctuation indicating that a seasonal head difference is not simulated. Weighted residuals at well USGS 25 (fig. 31C) show the head difference with time and indicates an underestimate of head difference and small difference in timing.

Analyses of Alternative Model Conceptualizations

The calibrated transient model was evaluated by examining the response of the model to alternative conceptualizations and quantities of inflow. In the analyses reported below, the effects of the alternative representations primarily are quantified in relation to the base case values of specific yield (SY) parameters. Results of simulations with alternative model conceptualizations or implementations are compared to the calibrated model simulation results to evaluate the selection of specific storage, uncertainty in boundary conditions, or possible shortcomings in the ability of the model to simulate some observed response.

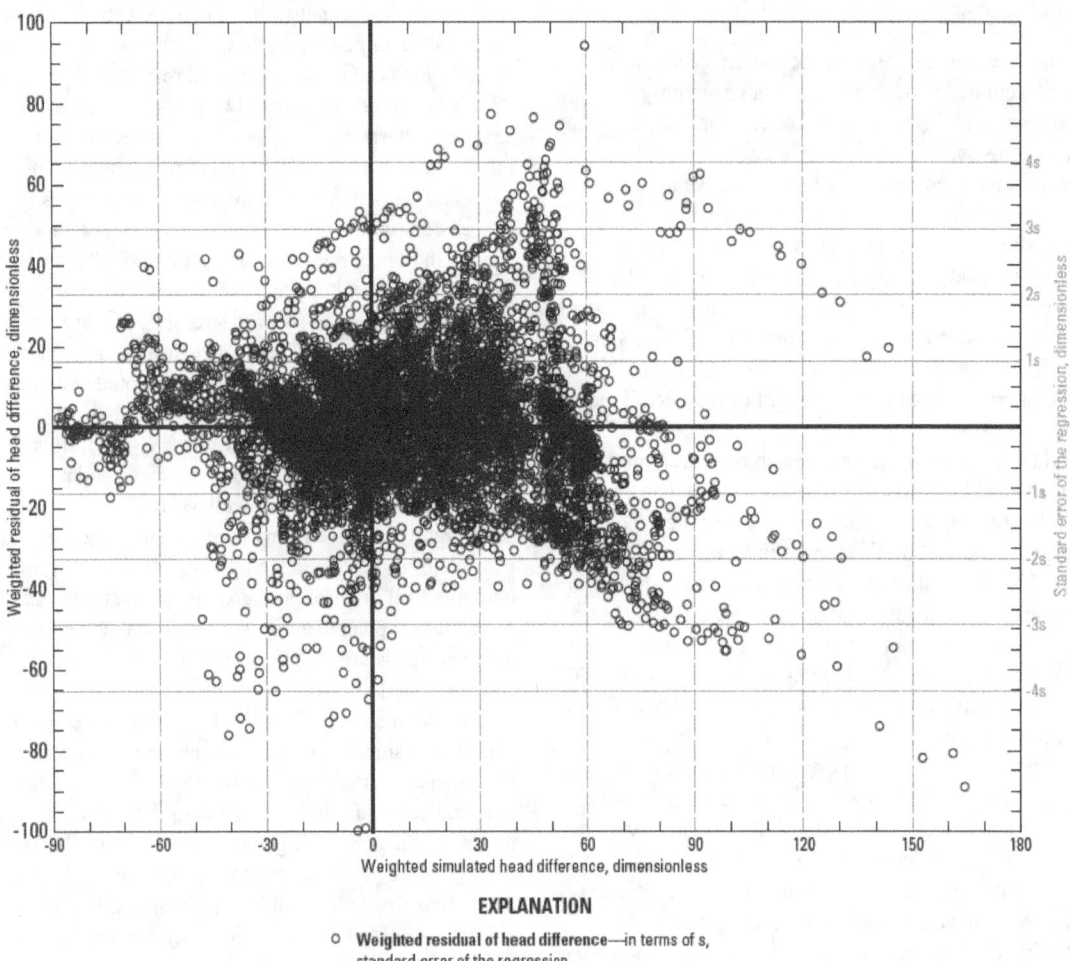

Figure 32.   Distribution of weighted residuals of head difference for model of transient groundwater flow, Idaho National Laboratory and vicinity, Idaho.

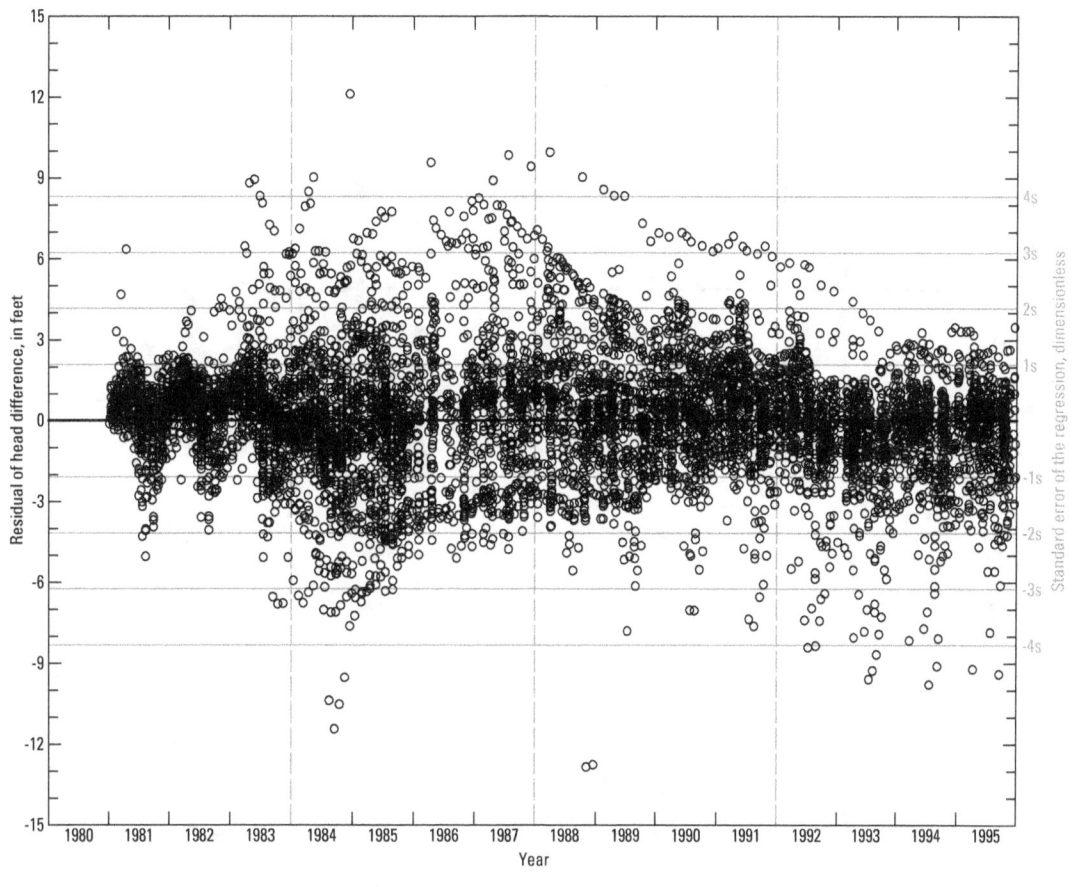

EXPLANATION

○   **Residual of head difference**—in terms of s, standard
    error of the regression

Note: Positive residual indicates that the observed value is
    greater than the simulated value.

Figure 33.   Distribution of weighted residuals of head difference over time for model of transient groundwater
flow, Idaho National Laboratory and vicinity, Idaho.

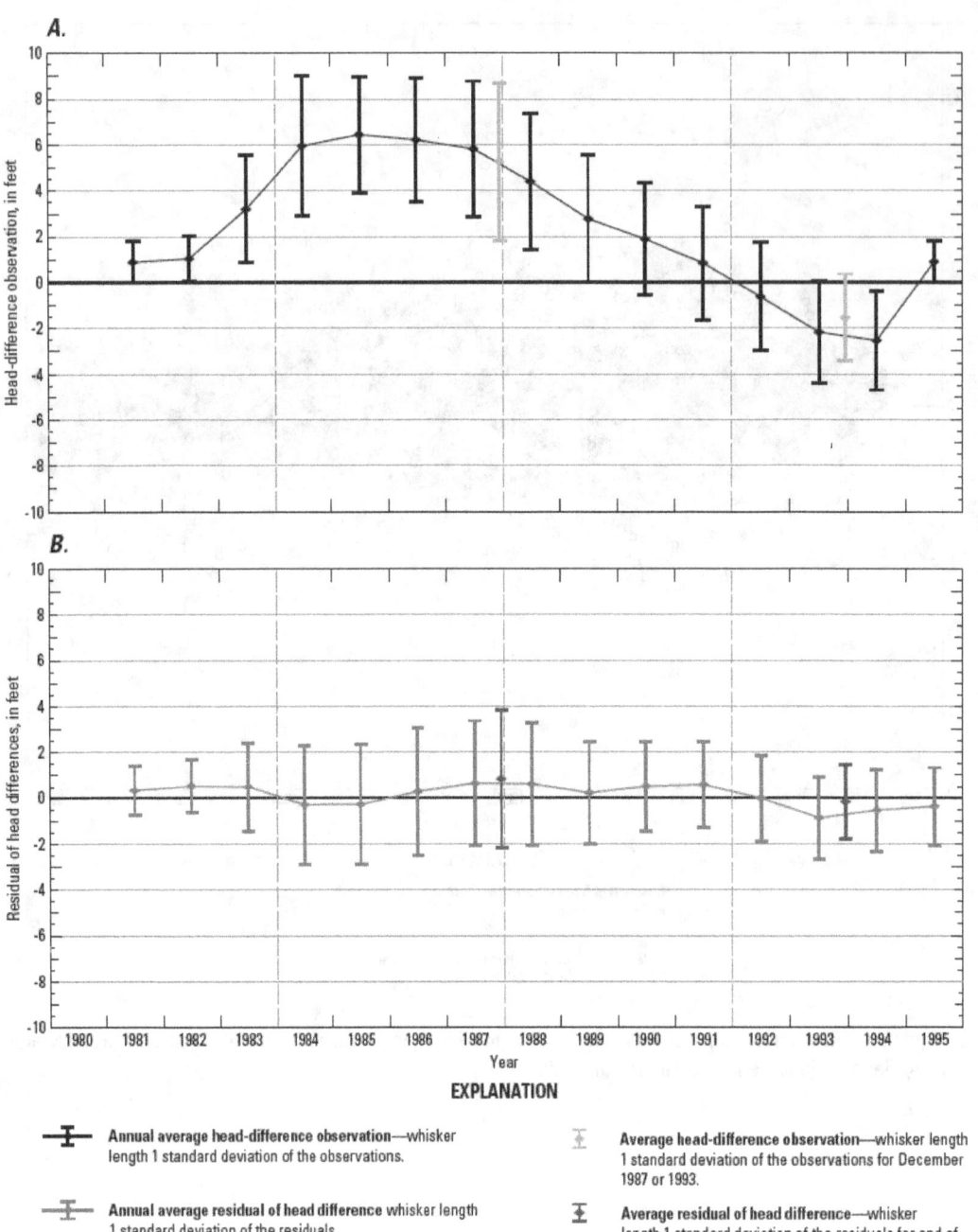

Figure 34. Distribution of annual and select monthly (*A*) head-difference observations and (*B*) residuals of head differences over time for model of transient groundwater flow, Idaho National Laboratory and vicinity, Idaho.

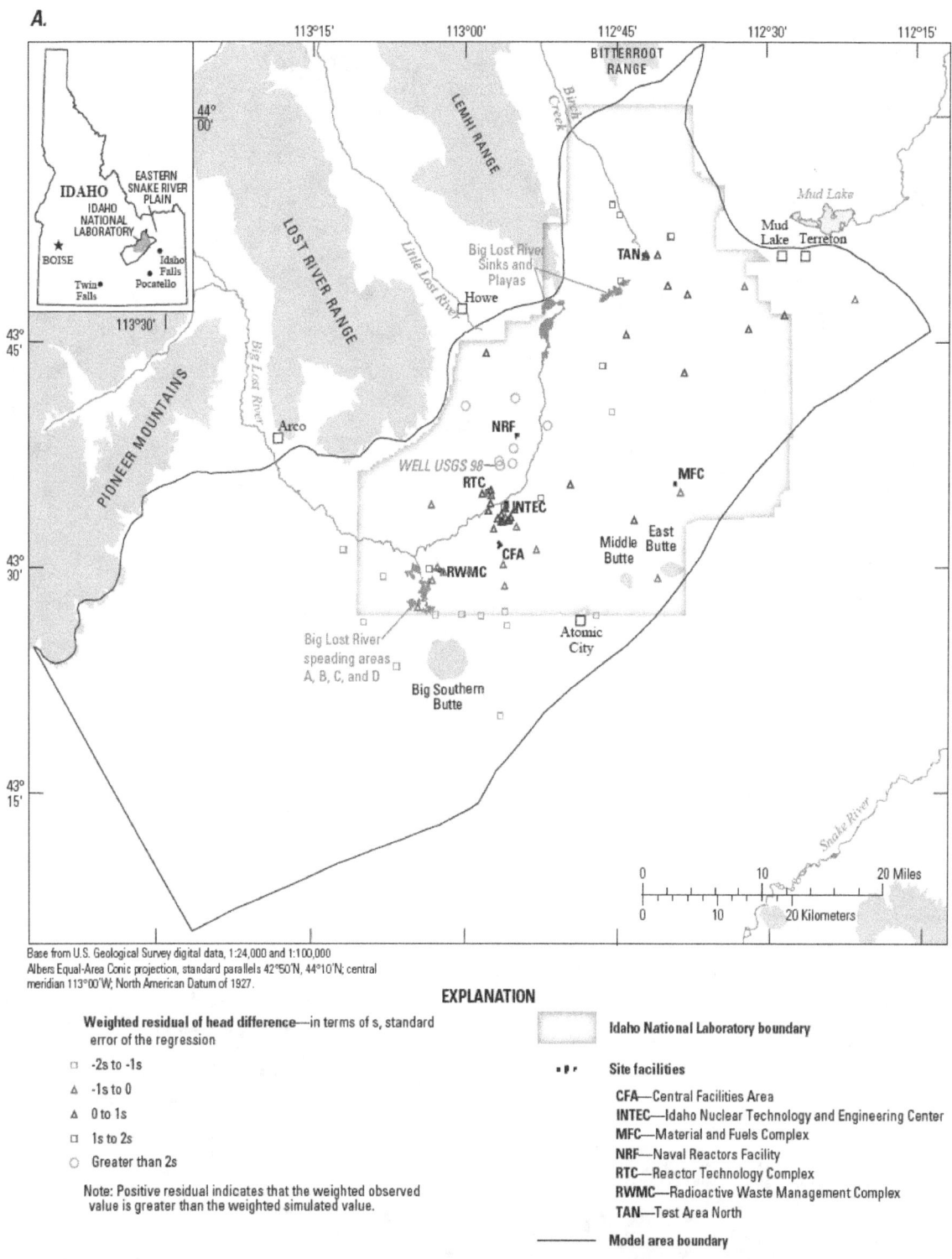

Figure 35. Distribution of weighted residuals of head difference for the end of (*A*) December 1987, stress period 22, and (*B*) December 1993, stress period 43, for model of transient groundwater flow, Idaho National Laboratory and vicinity, Idaho.

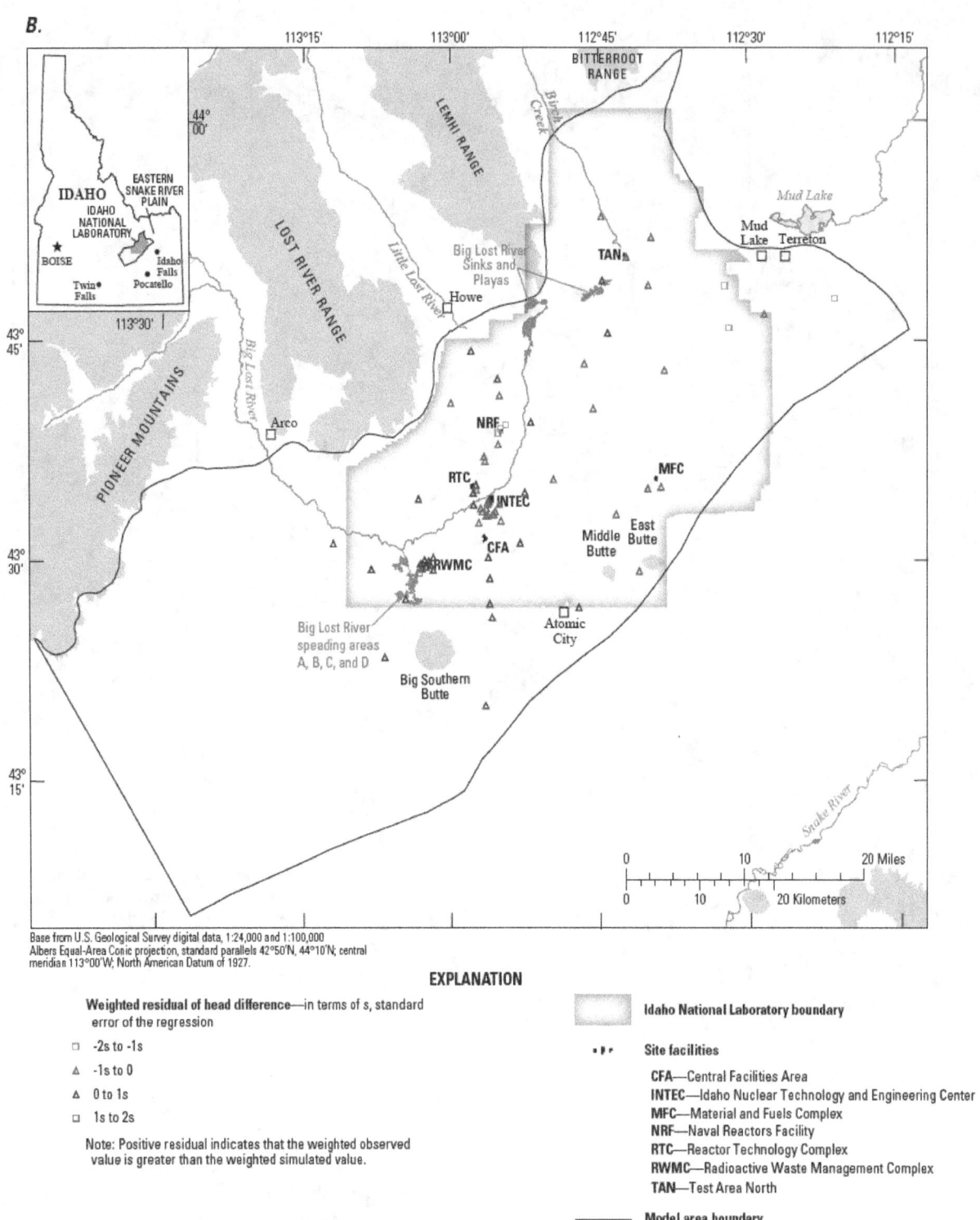

B.

Base from U.S. Geological Survey digital data, 1:24,000 and 1:100,000
Albers Equal-Area Conic projection, standard parallels 42°50'N, 44°10'N; central
meridian 113°00'W; North American Datum of 1927.

EXPLANATION

Weighted residual of head difference—in terms of s, standard
error of the regression

☐ -2s to -1s

△ -1s to 0

▲ 0 to 1s

☐ 1s to 2s

Note: Positive residual indicates that the weighted observed
value is greater than the weighted simulated value.

☐ **Idaho National Laboratory boundary**

• ▪ • **Site facilities**

CFA—Central Facilities Area
INTEC—Idaho Nuclear Technology and Engineering Center
MFC—Material and Fuels Complex
NRF—Naval Reactors Facility
RTC—Reactor Technology Complex
RWMC—Radioactive Waste Management Complex
TAN—Test Area North

——— **Model area boundary**

Figure 35.—Continued.

Early in the calibration process, two analyses were used to check numerical considerations for the implemented transient simulation. The calibrated model was reconfigured to allow three time steps per stress period with a time step multiplier of 1.2. The alternative model had similar estimated values of parameters and no improvement in the sum of squared weighted residuals and standard error of the regression when compared to the calibrated model. The results imply that if future use of the model would benefit from a change in time discretization by use of additional time steps, the change could be made without recalibration.

Another analysis was made to re-evaluate the areal discretization of the model. Although initial modeling was done at a 0.5 mi grid spacing, final modeling has been done at a 0.25 mi spacing. Results from initial model runs were not directly comparable due to coincident changes in sets of calibration observations, grid spacing, and stress period length. The calibrated model with 0.25 mi spacing was subsequently reconfigured to 0.5 mi spacing and recalibrated. The alternative model had similar estimated values of parameters and slight increases in the sum of squared weighted residuals and standard error of the regression. A slight decrease in average weighted residual was estimated when compared to the calibrated model. The configuration of the model is not strongly sensitive to doubling the cell spacing. The results imply that the model could be used with a variable spacing that is less fine in areas remote from simulated contaminant travel to improve run-time efficiency for transport models.

Seasonal Variation of Northeast Boundary Flux

One possible reason the calibrated model did not simulate the annual component of water-level change is that underflow across the northeast boundary varies seasonally. Flux could vary seasonally following the general trend of annual cycles of water levels in wells, which probably respond to seasonal irrigation pumpage upgradient of the northeast boundary.

The transient model was modified from the calibrated model implementation of flux simulating constant inflow across the northeast model boundary to an alternative model simulating seasonal variation in flux. Total flux remained constant from year to year at the same rate as the calibrated model, but was reduced to 60 percent of annual average during the simulated summer irrigation-pumping season for model layers 1 and 2 and 70 percent for layer 3. For the remainder of the year, flux increased to 160 percent of annual average flux for model layers 1 and 2 and 130 percent of annual average flux for layer 3. Flux into the model area across the boundary decreased 150 ft³/s during the simulated summer irrigation-pumping season and increased 75 ft³/s for the remainder of the year compared to the annual average.

The simulated values of head in the aquifer showed additional annual drawdown and recovery of as much as 17 ft adjacent to the northeast boundary. The effect of seasonal variation in flux dissipated quickly away from the boundary with additional annual drawdown and recovery of about 2 and 1 ft within 3 and 6 mi of the boundary.

Estimated parameter values for the simulation were close to those of the calibrated model except for SY44, which was 5 percent different, within 0.001 of the minimum of the confidence intervals of the calibrated model, and within the range of the confidence interval for the base case model (fig. 27; table 15). As an additional check, the model incorporating seasonal changes in northeast boundary flux was recalibrated with three time steps per stress period. The model with multiple time steps had similar estimated values of parameters to the alternative model and only a slight improvement in the sum of squared weighted residuals and standard error of the regression when compared to the alternative model. Perhaps further research and consideration of concepts of storativity for the layered basalts and better aquifer test procedures are needed to understand this problem and, if necessary, modify conceptual and numerical models of the aquifer. In summary, simulating seasonal changes in flux across the northeast boundary does not improve the calibration of the model nor substantially improve the simulation of annual fluctuations of head throughout the northeast quarter of the model area.

Evaporation of Big Lost River Streamflow Infiltration

The decision to include estimates of evaporation for transient model calibration was based on the possibility an overestimate of infiltration in sinks, playas, and spreading areas causing a bias in estimated values of specific yield that were not seen for hydraulic conductivity in the steady-state model. Because head differences are largest near the Big Lost River, the locus of changes in storage, the sensitivity of the transient model estimates of specific yield to streamflow infiltration might be greater than the sensitivity indicated for estimates of hydraulic conductivity in the steady-state model. Although the streamflow infiltration is relatively well known for most river reaches, neglecting small overestimates of infiltration that are a result of evaporation losses may bias estimates of specific yield.

The size of the possible bias was evaluated by analyzing model recalibration without reducing streamflow infiltration by estimates of evaporation. The model recalibration was only slightly sensitive to the increase in estimated infiltration from neglecting evaporation losses. The bias generally resulted in estimated parameter values well within the range of the confidence intervals of the calibrated model except those of SY11 and SY2, which were 2 and 4 percent different and within 0.001 and 0.004, respectively, of the maximums of the confidence intervals for the calibrated model (fig. 27; table 15). Evaporation of streamflow in channels, sinks, playas, and spreading areas, as implimented for transient flow modeling, is an addition to the conceptual model.

Analysis of Advective Flow and Transport

Particle-tracking simulations were performed to evaluate (1) how simulated groundwater flow paths and travel times differ between the steady-state and transient flow models, (2) the effects of wet- and dry-climate cycles on groundwater flow paths and travel times, (3) the effects of streamflow infiltration on advective transport, and (4) how well model predictions of the source and travel times of groundwater flow compare to estimates based on other indicators such as water chemistry, contaminant chemistry, and environmental tracers. These simulations did not include the effects of natural dispersive processes attributable to small-scale heterogeneity within individual hydrogeologic zones, lateral dispersion of pond infiltration through the unsaturated zone, and molecular and thermal diffusion.

Particle-tracking computations were made using MODPATH (Pollock, 1994), a post-processing program for MODFLOW-2000. MODPATH computes particle paths using a semi-analytical expression of the flow path of a particle within each cell and tracks the movement of the particle from one cell to the next until the particle is terminated at a boundary, an internal sink or source, or some other user-defined criterion (Pollock, 1994, p. 1-1). Particle movement was plotted to produce maps of 3-D flow paths in two dimensions.

The spatial and temporal distribution of particle releases, and the duration of the particle-tracking simulations, varied depending on the purpose of the particle-tracking simulation. The number of particles released, location of cells where particles were released, distribution of the particles throughout cells, time(s) of particle release, and length of time that particle movement was simulated are presented at the beginning of each section where particle-tracking results are discussed.

Effective porosity values used in these simulations (table 19) are not calibrated and were estimated from large-scale model-derived values of specific yield (table 15), small-scale measurements of bulk or total porosities on individual core samples (table 3), and literature derived estimates of porosity for similar rock types (Freeze and Cherry, 1979, p. 158).

Approximations in the particle-tracking computer code include how weak sink cells are treated. Weak sink cells are cells that contain sinks that do not discharge at a rate large enough to consume all the water entering the cell (Pollock, 1994, p. 2-17). Because there is no way to know whether particles entering a weak sink cell discharge to the sink or pass through the cell, an approximation of particle behavior

Table 19. Estimated effective porosity for groundwater model hydrogeologic zones, Idaho National Laboratory and vicinity, Idaho.

Hydrogeologic zone	Estimated effective porosity
1	0.07
2	.14
3	.03
4	.05
11	.07
22	.15
33	.05
44	.03
6	.05

in the cell is necessary. For the particle-tracking simulations presented here, particles were instructed to terminate upon entering cells in which discharge to sinks is larger than 0.5 of the total inflow to the cells.

Comparison of Steady-State and Transient Flow

Particle tracking, simulated with the steady-state and transient flow models, was used to evaluate the influence of transient stresses on groundwater flow directions and average linear velocities. Transient stresses evaluated included (1) groundwater withdrawals from irrigation wells and infiltration from irrigation return flows; (2) streamflow infiltration from the Big Lost River channel, sinks, playas, and spreading areas; and (3) industrial groundwater withdrawals and wastewater return flows at the RTC and the INTEC.

Water-level rises throughout the model area were observed and simulated during the early- to mid-1980s, with the largest rises centered near the sinks and spreading areas (figs. 12 and 36). Observed heads, simulated heads, and residuals of head difference for wells USGS 9 and USGS 87 (fig. 37) near the spreading areas (fig. 1) indicate that the transient flow model overestimated the maximum water-level rise during the wet cycle by about 10 ft at USGS 9 and by about 6 ft at USGS 87. Similarly, a contour map of water-level rises from July 1981 to July 1985 (fig. 12) indicates that the simulated water-table rise beneath the sinks also may be overestimated (fig. 36). Consequently, the influence of streamflow infiltration at the spreading areas and sinks on particle directions and velocities may be overestimated in the transient particle-tracking simulations.

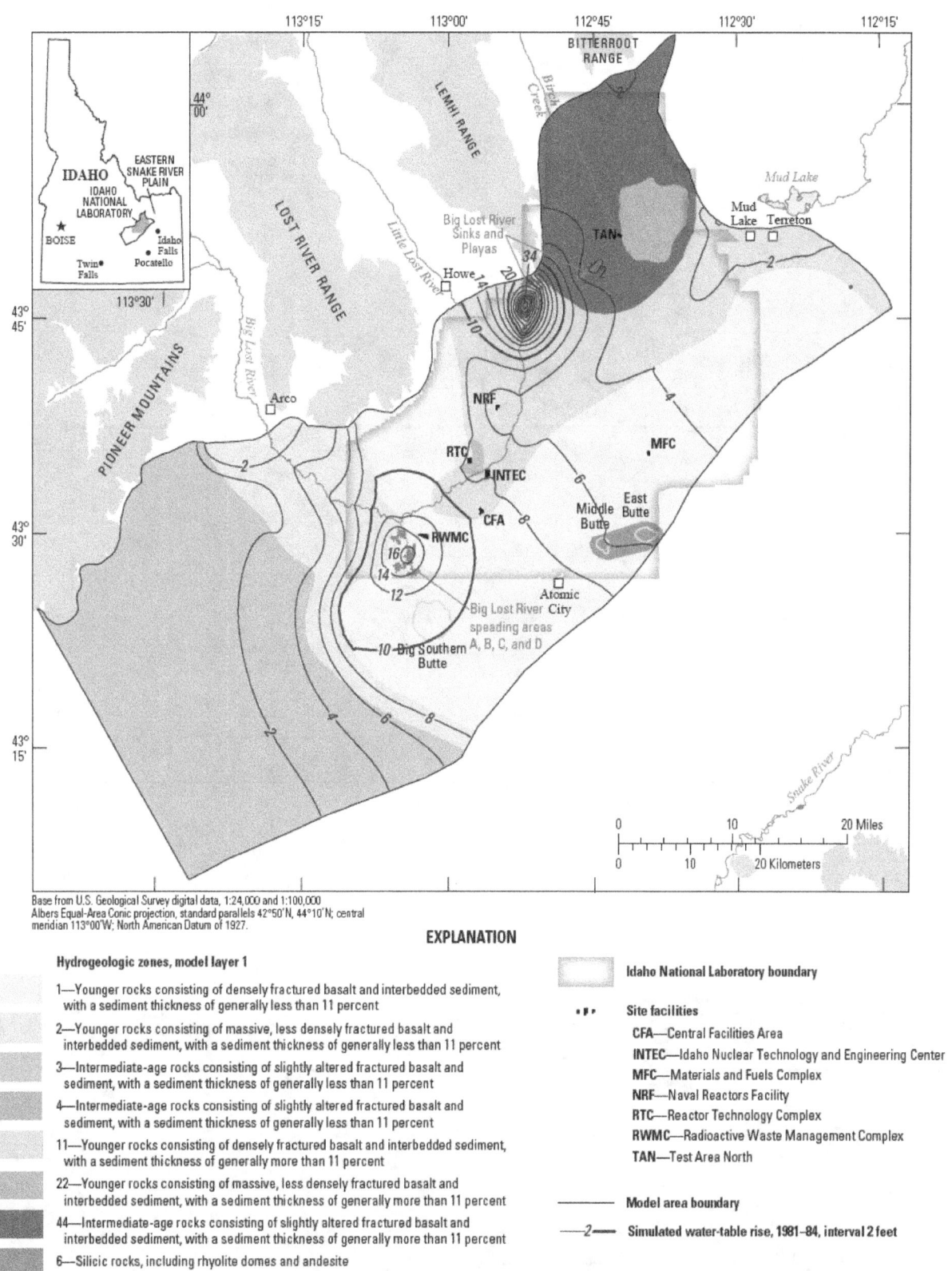

Figure 36.   Simulated water-table rise from January 1981 through December 1984, Idaho National Laboratory and vicinity, Idaho.

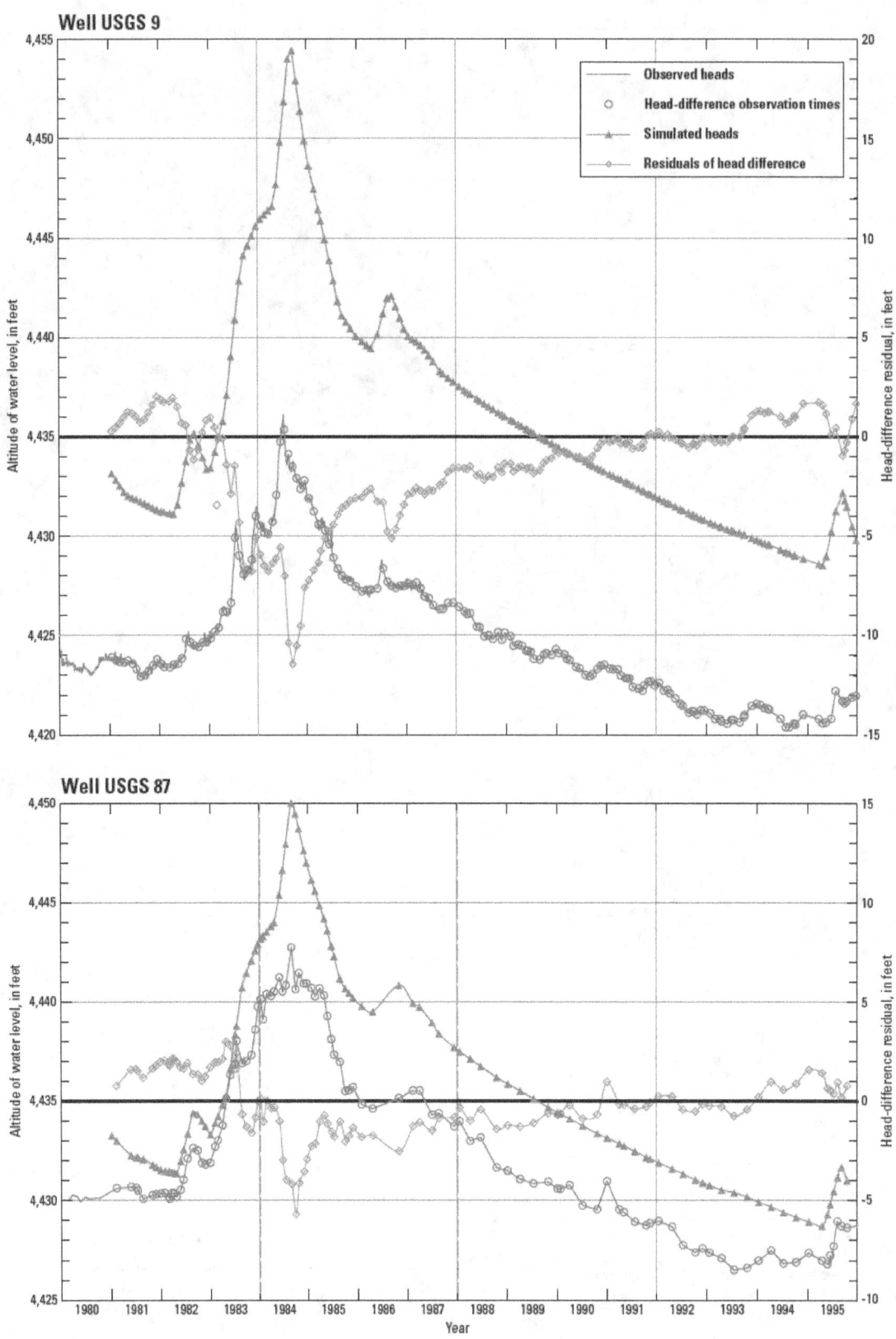

Figure 37.   Observed and simulated hydraulic head, head-difference observation times, and residuals of head difference for wells USGS 9 and USGS 87 from 1981 through 1995, Idaho National Laboratory, Idaho. Residuals of head difference are differences between simulated head difference and observed head difference.

Analysis of Steady-State and Transient Flow Paths

To evaluate how transient recharge influences groundwater flow, particles were released from identical locations in the steady-state and transient flow models and their paths were simulated for 10 years with the steady-state flow model and for 10 years, from 1983 through 1992, with the transient flow model. The transient simulation period included the last 4 years of an extreme wet-climate cycle (1983 through 1986) that also included the two largest recorded annual mean discharges for the Big Lost River onto the INL (1983 and 1984), and the first 6 years of an extreme dry-climate cycle (1987 through 1992) (fig. 10).

The particles, labeled A through O (fig. 38), were released in the steady-state and transient flow models at the center of 15 cells in model layer 1. The release points were (1) restricted to model layer 1 because infiltration from the Big Lost River, which is by far the largest transient stress, was applied to this model layer and (2) selected to provide a distribution of particle paths throughout the model area, with an emphasis on release points near transient stresses.

The A, B, and C particles originated near the northeast boundary of the model area within or near irrigated areas (figs. 18 and 38). The D, E, and F particles originated near the Big Lost River Playas and Sinks. The G, H, I, J, and K particles originated near the Big Lost River channel. The L and M particles originated east and northeast of the Big Lost River spreading areas, respectively. The N particles originated southwest of the RTC and the O particles originated south of the INTEC.

Comparisons of the steady-state and transient particle simulations are based on differences in the (1) horizontal flow direction,[1] (2) ending depth, (3) net distance traveled, and (4) average linear velocity (table 20) of the particles. In most cases, particles move along one general direction and the flow path is described by a single compass direction. Where the particles followed an irregular path the flow path is described with multiple compass directions. The net distance traveled was calculated as the straight-line distance between the starting and ending locations of the particles. The ending depth for all particles is model layer 1 unless indicated otherwise. Average linear velocity was calculated as the net distance traveled divided by travel time, in this case 10 years. In the following discussion, the values for net distance traveled and average linear velocities are presented first for particle simulations during steady-state flow followed by particle simulations during transient flow.

Recharge from Irrigation—Particles A, B, C

The steady-state and transient A particles traveled about 7.3 and 6.9 mi, respectively, southeast, south, and southwest, with the transient particle ending about 0.2 mi northwest of the steady-state pathline (fig. 38; table 20). The average linear velocities were 11 and 10 ft/d. The initial southeast flow direction was perpendicular to the regional groundwater flow direction and was caused by groundwater moving toward hydrogeologic zone 4 (fig. 38), a zone with a large hydraulic conductivity (HC). The shorter distance traveled by the transient particle resulted from the simulated decrease in hydraulic gradient northeast of the sinks during the early to mid 1980s. The gradient decrease was caused by a rise in water levels during the wet cycle of the early to mid 1980s (fig. 36). Water levels rose throughout the model area due to the large amount of streamflow infiltration from the Big Lost River, but the largest water-level rises were centered at the Big Lost River Sinks and spreading areas (fig. 36). This water-level rise caused a decrease in hydraulic gradient upgradient (north and east), and an increase in hydraulic gradient downgradient (south and west), of the sinks and spreading areas. The A particles moved downward to model layer 2, with most of the downward flow taking place in hydrogeologic zone 4. Downward flow in hydrogeologic zone 4 indicates that simulated heads in this zone decrease with depth, reflecting the downward displacement effects of inflow across the northeast boundary, underflow from the Birch Creek alluvial aquifer, streamflow infiltration from Birch Creek diversions (fig. 14), and the larger HC of hydrogeologic zone 4 relative to that of hydrogeologic zone 44 (fig. 21).

The steady-state and transient B particles traveled south about 6.6 and 5.7 mi, with the transient particle ending about 0.2 mi east of the steady-state pathline. The average linear velocities were 9.5 and 8.2 ft/d. The transient particle traveled a shorter distance because of the simulated decrease in hydraulic gradient northeast of the Big Lost River Sinks during the early to mid 1980s. The B particles, although originating in hydrogeologic zone 4 at the center of model layer 1, moved downward to the bottom of model layer 1 reflecting the downward displacement effects of upgradient inflow across the northeast boundary.

The steady-state and transient C particles traveled about 46.6 and 43.4 mi southwest, with the transient particle ending about 0.4 mi northwest of the steady-state pathline. The average linear velocities were 67.4 and 62.7 ft/d. The long distance these particles traveled reflects the large HC for hydrogeologic zone 1 (fig. 21). The transient particle traveled a shorter distance because of the simulated decrease in hydraulic gradient east of the sinks and northeast of the spreading areas during the early to mid 1980s.

---

[1]Difference in horizontal flow direction, also defined as the particle divergence, is the perpendicular offset distance between the endpoint of the slowest particle and the pathline of the fastest particle

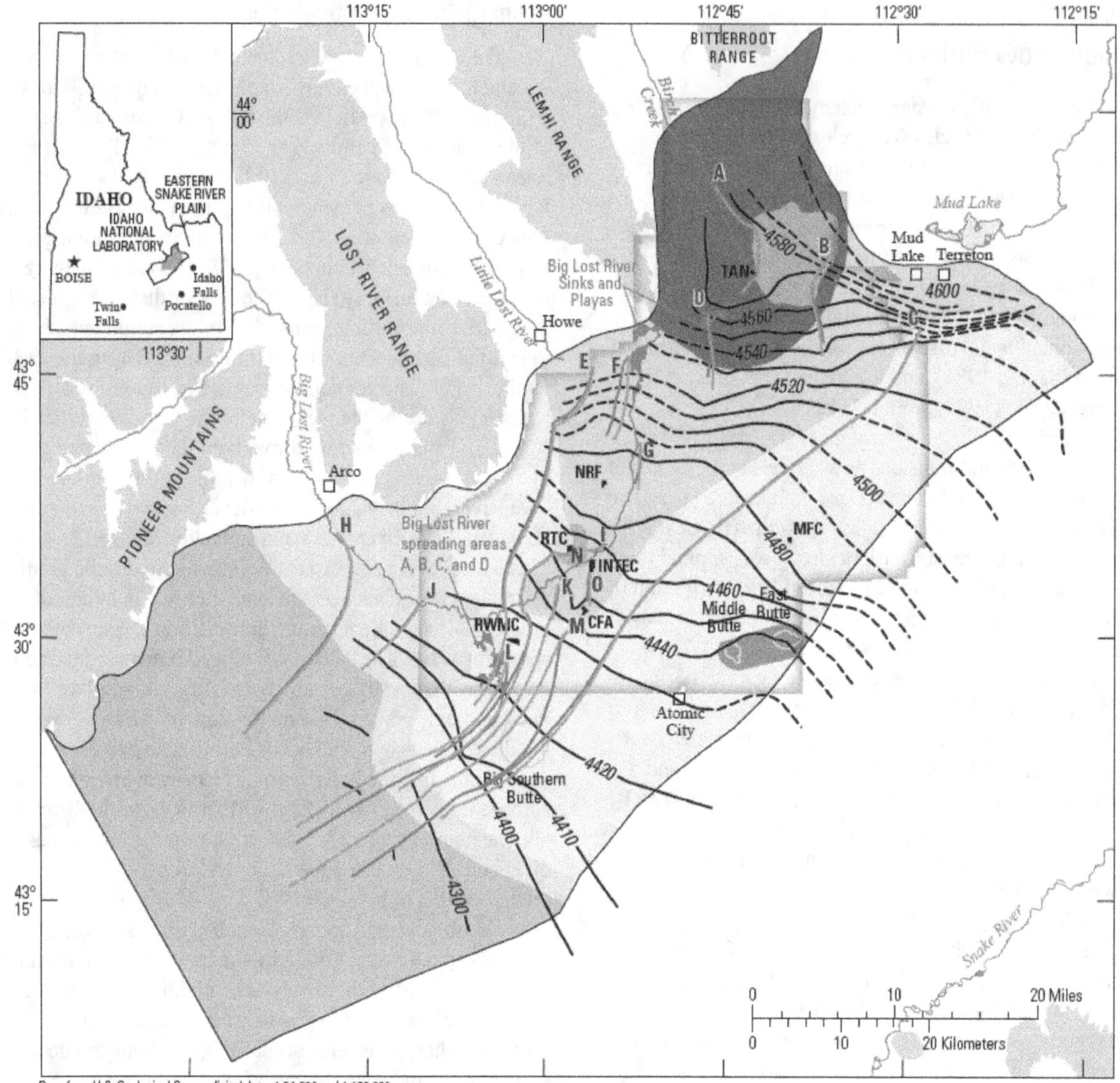

EXPLANATION

Hydrogeologic zones, model layer 1

1—Younger rocks consisting of densely fractured basalt and interbedded sediment, with a sediment thickness of generally less than 11 percent

2—Younger rocks consisting of massive, less densely fractured basalt and interbedded sediment, with a sediment thickness of generally less than 11 percent

3—Intermediate-age rocks consisting of slightly altered fractured basalt and sediment, with a sediment thickness of generally less than 11 percent

4—Intermediate-age rocks consisting of slightly altered fractured basalt and sediment, with a sediment thickness of generally less than 11 percent

11—Younger rocks consisting of densely fractured basalt and interbedded sediment, with a sediment thickness of generally more than 11 percent

22—Younger rocks consisting of massive, less densely fractured basalt and interbedded sediment, with a sediment thickness of generally more than 11 percent

44—Intermediate-age rocks consisting of slightly altered fractured basalt and interbedded sediment, with a sediment thickness of generally more than 11 percent

6—Silicic rocks, including rhyolite domes and andesite

Idaho National Laboratory boundary

Site facilities
CFA—Central Facilities Area
INTEC—Idaho Nuclear Technology and Engineering Center
MFC—Materials and Fuels Complex
NRF—Naval Reactors Facility
RTC—Reactor Technology Complex
RWMC—Radioactive Waste Management Complex
TAN—Test Area North

A———— 10-year steady-flow pathlines and identifier

A———— 1983-1992 transient-flow pathlines and identifier

———— Model area boundary

—4500—— Water-table contour—Shows altitude of water table, March–May, 1995. Intervals 10 and 100 feet. Dashed where approximately located. From Bartholomay and others (1997, figure 9). Datum is NGVD27.

Base from U.S. Geological Survey digital data, 1:24,000 and 1:100,000
Albers Equal-Area Conic projection, standard parallels 42°50'N, 44°10'N;
central meridian 113°00'W; North American Datum of 1927.

Figure 38.   Pathlines representing 10-year particle-tracking simulations under steady-state and transient flow conditions, Idaho National Laboratory and vicinity, Idaho. Particles are labeled A through O.

Table 20. Summary comparison of particle divergence, net distances traveled, and average linear velocities at the end of a 10-year simulation period under 1980 steady-state flow conditions and transient flow conditions from 1983 through 1992, Idaho National Laboratory and vicinity, Idaho.

[**Particle divergence:** The perpendicular offset distance between the endpoint of the slowest particle and the pathline of the fastest particle Net distance traveled is the straight-line distance between a particle's starting and ending locations Percentages are (1) the particle divergence divided by the smaller net distance traveled and (2) the difference in net distance traveled divided by the larger net distance traveled **Average linear velocity:** The net distance traveled in feet divided by travel time in days Pathlines are shown in figure 38 **Abbreviations**: SS, steady-state; T, transient; mi, mile; ft/d, foot per day; BLR, Big Lost River; sa, Big Lost River spreading areas]

Particle	Transient stress	SS and T particle divergence (mi)	SS and T particle divergence (percent)	Ending model layer (SS/T)	Net distance traveled (SS/T) (mi)	Difference in net distance traveled (percent)	Average linear velocity (SS/T) (ft/d)
A	Irrigation	0.2	3	2/2	7.3/6.9	5	11/10
B	Irrigation	.2	4	1/1	6.6/5.7	14	9.5/8.2
C	Irrigation	.4	1	1/1	46.6/43.4	7	67.4/62.7
D	BLR Playa	.4	10	2/2	5.0/4.2	16	7.2/6.1
E	BLR Sinks	.2	1	1/1	23.9/17.0	29	34.5/24.6
F	BLR Sinks	.6	11	2/2	5.7/5.5	4	8.2/8.0
G	BLR channel	.3	11	1/1	3.1/2.7	13	4.5/3.9
H	BLR channel	.1	1	4/4	14.7/14.2	3	21.3/20.5
I	BLR channel	.2	20	1/1	1.3/1.0	23	1.9/1.4
J	BLR channel	.1	4	2/2	6.4/2.6	59	9.3/3.8
K	BLR channel	.1	4	1/1	12.2/2.5	80	17.6/3.6
L	BLR sa	.9	5	3/4	18.6/19.2	3	26.9/27.8
M	BLR sa	1.7	8	2/2	26.1/22.3	15	37.7/32.2
N	Wastewater	.6	4	1/1	19.8/16.3	18	28.6/23.6
O	Wastewater	.2	11	1/1	2.0/1.8	10	2.9/2.6

Recharge from Big Lost River Playas and Sinks—Particles D, E, F

The steady-state and transient D particles traveled about 5.0 and 4.2 mi south, with the transient particle ending about 0.4 mi west of the steady-state pathline (fig. 38; table 20). The average linear velocities were 7.2 and 6.1 ft/d. The differences in the net distance and direction traveled were caused by the transient particle initially traveling northwest. The northwest flow direction was caused by a short-term reversal in gradient from streamflow infiltration at Big Lost River Playa 3 in 1983. The D particles moved downward to model layer 2 reflecting the downward displacement effects of streamflow infiltration across the top of model layer 1 beneath Big Lost River Playa 3 (fig. 13B).

The steady-state and transient E particles traveled about 23.9 and 17.0 mi south-southwest, with the transient particle ending about 0.2 mi west of the steady-state pathline (table 20). The average linear velocities were 34.5 and 24.6 ft/d. The transient particle traveled a shorter distance because of the simulated decrease in hydraulic gradient north of the spreading areas resulting from the large amount of streamflow infiltration at the Big Lost River spreading areas during the early to mid 1980s (fig. 36).

The steady-state and transient F particles traveled about 5.7 and 5.5 mi south-southwest, with the transient particle ending about 0.6 mi west of the steady-state pathline (table 20). The average linear velocities were 8.2 and 8.0 ft/d. A more westerly orientation of the hydraulic gradient during the early to mid 1980s, caused by the large influx of streamflow infiltration at the Big Lost River Sinks and Playas (fig. 36), was responsible for the transient particle ending west of the steady-state particle. The downward displacement effects of recharge at the sinks also caused the F particles to move downward to model layer 2.

Recharge from Big Lost River Channel—Particles G, H, I, J, K

The steady-state and transient G particles traveled about 3.1 and 2.7 mi south, with the transient particle ending about 0.3 mi west of the steady-state pathline (fig. 38; table 20). The average linear velocities were 4.5 and 3.9 ft/d. The westerly movement of the transient particle, relative to the steady-state particle, was caused by a lack of streamflow infiltration from the Big Lost River channel from 1987 through 1992. The reduction in recharge resulted in a 1992 simulated water table beneath the Big Lost River south of the sinks that was about 5 to 10 ft lower than initial conditions in 1980.

The steady-state and transient H particles initially traveled south-southeast, a direction that was influenced by underflow from the Big Lost River alluvial aquifer, before traveling south and southwest. The H particles traveled about 14.7 and 14.2 mi southwest, with the transient particle ending about 0.1 mi northwest of the steady-state pathline (table 20). The average linear velocities were 21.3 and 20.5 ft/d. The shorter distance traveled by the transient particle was caused by the reduced gradient in this area during the dry-climate cycle. The H particles moved downward to model layer 4 reflecting the downward displacement effects of underflow from the Big Lost River alluvial aquifer and streamflow infiltration from the Big Lost River (fig. 14).

The steady-state and transient I particles traveled about 1.3 and 1.0 mi southwest, with the transient particle ending about 0.2 mi southeast of the steady-state pathline (table 20). The average linear velocities were 1.9 and 1.4 ft/d. Streamflow infiltration at the Big Lost River spreading areas during 1983 and 1984 produced a decrease in the simulated hydraulic gradient northeast of the spreading areas (fig. 36) and caused the transient particle to travel a shorter distance than the steady-state particle. The transient particle traveled south of the steady-state particle because the small amount of recharge from the Big Lost River channel between 1987 and 1992 resulted in a 1992 simulated water table beneath the INTEC that was about 3–4 ft lower than in the steady-state simulation and resulted in a shift of the water-table gradient toward a more southerly direction.

The steady-state and transient J particles traveled about 6.4 and 2.6 mi southwest, with the transient particle ending about 0.1 mi northwest of the steady-state pathline (table 20). The average linear velocities were 9.3 and 3.8 ft/d. The distance the transient particle traveled was influenced by streamflow infiltration from the Big Lost River during the early to mid 1980s and a lack of recharge during the mid 1980s to early 1990s. Recharge produced a decrease in the hydraulic gradient upgradient of the river, and the lack of recharge produced a decrease in hydraulic gradient downgradient of the river. The decreases in hydraulic gradient influenced the velocity of the transient particle. The J particles from both simulations moved downward to model layer 2 reflecting the downward displacement effects of streamflow infiltration from the Big Lost River (fig. 14).

The K steady-state particle traveled south-southwest and southwest about 12.2 mi and the transient particle traveled about 2.5 mi south-southwest (table 20). The average linear velocities were 17.6 and 3.6 ft/d. The extreme difference in distance traveled was caused by the (1) simulated decrease in hydraulic gradient northeast of the spreading areas during the early to mid 1980s (fig. 36), which slowed the velocity of the transient particle; and (2) earlier arrival of the steady-state particle in hydrogeologic zone 1, which has an HC about 50 times larger than that of hydrogeologic zone 11 (fig. 21).

Recharge from Big Lost River Spreading Areas— Particles L, M

The steady-state and transient L particles traveled about 18.6 and 19.2 mi southwest, with the steady-state particle ending about 0.9 mi northwest of the transient pathline (fig. 38; table 20). The average linear velocities were 26.9 and 27.8 ft/d. The 14 to 16 ft increase in hydraulic head beneath the spreading areas from 1981 through 1984 (fig. 36) caused the transient particle to travel farther south than the steady-state particle. The steady-state and transient particles moved downward to model layers 3 and 4, respectively, reflecting the downward displacement effects of streamflow infiltration across the top of model layer 1 beneath the spreading areas. The large amount of recharge from the spreading areas in 1983 and 1984, with a subsequent increase in simulated vertical head gradients, increased downward flow in the aquifer beneath the spreading areas and resulted in the transient particle ending at a greater depth than the steady-state particle.

The steady-state and transient M particles traveled about 26.1 and 22.3 mi southwest, with the transient particle ending about 1.7 mi east of the steady-state pathline (table 20). The average linear velocities were 37.7 and 32.2 ft/d. The differences in distance and direction resulted from the 10 to 12 ft increase in head east and south of the spreading areas from 1981 through 1984 (fig. 36). The increased head, and the resulting decrease in the hydraulic gradient east of the spreading areas, caused the transient particle to travel slower than, and farther east of the steady-state particle. The M particles from both simulations moved downward to model layer 2 reflecting the downward displacement effects of streamflow infiltration from the Big Lost River and spreading areas.

Wastewater Discharge at RTC and INTEC—Particles N, O

The steady-state and transient N particles originated about 0.25 mi southwest of the RTC infiltration ponds and traveled west, southwest, south, southeast, and then southwest 19.8 and 16.3 mi (fig. 38; table 20). The average linear velocities were 28.6 and 23.6 ft/d. The initial westerly flow direction of these particles was caused by a gradient that was influenced by wastewater discharge at the RTC infiltration ponds and the large HC of hydrogeologic zone 1 to the west. The southwest, south, and southeast flow directions resulted from the particles moving around the perimeter of hydrogeologic zone 11, a zone with a small HC. The particles moved along the perimeter of zone 11 because little flow exited zone 11 relative to the volume of flow within zone 1. The steady-state particle traveled about 3.5 mi farther than the transient particle because the particles traveled through the area northeast of the Big Lost River spreading areas during the simulated hydrologic conditions of the early 1980s. The hydraulic head in this area increased by about 8–10 feet between 1981 and 1984 (fig. 36), decreasing the hydraulic gradient, and causing the transient particle to move slower than the steady-state particle. The transient particle ended 0.6 mi northwest of the steady-state pathline because the particles traveled south of the spreading areas while water levels were declining beneath the spreading areas during the late 1980s and early 1990s (fig. 38). The declining water table beneath the spreading areas shifted the gradient in this area toward a more northerly direction.

The steady-state and transient O particles originated about 0.5 mi south of the INTEC disposal well (CPP 3) and traveled about 2.0 and 1.8 mi south, with the transient particle ending about 0.2 mi west of the steady-state pathline (table 20). The average linear velocities were 2.9 and 2.6 ft/d. The transient particle traveled slower than the steady-state particle because the hydraulic gradient decreased due to recharge at the spreading areas during the early to mid 1980s. The transient particle traveled slightly west of the steady-state particle because of wastewater discharge, beginning in 1984, at the INTEC infiltration ponds, about 0.1 mi east of the of the O particle origination point.

Evaluation of Transient Flow Influence on Particle Paths

The 1 to 4 percent divergence of the A, B, and C steady-state and transient pathlines and the relatively small differences in net distance traveled, 5 to 14 percent, (table 20) indicates that irrigation withdrawals and return flows do not have a large influence on particle flow directions and velocities. Particle flow directions and net distance traveled were not influenced by irrigation because irrigation withdrawals and return flows were (1) about equal, so the net recharge from irrigation was about zero; (2) seasonal, but relatively constant on an annual basis from 1981 through 1995; and (3) distributed fairly uniformly over a large area, which resulted in a small average irrigation-infiltration flux of about 0.4 (ft³/s)/mi² (Ackerman and others, 2006, p. 34), compared to the average infiltration flux of 20 (ft³/s)/mi² (Ackerman and others 2006, p. 34) from the Big Lost River. Consequently, head differences with time resulting from irrigation withdrawals and return flows were small and broadly distributed.

The large, concentrated fluxes of streamflow infiltration from the Big Lost River Sinks, Playas, and spreading areas influenced the movement of particles across a large part of the INL. Particles that traveled near these recharge centers, such as the D, F, L, M, and N particles, were influenced more by streamflow infiltration than particles that remained far from the recharge centers, such as the A, B, and H particles. The relatively smaller concentrated fluxes of streamflow infiltration from the Big Lost River channel only influenced the movement of particles near the channel. The influence of this recharge on particle flow direction and travel distance,

however, was large as shown by the 11–20 percent divergence for particle pairs G and I and the 13 to 59 percent difference in net distance traveled for particle pairs G, I, and J (table 20). The effect of episodic streamflow infiltration on particle paths was partly masked in these simulations because the transient simulation included wet- and dry-climate cycles. During the early to mid 1980s, wet cycle recharge from the Big Lost River caused rises in the water table, with the largest rises centered at the recharge locations. During the mid 1980s to early 1990s dry cycle the water table declined, with the largest declines centered at the recharge locations. Consequently, changes in particle directions and velocities that were a result of the persistent rise in the water table were partly counteracted by the subsequent decline in the water table.

The concentrated flux of wastewater discharge at the RTC and the INTEC influenced the movement of particles within about 1 mi of these facilities. This small area of influence may result from the (1) complex hydraulic gradients near the facilities because of large groundwater withdrawal fluxes adjacent to large wastewater discharge fluxes and (2) larger annual groundwater withdrawal than annual wastewater discharge at these facilities. Local influence of wastewater discharge on particle movement was evident from the initial westerly movement of the N particles. This westerly movement was partly influenced by wastewater discharge at the RTC, and was different from the regional southwest direction of groundwater flow. However, the initial movement of the steady-state and transient particles was similar because of the relatively constant wastewater discharge rate at the RTC from 1980 through 1995. The 11 percent divergence of the steady-state and transient O particles (table 20) near the INTEC resulted from the relocation in 1984 of wastewater discharge from the INTEC disposal well (CPP 3) to the INTEC infiltration ponds, 0.5 mi south of the disposal well. This relocation affected the transient particle, but did not affect the steady-state particle because the steady-state particle moved in response to simulated 1980 hydrologic conditions. Differences in net distance traveled were 18 and 10 percent for particle pairs N and O, respectively. These differences resulted from streamflow infiltration at the Big Lost River spreading areas and, in the case of particle O, from relocation of wastewater disposal from CPP 3 to the INTEC infiltration ponds in 1984.

Transient and steady-state average linear particle velocities ranged from 1.4 to 67.4 ft/d (table 20). This large variability in particle velocities was caused mainly by geographic variability in hydraulic properties, however, and not by variability of recharge. The primary reason particle velocities varied so much was the 1.7 order-of-magnitude range of simulated HC values for the hydrogeologic zones (table 9). The factor of five range of model values for effective porosity (0.03 to 0.15, table 19) and the 1.3 order-of-magnitude range of simulated hydraulic gradients (from about 2.3 to 80 ft/mi; fig. 23) also influenced particle velocities. For example, velocities of 4.5 ft/d or slower were calculated for particles G, I, and O (table 20). These particles remained entirely within the small-HC area of hydrogeologic zone 11. The F particles also remained entirely within hydrogeologic zone 11, but the slightly faster F particle velocities (8.0 and 8.2 ft/d) were attributed to recharge at the Big Lost River Sinks that increased the hydraulic gradient (fig. 38). The J particle velocities were 3.8 and 9.3 ft/d, and these particles traveled entirely within hydrogeologic zone 2, a zone with a small HC. However, the large difference in velocity between the J particles was attributed to variable streamflow infiltration. Velocities ranging from 6.1 to 11 ft/d were calculated for particles A, B, and D. These particles traveled primarily within hydrogeologic zone 44, a zone with a small HC, but particles A and B also traveled partly through hydrogeologic zone 4, which has a large HC. The velocities calculated for the K particles were 3.6 and 17.6 ft/d. The slower velocity was calculated from a K particle that remained entirely within zone 11, and the faster velocity was calculated from a K particle that traveled a long distance in zone 1. The H particles traveled through hydrogeologic zones 2 and 3, which have a small HC. The velocities calculated for the H particles, 21.3 and 20.5 ft/d, were faster than expected for travel through these zones and reflect the steeper hydraulic gradient in the ESRP aquifer at the mouth of the Big Lost River (fig. 22), a result of the large amount of underflow recharge from this tributary valley. Velocities of 24 ft/d or faster were calculated for particles C, E, L, M, and N. These particles traveled long distances through hydrogeologic zone 1, with the fastest velocities (62.7–67.4 ft/d) calculated for the C particles, which traveled almost exclusively in hydrogeologic zone 1.

Representativeness of Flow Directions and Average Linear Particle Velocities

The representativeness of the simulated particle flow directions and average linear velocities was evaluated by comparing (1) steady-state flow directions to a contour map of 1995 water-table altitudes and (2) steady-state average linear particle velocities to independent average linear groundwater velocity estimates based on the tritium/helium-3 (^{3}H/^{3}He) and chlorofluorocarbon (CFC) model ages of the young fraction of groundwater and assumed first and peak arrivals of chlorine-36 (^{36}Cl) at selected wells in geographic proximity to the simulated pathlines (table 21; fig. 39).

The 1995 water-table contours used in this comparison depict the shape of the water table at the end of a long dry cycle. Even though the water table rises and falls over time in response to variable recharge, these contours are believed to be a reasonable representation of the water table because the general shape of the water table is relatively stable (Ackerman and others, 2006, p. 39). Particle velocities derived from the steady-state flow simulation were used in this comparison because these particles were not influenced by episodic streamflow infiltration and therefore, were more representative of long-term aquifer conditions. Independent velocity estimates used in this comparison were calculated

based on the straight-line or minimum distance from the point of origination to the point of observation and represent minimum velocities that do not take into account the tortuosity of the pathway. The reliability of velocity estimates based on assumed peak and first arrivals is limited by the sensitivity of the analytical detection limit and the frequency of sampling prior to a confirmed detection. In almost all cases, first-arrival estimates likely occur after the actual first arrival; therefore, velocity estimates based on assumed first arrival times should be viewed as minimum velocities.

Most simulated particle paths seem to be reasonable because the particles moved in directions perpendicular or nearly perpendicular to the 1995 water-table contours (fig. 38). The only particle paths that seem inconsistent are those associated with the A particles. These particles moved in a direction parallel to the 4,580 ft water-table contour before tracking southward across hydrogeologic zone 4 and in a direction parallel to the regional direction of groundwater flow. As noted in the section, Comparison of Simulated and Observed Steady-State Heads head observations are sparse in the northwestern part of the model area near the Birch Creek alluvial aquifer underflow boundary and are insufficient to result in improved agreement between simulated and observed heads in this area. The orientation of the simulated 4,600 ft head contour (fig. 23) reflects a complex interplay of inflows from multiple sources that does not reflect the general shape of the water table in this area as defined by the limited data available.

The steady-state particle velocities and independently derived velocity estimates ranged from 1.9 to 67.4 ft/d (table 20; fig. 38) and 1.6 to 33 ft/d (table 21; fig. 39), respectively. A geographic comparison indicates that particle velocities (1) within hydrogeologic zone 11 were slightly slower than independent estimates, (2) within hydrogeologic zone 1 were much faster than independent estimates, and (3) within other hydrogeologic zones were faster than independent estimates.

For example, steady-state velocities for particles F and G, located entirely within hydrogeologic zone 11, were 8.2 and 4.5 ft/d, respectively (table 20), and were slightly slower than the 10 to 14 ft/d velocity estimates to well USGS 17 near and south of the Big Lost River Sinks (fig. 39). The velocities for particles C and M, located entirely within hydrogeologic zone 1, were 67.4 and 37.7 ft/d, respectively. The velocity for particle C was 2 to 4 times the 16 and 33 ft/d velocity estimates to wells USGS 1 and USGS 100 for the southeast part of the model area, and the velocity for particle M was about 2 to 3 times the maximum velocity estimates of 15.9 and 11.9 ft/d to wells USGS 11 and USGS 14 south of the INL. The velocity for particle D, mostly within hydrogeologic zone 44, was 7.2 ft/d, 3 times the estimated velocity of 2.3 ft/d to well PSTF Test near TAN. Similarly, the 9.3 ft/d velocity for particle J, within hydrogeologic zone 2, was nearly 3 times the 3.3 and 3.6 ft/d estimated velocity to wells USGS 7 and USGS 86 near the west INL boundary.

Table 21. Average linear groundwater velocities calculated based on tritium/helium-3 and chlorofluorocarbon model ages of the young fraction of groundwater and assumed first and peak arrivals of chlorine-36, between selected wells, Idaho National Laboratory and vicinity, Idaho.

[Velocity traces are shown in figure 39 **Method:** ^{3}H/^{3}He, Busenberg and others (2001); **CFC**, Busenberg and others (2001); ^{36}Cl, Cecil and others (2000) Abbreviations: ^{3}H/^{3}He, tritium/helium-3; CFC, chlorofluorocarbon; ^{36}Cl, chlorine-36; ^{18}O, oxygen-18; BLR, Big Lost River; INTEC, Idaho Nuclear Technology and Engineering Center; USGS, U S Geological Survey; LLR, Little Lost River; NRF, Naval Reactors Facility; yr, year; σ, sigma; ft/d, foot per day; ±, plus or minus; –, no information]

Well	^{3}H/^{3}He Age (yr)	±1σ	CFC Age (yr)	^{36}Cl First arrival Time (yr)	^{36}Cl Peak arrival Time (yr)	Source	Velocity (ft/d)	Young fraction from ^{18}O (percent ±1σ)
USGS 1	–	–	26–28	–	–	Mud Lake-Terreton	33	43 ± 7
USGS 5	16.5	0.5		–	–	BLR Sinks	10.8	47 ± 8
	16.3	.3		–	–	BLR Sinks	10.8	
USGS 8	8.4	.2		–	–	BLR channel	3.3	44 ± 7
USGS 9	21.3	.6		–	–	INTEC disposal well	6.9	38 ± 6
	22.7	.4		–	–	INTEC disposal well	6.2	
USGS 11	17.3	.3		–	–	INTEC disposal well	12.5	37 ± 6
				13.6[1]	–	INTEC disposal well	15.9	
					25.7	INTEC disposal well	8.4	
USGS 12	2.9	.4		–	–	LLR underflow	4.9	62 ± 10
	4.5	.4		–	–	LLR underflow	3.9	
USGS 14	27.3	.5		–	–	INTEC disposal well	9.2	53 ± 9
				19.3[1]		INTEC disposal well	11.9	
				–	28.5	INTEC disposal well	8.1	
USGS 17	16.1	.3		–	–	BLR Sinks	9.8	54 ± 9
	11.1	.3		–	–	BLR Sinks	14.4	
USGS 86	12.1	.5		–	–	BLR channel	3.6	26 ± 6
USGS 97	6.3	.5		–	–	NRF	3.3	56 ± 9
USGS 98	6.7	1.3		–	–	NRF	6.6	29 ± 5
USGS 99	3.9	.2		–	–	NRF	8.2	33 ± 6
USGS 100			23–24	–	–	Mud Lake-Terreton	16	44 ± 7
USGS 102	5.7	.2		–	–	NRF ditch	1.6	59 ± 10
USGS 103	26.1	.4		–	–	INTEC disposal well	4.3	41 ± 7
USGS 105[2]				–	–	INTEC disposal well	6.2	41 ± 7
USGS 109	20.0	.4		–	–	INTEC disposal well	6.2	45 ± 8
	17.7	.4		–	–	INTEC disposal well	7.2	
USGS 121	15.5	.6		–	–	NRF	5.2	48 ± 8
USGS 124	23.7	.1		–	–	INTEC disposal well	7.5	36 ± 6
	23.6	.5		–	–	INTEC disposal well	7.5	
USGS 125	17.0	.3		–	–	INTEC disposal well	10.2	43 ± 7
Cross Road[3]	13.1	.4		–	–	BLR spreading areas	9.8	
NPR Test	13.9	.4		–	–	BLR Sinks	14.1	47 ± 8
PSTF	9.3	2.2		–	–	Birch Creek Playa	2.3	51 ± 9

[1] USGS 11 ^{36}Cl greater than 1 5 times background; USGS 14 ^{36}Cl greater than 1 8 times background
[2] Busenburg and others (2001, fig 25)
[3] Plummer and others (2000)

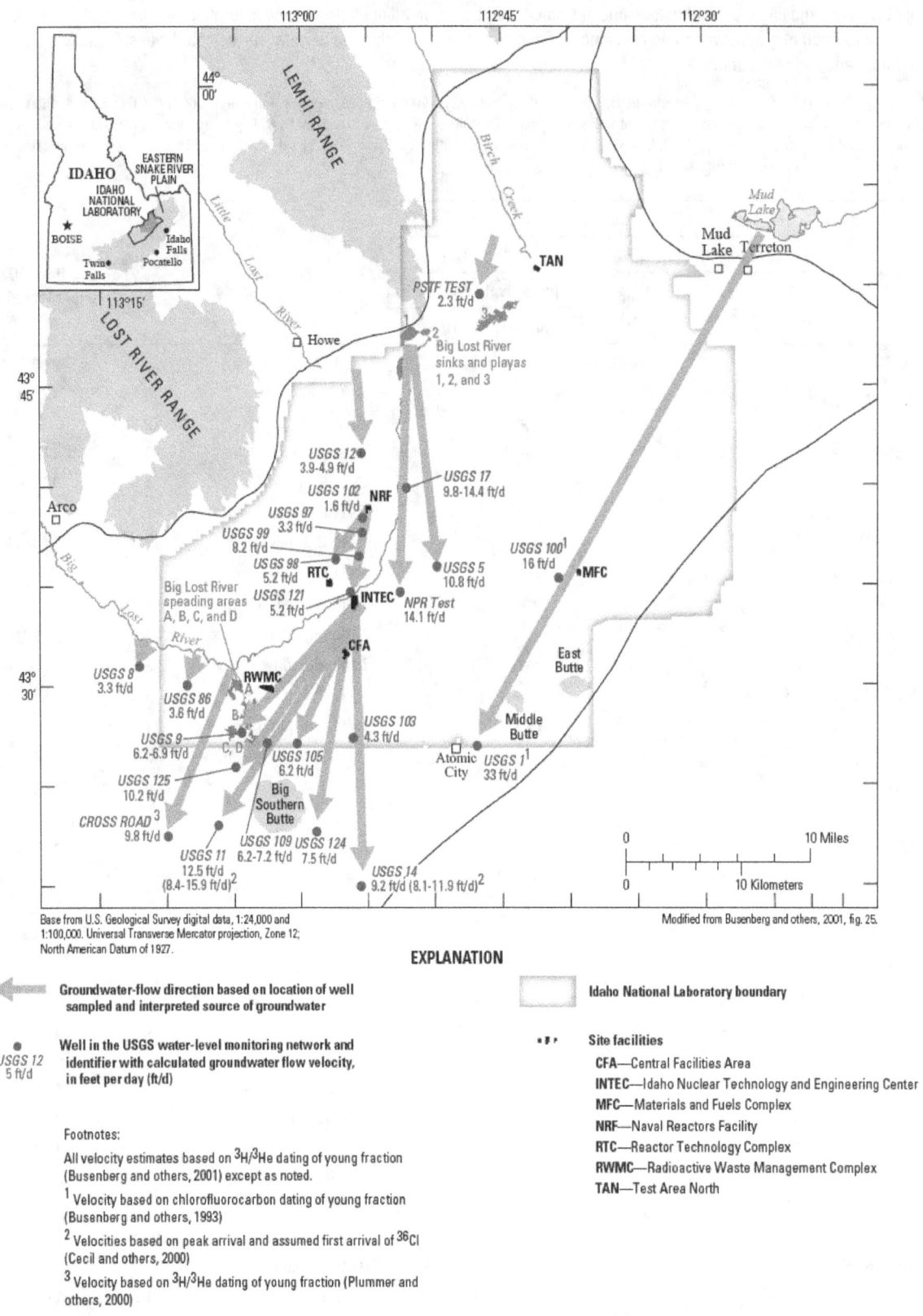

Base from U.S. Geological Survey digital data, 1:24,000 and
1:100,000. Universal Transverse Mercator projection, Zone 12;
North American Datum of 1927.

Modified from Busenberg and others, 2001, fig. 25.

EXPLANATION

Groundwater-flow direction based on location of well
sampled and interpreted source of groundwater

Idaho National Laboratory boundary

Well in the USGS water-level monitoring network and
identifier with calculated groundwater flow velocity,
in feet per day (ft/d)

USGS 12
5 ft/d

Site facilities

CFA—Central Facilities Area
INTEC—Idaho Nuclear Technology and Engineering Center
MFC—Materials and Fuels Complex
NRF—Naval Reactors Facility
RTC—Reactor Technology Complex
RWMC—Radioactive Waste Management Complex
TAN—Test Area North

Footnotes:

All velocity estimates based on ^{3}H/^{3}He dating of young fraction
(Busenberg and others, 2001) except as noted.

[1] Velocity based on chlorofluorocarbon dating of young fraction
(Busenberg and others, 1993)

[2] Velocities based on peak arrival and assumed first arrival of ^{36}Cl
(Cecil and others, 2000)

[3] Velocity based on ^{3}H/^{3}He dating of young fraction (Plummer and
others, 2000)

Figure 39.   Flow directions and average linear groundwater velocities based on tritium/helium-3 and chlorofluorocarbon
model ages of the young fraction of groundwater and assumed first and peak arrivals of chlorine-36 at selected wells,
Idaho National Laboratory and vicinity, Idaho.

Comparison of Flow During a Wet- and a Dry-Climate Cycle

The extended wet- and dry-climate cycles of the 1980s and 1990s produced large differences in streamflow infiltration to the ESRP aquifer in the model area, and provided an opportunity to examine the influence of this climate-controlled recharge on the movement of groundwater beneath the INL. To investigate this influence, particle tracking was used to simulate groundwater movement under transient flow conditions for 4 years during a wet cycle and for 4 years during a dry cycle. The wet-cycle particle-tracking simulations were run from 1983 through 1986. This period included the last 4 years of a 5-year wet cycle (1982 through 1986) and included the 2-year period (1983 and 1984) with the largest recorded annual mean discharges onto the INL (fig. 10). The starting time for this simulation was selected to emphasize the influence of streamflow infiltration on particle flow directions, velocities, and depths. The dry-cycle particle-tracking simulations were run from 1989 through 1992. This period was part of an 8-year dry cycle that lasted from 1987 through 1994 and included a 4-year period when the Big Lost River did not flow onto the INL (fig. 10). The starting time for this simulation, 2 years following the wet-cycle simulation, was selected to minimize residual (or antecedent) effects from the previous wet-cycle simulation.

Analysis of Flow Paths During a Wet- and a Dry-Climate Cycle

Particles were released at the beginning of the wet-cycle and dry-cycle simulations at the same particle release locations as described in the section Analysis of Steady-State and Transient Flow Paths. The particles were tracked and the resulting particle paths were plotted as pathlines (fig. 40). Although these particle-tracking simulations were designed to evaluate the influence of streamflow infiltration on particle movement, the discussion of the particle paths follows the same organization as the previous section, where particles were grouped together based on the proximity of particle starting locations to a particular transient stress. The same features of the particle pathlines were evaluated and the same calculations (table 22) made for net distance traveled and average linear particle velocity as described in the previous section. In the following discussion the (1) ending depth for all particles is model layer 1 unless indicated otherwise and (2) values for net distance traveled and average linear velocity are presented first for the wet-cycle simulation followed by the dry-cycle simulation.

Effects of Irrigation during a Wet- and a Dry-Climate Cycle—Particles A, B, C

The A particles traveled about 2.5 mi southeast, with the wet-cycle particle ending about 0.1 mi northeast of the dry-cycle pathline (fig. 40; table 22). Both average linear velocities were 9.0 ft/d. The small change in the flow directions of the A particles was caused by the different amounts of recharge at the sinks during the wet and dry cycles and the resulting differences in hydraulic head and gradient in the northern part of the INL (fig. 36).

The B particles traveled about 4.2 and 4.4 mi south-southwest, with the wet-cycle particle ending about 0.3 mi east of the dry-cycle pathline. The average linear velocities were 15 and 16 ft/d. The small change in the flow directions and distances of the B particles was caused by the different amounts of recharge at the sinks during the wet and dry cycles and the resulting differences in hydraulic head and gradient in the northeastern part of the INL (fig. 36). The C particles traveled along nearly identical paths about 15.1 and 15.6 mi southwest. The average linear velocities were 54.6 and 56.4 ft/d. The shorter travel distance of the C wet-cycle particle was caused by the large amount of recharge at the spreading areas and sinks during the wet-climate cycle, which produced a slight rise in water levels and a coincident decrease in hydraulic gradient in the eastern part of the INL (fig. 36).

Effects of Recharge at Big Lost River Playas and Sinks during a Wet- and a Dry-Climate Cycle—Particles D, E, F

The D wet-cycle particle initially traveled northwest in response to a pulse of recharge and rising water levels at Big Lost River Playa 3 in 1983, before traveling south-southeast and south. The dry-cycle particle traveled south. The wet-cycle particle ended about 0.2 mi east of the dry-cycle pathline. The net travel distance of the wet-cycle particle was about 1.3 mi, or about 0.9 mi less than the dry-cycle particle. The average linear velocities were 4.7 and 8.0 ft/d.

The E particles traveled about 3.5 and 2.8 mi south-southwest, with the dry-cycle particle ending about 0.5 mi southeast of the wet-cycle pathline. The average linear velocities were 13 and 10 ft/d. The different flow direction and the longer travel distance of the wet-cycle particle were caused by an increase in hydraulic head and gradient in response to recharge at the sinks (fig. 36).

The F particles traveled about 3.0 mi south-southwest and 1.8 mi south-southeast, about a 45-degree difference in flow directions. The dry-cycle particle ended about 1.5 mi east of the wet-cycle pathline, and traveled south-southeast in response to a local hydraulic gradient influenced by alluvial aquifer underflow from the Little Lost River valley. Recharge at the sinks was responsible for the south-southwest flow direction and the longer distance traveled by the wet-cycle particle. The average linear velocities were 11 and 6.5 ft/d.

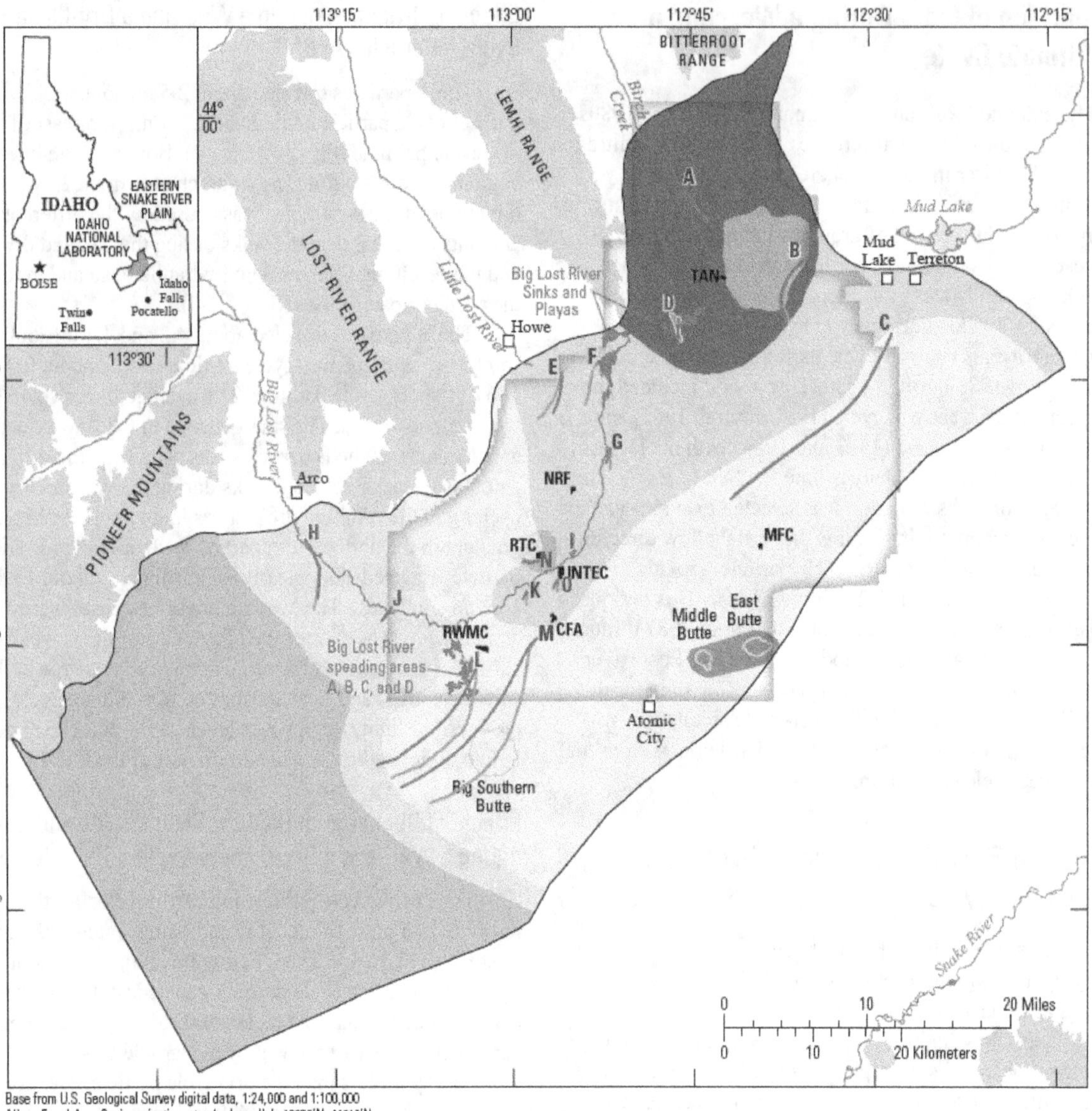

Figure 40.   Pathlines representing particle-tracking simulations for a 4-year wet cycle (1983 through 1986) and a 4-year dry cycle (1989 through 1992) under transient flow conditions, Idaho National Laboratory and vicinity, Idaho. Particles marked A through O.

Table 22. Summary comparison of particle divergence, distances traveled, and average linear velocities at the end of simulations for a wet cycle (1983 through 1986) and a dry cycle (1989 through 1992), Idaho National Laboratory and vicinity, Idaho.

[**Particle divergence:** The perpendicular offset distance between the endpoint of the slowest particle and the pathline of the fastest particle **Net distance traveled:** The straight-line distance between the particle's starting and ending locations Percentages are the particle divergence divided by the smaller net distance traveled and the difference in net distance traveled divided by the larger net distance traveled **Average linear velocity:** The net distance traveled in feet divided by travel time in days Pathlines are shown in figure 40 **Abbreviations:** W, wet climate cycle; D, dry climate cycle; mi, mile; ft/d, foot per day; BLR, Big Lost River; sa, Big Lost River spreading areas]

Particle	Transient stress	W and D particle divergence		Ending model layer (W/D)	Net distance traveled (W/D) (mi)	Difference in net distance traveled (percent)	Average linear velocity (W/D) (ft/d)
		(mi)	(percent)				
A	Irrigation	0.1	4	1/1	2.5/2.5	0	9.0/9.0
B	Irrigation	.3	7	1/1	4.2/4.4	5	15/16
C	Irrigation	.1	1	1/1	15.1/15.6	3	54.6/56.4
D	BLR Playa	.2	15	1/1	1.3/2.2	41	4.7/8.0
E	BLR Sinks	.5	18	1/1	3.5/2.8	20	13/10
F	BLR Sinks	1.5	83	1/1	3.0/1.8	40	11/6.5
G	BLR channel	.3	27	1/1	1.5/1.1	27	5.4/4.0
H	BLR channel	.1	3	2/2	4.6/4.0	13	17/14
I	BLR channel	.2	40	1/1	0.5/0.5	0	1.8/1.8
J	BLR channel	.1	10	1/1	1.1/1.0	9	4.0/3.6
K	BLR channel	.2	33	1/1	0.8/0.6	25	2.9/2.2
L	BLR sa	1.1	12	3/2	9.3/9.4	1	34/34
M	BLR sa	1.8	14	2/1	13.1/13.6	4	47.3/49.1
N	Wastewater	.2	18	1/1	1.1/1.2	8	4.0/4.3
O	Wastewater	.3	50	1/1	0.8/0.6	25	2.9/2.2

Effects of Recharge along Big Lost River during a Wet- and a Dry-Climate Cycle—Particles G, H, I, J, K

The G wet-cycle particle traveled about 1.5 mi south and the dry-cycle particle traveled about 1.1 mi south-southwest, with the dry-cycle particle ending about 0.3 mi west of the wet-cycle pathline. The average linear velocities were 5.4 and 4.0 ft/d. The difference in flow direction was caused by more recharge from the Big Lost River channel during the wet cycle, which caused the hydraulic gradient east of the channel to be slightly more eastward than during the dry cycle. The difference in distance traveled was caused by more recharge from the Big Lost River Sinks during the wet cycle, which increased the hydraulic gradient south of the sinks (fig. 36).

The H particles traveled about 4.6 and 4.0 mi south-southeast and south, with the dry-cycle particle ending about 0.1 mi east of the wet-cycle pathline. The average linear velocities were 17 and 14 ft/d. The wet-cycle particle traveled farther because of a slight increase in hydraulic gradient caused by the large amount of recharge at the spreading areas (fig. 36). The H particles moved downward to model layer 2 in response to the downward displacement effects of underflow from the Big Lost River alluvial aquifer and streamflow infiltration from the Big Lost River (fig. 14).

The I particles traveled about 0.5 mi southwest, with the dry-cycle particle ending about 0.2 mi southeast of the wet-cycle pathline. Both average linear velocities were 1.8 ft/d. The more westerly flow direction of the wet-cycle particle was caused by a slight directional change in gradient due to increased recharge from the Big Lost River channel southeast of the particles.

The J particles traveled about 1.1 and 1.0 mi southwest, with the dry-cycle particle ending about 0.1 mi southeast of the wet-cycle pathline. The average linear velocities were 4.0 and 3.6 ft/d. The wet-cycle particle traveled farther and slightly more west than the dry-cycle particle because of the large amount of recharge at the spreading areas and the slight directional change in gradient produced by the recharge.

The K particles traveled about 0.8 mi south and 0.6 mi southwest, with the dry-cycle particle ending about 0.2 mi west of the wet-cycle pathline. The average linear velocities were 2.9 and 2.2 ft/d. The wet-cycle particle traveled farther and slightly more east than the dry-cycle particle because of increased recharge from the Big Lost River channel about 1 mi to the northwest and a slight directional change in gradient produced by the recharge.

Effects of Recharge at Big Lost River Spreading Areas during a Wet- and a Dry-Climate Cycle—Particles L, M

The L particles traveled about 9.3 mi southwest, with the wet-cycle particle ending about 1.1 mi southeast of the dry-cycle pathline. The average linear velocities were 34 ft/d. The particle directions diverged because the direction of the local hydraulic gradient varied in response to streamflow infiltration to the aquifer from the spreading areas (fig. 36). The wet- and dry-cycle particles traveled downward to model layers 3 and 2, respectively, reflecting the downward displacement effects of streamflow infiltration beneath the spreading areas across the top of model layer 1. The deeper movement of the wet-cycle particle was caused by the increased recharge at the spreading areas during the wet-cycle.

The M particles traveled about 13.1 and 13.6 mi southwest, with the wet-cycle particle ending about 1.8 mi southeast of the dry-cycle pathline. The average linear velocities were 47.3 and 49.1 ft/d. The wet-cycle particle traveled farther southeast than the dry-cycle particle because of the increased recharge at the spreading areas during the wet cycle. The wet-cycle particle moved downward to model layer 2 and the dry-cycle particle remained in model layer 1. The downward movement of the wet-cycle particle was caused by the increased recharge at the spreading areas during the wet cycle.

Effects of Wastewater Discharge at RTC and INTEC during a Wet- and a Dry-Climate Cycle—Particles N, O

The N particles traveled about 1.1 and 1.2 mi west, with the wet-cycle particle ending about 0.2 mi north of the dry-cycle pathline. The average linear velocities were 4.0 and 4.3 ft/d. The westerly flow direction was not coincident with the regional flow direction and reflected a local hydraulic gradient controlled by (1) wastewater discharge about 0.3 mi to the northeast at the RTC waste infiltration ponds and (2) and an increase in HC. Hydrogeologic zone 1, about 1.1 mi west of the particle release point has a much larger HC than hydrogeologic zone 11, in which the particles were released (fig. 40). The wet-cycle particle traveled north of the dry-cycle particle because of a slightly more northerly orientation to the local hydraulic gradient resulting from streamflow infiltration in the Big Lost River channel about 1 mi to the southeast.

The O particles traveled about 0.8 mi south and 0.6 mi southwest, with the dry-cycle particle ending about 0.3 mi west of the wet-cycle pathline. The average linear velocities were 2.9 and 2.2 ft/d. The difference in flow direction and the longer distance traveled by the wet-cycle particle was caused by a shift in the direction of the hydraulic gradient and an increase in the hydraulic gradient resulting from the larger amount of streamflow infiltration in the Big Lost River channel during the wet-cycle.

Evaluation of Short-Term Climate Cycles Influence on Particle Paths

Irrigation had minimal influence on the travel direction and distance of the A, B, and C particles. The small particle divergences of 1 to 7 percent and differences in net distance traveled of 0 to 5 percent (table 22) for these particles were caused by differences in the amount of streamflow infiltration at the Big Lost River Sinks and spreading areas during the wet- and dry-climate cycles. Because these particles are 12–13 mi distant from the sinks and more than 20 mi from the spreading areas, these results may indicate that episodic recharge at the sinks or spreading areas can influence groundwater movement over a fairly large area, a simulation result that seems to be supported by observed water-level rises and declines between 1981 and 1998 (fig. 12).

Differences in recharge at the Big Lost River Playas and Sinks corresponding to wet- and dry-climate cycles substantially affected the travel directions and distances of the D, E, and F particles. The divergence of these particles ranged from 15 to 83 percent, and the difference in net distance traveled ranged from 20 to 41 percent. Differences in recharge at the sinks and playas affected the F particles the most, because these particles originated closest to the sinks (fig. 40).

Variable recharge from the Big Lost River channel caused large divergences (27–40 percent) for the G, I, and K particles and a large difference in net distance traveled (25 percent) for the G and K particles. Even though the influence of recharge from the channel is probably limited to a distance of a few miles from the channel, the influence of recharge on these particles was large because they traveled short distances (1.5 mi or less; table 22) and remained within about 1 mi of the channel and in close proximity to the source of the transient stress throughout the simulation period.

Differences in recharge at the spreading areas caused a moderate divergence of 12–14 percent, but only a 1–4 percent net difference in distance traveled, for the L and M particles. Recharge was expected to have a greater influence on particle movement for particles traveling near such a concentrated recharge center as the spreading areas. The lesser influence was caused by the large HC of hydrogeologic zone 1, the zone underlying the spreading areas. Divergence of the particles was relatively small because the large HC allowed for rapid movement of water beneath the spreading areas (L and M particle velocities ranged from 34 to 49 ft/d; table 22), relative to the movement of water beneath the sinks (E and F particle velocities ranged from 6.5 to 13 ft/d). As a result, the water table beneath the spreading areas rose much less than the water table beneath the sinks during the wet cycle (fig. 36). The large HC for hydrogeologic zone 1 also influenced the small percentage differences in net distance traveled, because these percentage differences were minimized by the long distances the particles traveled (9.3–13.6 mi; table 22).

Flow directions and velocities for the N and O particles were influenced by (1) wastewater discharged at the RTC and INTEC, (2) large differences in HC between adjacent hydrogeologic zones, and (3) variable streamflow infiltration from the Big Lost River channel. These influences on particle movement illustrate the complexity of flow beneath the RTC and the INTEC. The effects of short-term climate cycles on N and O particle paths were subdued by the large HC of hydrogeologic zone 1 and by the volume of wastewater discharged at the RTC and the INTEC. Consequently, only episodic recharge from the river channel caused divergence or differences in net distance traveled for the N and O particles. The divergence (50 percent; table 22) and difference in net distance traveled (25 percent) for the O particles were relatively large because of the short distance traveled by these particles and the close proximity of the particles to the river channel (fig. 40).

The wet- and dry-cycle particle paths (table 22) were compared to the steady-state and transient particle paths (table 20) to evaluate how episodic recharge affects short-term (one wet- or dry-climate cycle) and long-term (multiple wet- and dry-climate cycles) advective transport. Considering just the K, L, M, N, and O particles, the particles that traveled in the southwestern part of the INL where contaminants are most prevalent in the aquifer, the percent divergences ranged from 12 to 50 percent for the wet-cycle/dry-cycle particles (table 22) and from 4 to 11 percent for the steady-state/transient particles (table 20). Differences in net distance traveled between wet-cycle/dry-cycle and steady-state/transient particles were compared only for the O particles. For the other starting locations, particles simulated under different conditions traveled through several HC zones, complicating the evaluation of the effects of episodic recharge. The differences in net distance traveled for the wet-cycle/dry-cycle and steady-state/transient O particles were 25 and 10 percent, respectively. These differences in particle divergence and net distance traveled illustrate how the proximity of the particles to recharge areas influenced the results. The wet- and dry-cycle particles typically traveled one-half the distance or less of the steady-state and transient particles, which meant that the wet- and dry-cycle particles remained closer to the recharge areas that were the source of change in the local hydraulic gradient. The farther away from recharge areas that particles moved, the less influence those recharge areas had on particle flow directions and velocities. The length of time that particle paths were simulated also influenced the results. The effect of episodic streamflow infiltration on these particles was large during short simulation periods that coincide with wet- or dry-climate cycles, but this effect was diminished over longer simulation periods that included both wet- and dry-climate cycles.

Evaluation of Flow and Transport in the Southwestern Part of Idaho National Laboratory

Particle tracking was used to simulate the growth of tritium (^{3}H) plumes at the INTEC and RTC over a 16-year period under steady-state conditions (1980) and over a 16-year period under transient conditions (1953 through 1968). These simulations were used to evaluate the representativeness of simulated groundwater flow directions and average linear velocities in that part of the aquifer most affected by contamination.

The 2-D shape, dimensions, and areal extent of these simulated plumes are compared to maps of the 1968 ^{3}H plumes in the aquifer (fig. 41) that originated at the INTEC in 1953 and at the RTC in 1952. In this analysis, model-derived particle velocities are compared to groundwater velocity estimates that are based on (1) the assumed position and an extended definition of the position of the leading edge of the ^{3}H plumes in 1968 (fig. 41; table 23), (2) the peak and assumed first arrivals of ^{3}H and ^{36}Cl at downgradient wells, and (3) the ^{3}H/^{3}He model ages of the young fraction of groundwater at downgradient wells.

The assumed position of the leading edge of the ^{3}H plume is defined by ^{3}H concentrations greater than 2,000 pCi/L in 1968. The extended definition of the position of the leading edge is based on a comparison of the shape of the concentration gradients used to define the leading edge of the 1968 and 1985 ^{3}H plumes. Peak-arrival velocity estimates are based on a time-lagged correlation of monthly ^{3}H releases at the INTEC disposal well with semi-annual concentrations of ^{3}H at USGS 36. First-arrival velocity estimates are based on ^{3}H detections at monitoring wells CFA 2, USGS 90, 103, 105, and 108; and ^{36}Cl detections at USGS 11 and 14. Velocity estimates for ^{3}H/^{3}He are based on the model ages of the young fraction of groundwater at USGS 103, 105, 108, 109, 124, and 125 (fig. 41).

Model-derived average linear groundwater velocities are based on the (1) transient travel times of multiple particles released in model layer 1 and 2 cells at the location of the INTEC disposal well (CPP 3) and in model layer 1 cells at the location of the INTEC disposal pit and the RTC warm-waste ponds (fig. 41), and (2) steady-state travel times of a single particle released in the center of a model layer 2 cell at the location of the INTEC disposal well.

Several assumptions, approximations, and simplifications were made for this analysis, including:

- [assumption] The ^{3}H plumes represent a continuum wherein ^{3}H is assumed to be dispersed throughout the water-filled pore space encompassed within the boundaries of the plume.

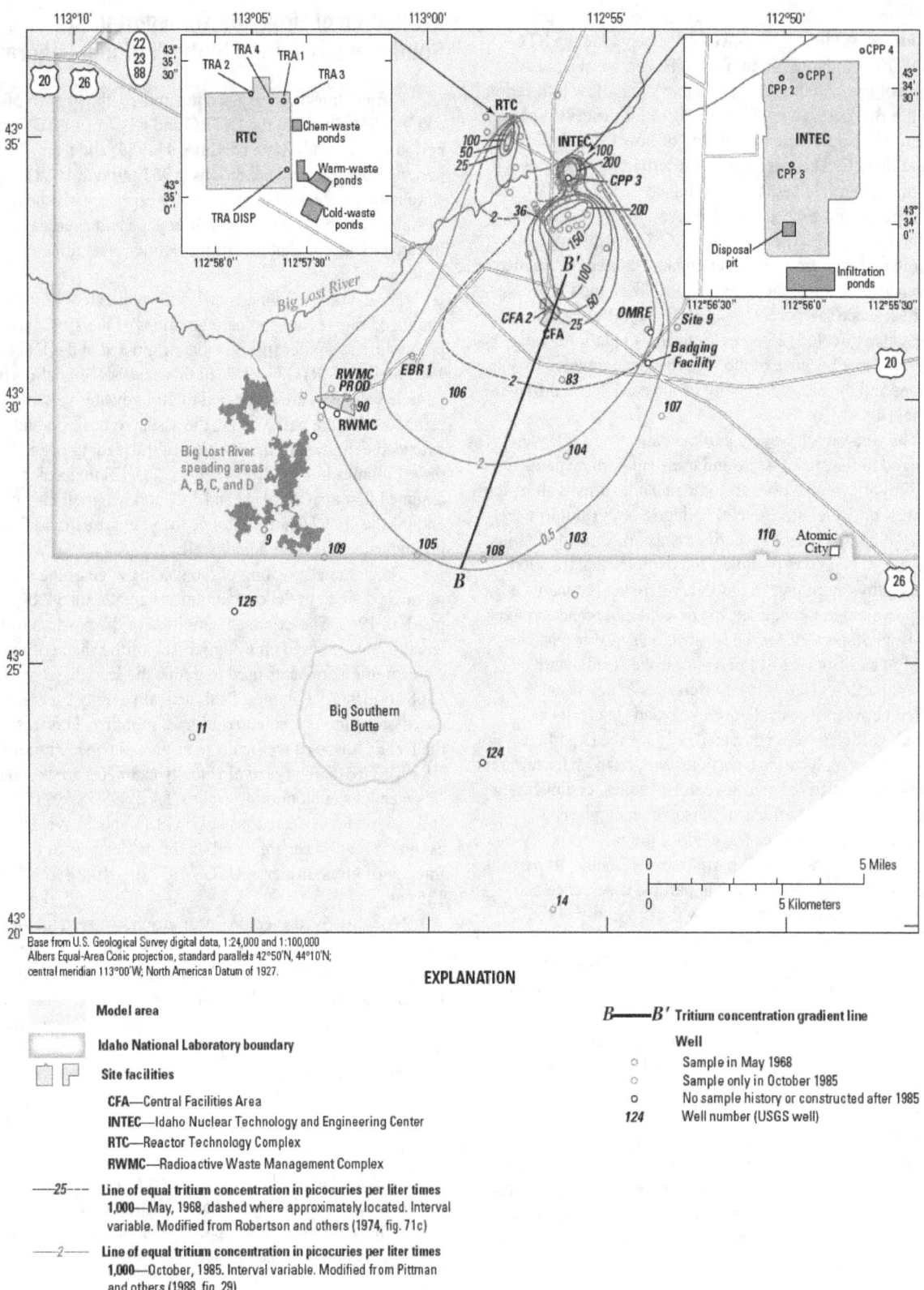

Figure 41. Location of tritium concentrations greater than 2,000 and 500 picocuries per liter, in 1968 and 1985, respectively, and monitoring wells used to compare model-derived average linear groundwater velocities to independently derived velocities, Idaho National Laboratory and vicinity, Idaho.

Table 23. Measured tritium concentrations used to define the leading edge of the tritium plume in 1968 and 1985, Idaho National Laboratory and vicinity, Idaho.

[Locations of wells are shown in figure 41 and appendix B Analytical detection limit for 1968 is 2 pCi/mL and for 1985 is 0 5 pCi/mL Values in **bold** indicate reportable detection **Abbreviations:** pCi/mL, picocurie per milliliter, σ, sigma; <, less than]

Well	May 1968		October 1985	
	Tritium (pCi/mL)	Counting error (2σ)	Tritium (pCi/mL)	Counting error (2σ)
USGS 83	**9**	4	0.0	0.6
USGS 90[1]			**1.6**	**.3**
USGS 103			.4	.6
USGS 104			.9	.6
USGS 105			.0	.6
USGS 106			**2.7**	**.8**
USGS 107			.1	.6
USGS 108			**.8**	**.6**
USGS 109			.2	.6
EBR 1	< 2		.1	.6
OMRE	**4**	2	**2.6**	**.8**
RWMC PROD			**1.4**	**.3**
SITE 9	< 2		.0	.6

[1] January 1985

- [approximation] The true areal extent, shape, and dimensions of the ^{3}H plumes at any point in time are known only to the extent that well coverage and analytical detection limits in use at the time will allow.

- [assumption] The 2-D areal distribution of ^{3}H and the boundaries of the ^{3}H plumes are defined by ^{3}H concentrations that represent mixtures of water, derived from various depths in open wells (in some cases pumped samples and in other cases thief or submerged bailer samples) that are assumed to be uniform with depth.

- [approximation] The transition from facility-derived ^{3}H to background concentrations is represented by a sharp boundary or concentration isopleth even though the true boundary, particularly along the forward or leading edge of the plume, likely lies within a diffuse transition zone where facility-derived ^{3}H concentrations gradually merge with and become indistinguishable from background[2] levels. The shape of this boundary, although depicted as smooth and continuous, is likely to be very irregular and even may be discontinuous, reflecting the effects of preferential flow in an aquifer characterized by extreme heterogeneity and anisotropy.

- [assumption] The reliability of assumed first arrival times, upon which independent velocity estimates are based, is limited by the sensitivity of the analytical detection limit and its ability to distinguish background concentrations from facility-derived concentrations, and the frequency of sampling prior to a confirmed detection. In almost all cases, first arrival estimates likely are after the actual first arrival; therefore, first arrival time estimates should be viewed as maximum travel times and velocity estimates as minimum velocities.

- [simplification] Velocity estimates, based on simulated travel times and the assumed first arrivals of ^{3}H and ^{36}Cl at downgradient wells, are calculated on the basis of the straight-line or minimum distance from point of origination to point of observation and represent minimum velocities that do not take into account the tortuosity of the pathway.

- [approximation] Particles were released within cell blocks that are 0.25 × 0.25 mi (width by length) and dimensionally much larger than the physical dimensions of the disposal pits, ponds, and wells that were used to dispose of wastewater. Modeling ^{3}H releases in this manner was a restriction imposed from using the uniform cell discretization of the steady-state and transient flow models. Substantial evidence suggests, however, that because of lateral flow in the unsaturated zone, inflows to the aquifer from the INTEC disposal pit and RTC warm-waste ponds probably occur over a substantially larger area than indicated by the surface expression of these disposal sites (Robertson and others, 1974; Wood and Norrell, 1996; Nimmo and others, 2002). In a study to explain the processes controlling the longitudinal and transverse dimensions of the tritium plume downgradient of the INTEC, Duffy and Harrison (1987, p. 900) concluded that the radial spreading effect of wastewater injection into the INTEC disposal well probably extended about 450 m (0.28 mi) outward from the well.

---

[2]In 1968, elevated background concentrations of ^{3}H (half life = 12 26 years) in the aquifer beneath the spreading areas, the Big Lost River channel, sinks, and playas, derived mostly from above-ground testing of nuclear weapons in the late 1950s and early 1960s, (^{3}H concentrations in the atmosphere peaked in 1964) may have been as high as 1,900–6,300 pCi/L from streamflow infiltration in 1964 (6,300 pCi/L), 1965 (5,800 pCi/L), and 1967 (1,900 pCi/L) Away from areas of rapid focused recharge, these background levels probably ranged from about 160 to 320 pCi/L, representing recharge prior to the 1964 peak These estimates are based on (1) a model of ^{3}H concentrations that assumes the residence time of water in streams recharging the aquifer is 1 year and the age of the young fraction is "... greater than 5 years but less than 40 years" and (2) measured concentrations of ^{3}H in wells at the INL from 1994 to 1996 that were shown not to be contaminated with facility-derived ^{3}H (Busenberg and others, 2001, p 25, figs 15, 16) In 1989, ^{3}H concentrations in water from 19 springs near Twin Falls, Idaho (fig 1), were less than about 0 2 pCi/mL and from 12 irrigation and domestic wells, 65 mi south of the INL, were less than the reporting level to 0 1 pCi/mL (or 615–307 pCi/L when decay corrected to 1968) (Mann and Cecil, 1990, p 11)

Although the hydrogeologic zones used to represent the aquifer are assumed to be adequately represented as homogeneous, anisotropic porous media at the scale of this analysis, small-scale heterogeneity within individual hydrogeologic zones is not represented and likely has an appreciable if not a large influence on dispersion of contaminants released at the INTEC and RTC. Because of this limitation, the particle plume simulations do not incorporate the effects of small-scale tortuosity and likely underestimate the full extent of dispersion.

Particle-Tracking Simulation Scenarios for Tritium Releases at INTEC and RTC, 1953–68

Background information used for the steady-state and transient particle-plume simulations is summarized in table 24. Major differences between the 1953–68 historical ^{3}H release scenario and the simulated particle-plume scenarios included simulating the (1) temporally variable ^{3}H releases during the historical period as constant discrete particle releases at the beginning of each simulation year and (2) downward percolation of ^{3}H in wastewater discharged to the INTEC disposal pit and RTC warm-waste ponds during the historical period as instantaneous particle releases to the aquifer.

The steady-state particle-tracking simulations are based on groundwater fluxes derived from calibration of the steady-state model. These fluxes are constant for each steady-state stress period. Groundwater fluxes are variable in the transient simulation and reflect the effects of temporal variations in wastewater disposal, groundwater withdrawals, and streamflow infiltration from one stress period to another. To facilitate comparisons of the particle tracking results the timing, location, and number of particle releases were identical in the steady-state and transient simulations.

Comparison of Historical Tritium Releases (1953–1968) and Simulated Particle Releases at the INTEC and RTC

From 1952 through 1968, an estimated 21,000 curies (Ci) of ^{3}H were discharged in wastewater at the RTC and INTEC (figs. 42A, B, and D; table 11). At the INTEC, 16,000 Ci were disposed of in a 20-in. diameter, 600-ft deep, gravel-packed well (CPP 3) (fig. 41) at inflow depths of 412–452 and 489–592 ft below land surface and 400 curies in a disposal pit (INTEC disposal pit), 0.3 mi south of CPP 3 and about 450 ft above the water table. At the RTC, 4,400 Ci were disposed of in wastewater infiltration ponds (warm-waste ponds) also about 450 ft above the water table. Annual releases at the INTEC during this period were highly variable, ranged from 100 to 3,500 Ci/yr, and averaged 1,000 Ci/yr (Robertson and others, 1974). Peak releases at the INTEC occurred in

1958 (3,500 Ci) and 1959 (2,500 Ci). Annual releases at the RTC over this period were fairly uniform, ranged from 20 to 500 Ci/yr, and averaged 280 Ci/yr (table 11).

In the models, particles were released at the INTEC disposal well, the INTEC disposal pit, and the RTC warm-waste ponds to simulate ^{3}H releases at these two facilities from 1953 through 1968. Small amounts of tritium were released at the RTC in 1952, but were not considered important enough to extend the duration of the simulation period to 17 years.

Particles were released at equally spaced points defined by a 3×3 array (440 ft between release points) across the four vertical faces of model cell blocks in model layers 1 and 2 to simulate the inflow depths of wastewater injected into the INTEC disposal well. Particles also were released across the four vertical faces of model cell blocks in model layer 1 at the INTEC disposal pit and the RTC warm-waste ponds to simulate inflow of wastewater from the ground surface. In the steady-state and transient particle-tracking simulations, 36 particles (9 particles across each cell face) were released in each model cell block at the beginning of each year in which tritium was released to the aquifer (fig. 42). During the 16-year simulation period, 2,124 particles were released in the steady-state and transient simulations.

Releasing particles within the vertical cell faces rather than distributing the particles within a model cell allows the plume of particles to reach its full extent while minimizing the number of particles released in a simulation. For example, a particle released from a cell face has a greater potential for describing the outer range of plume movement compared with a particle released from within a model cell. Although an increase in the number of particles provides greater accuracy in defining particle plume movement, the tradeoff is an increase in the computational burden associated with the use of additional particles.

Because flow through the unsaturated zone is treated implicitly in the numerical flow models, particles released at INTEC disposal pit and the RTC warm-waste ponds entered the flow field at the water table instantaneously with no provision to account for the delay time for wastewater to infiltrate through the 450-ft thick unsaturated zone at these facilities. This limitation probably has minimal effect on the representativeness of simulation results. Field experiments in 1999 at the spreading areas and in 1994 at the Large-Scale Infiltration Test (LSIT) site, about 1 mi east of the spreading areas and 1 mi south of the RWMC (fig. 41), indicate that water infiltration rates beneath surface-water impoundments that receive substantial amounts of inflow ranged from 16 ft/d (Wood and Norrell, 1996) at the LSIT site to as much as 72 ft/d (Nimmo and others, 2002, p. 95) at the spreading areas. These experiments indicate that under ponded-infiltration conditions the sedimentary interbeds and any other stratigraphic units expected to impede vertical flow in the unsaturated zone are not an effective barrier.

Table 24. Summary of tritium disposal, wastewater disposal, and groundwater withdrawals at the Idaho Nuclear Technology and Engineering Center and the Reactor Technology Complex used to simulate the growth of tritium plumes in the eastern Snake River Plain aquifer under steady-state conditions from 1980 and transient flow conditions from 1953 to 1968, Idaho National Laboratory, Idaho.

[Locations of facilities, disposal wells, production wells, disposal pit, and warm-waste ponds are shown in figure 41 Tritium disposal, wastewater discharge, and groundwater withdrawal estimates are given in appendix A **Abbreviations**: INTEC, Idaho Nuclear Technology and Engineering Center; RTC, Reactor Technology Complex; Ci, Curies; Ci/yr, Curies per year; L, model layer; ft, foot; bls; below land surface; ft³/s, cubic foot per second]

Tritium and particle releases

Historical tritium releases (1952–68)

Facility	Total tritium releases (Ci)	Annual release rates (Ci/yr)		Release location	Release depths (ft-bls) or (model layer)	Calendar years
		Range	Average			
INTEC	16,000	100–3,500	1,000	CPP 3	412–452	1953–68
				CPP 3	489–592	1953–68
	400	12–190	65	Disposal pit	Ground surface	1954–60, 1962–66
RTC	4,000	20–500	280	Warm-waste ponds	Ground surface	1952–68

Simulated particle releases

Steady-state and transient simulations

Facility	Total particle releases	Particles per release	Number of particle releases			
INTEC	576	36	16	CPP 3	L1	
	540	36	15	CPP 3	L2	
	432	36	12	Disposal pit	L1	
RTC	576	36	16	Warm-waste ponds	LI	

Wastewater disposal

Historical (1952–68)

Facility	Annual discharge rates (ft³/s)		Discharge location	Discharge depths (ft-bls) or (model layer)	Calendar or number of years
	Range	Average			
INTEC	0.8–1.8	1.3	CPP 3	412–452	1953–68
			CPP 3	489–592	1953–68
	0.002–0.034	.017	Disposal pit	Ground surface	1954–60, 1962–66
RTC	0.019–0.757	.421	TRA DISP	512–697	1964–68
				935–1,005	1964–68
				1,045–1,070	1964–68
	0.064–1.2	.692	Warm-waste ponds	Ground surface	1952–68

Simulated

Steady-state (average for 1966–80) simulation

INTEC	1.694	CPP 3	L1,2	16 yrs	L1 = 82 percent; L2 = 18 percent
	0	Disposal pit			
RTC	2.008	TRA DISP	L1,2,4	16 yrs	L1 = 63 percent; L2 = 32 percent; L4 = 5 percent
	.241	Warm-waste ponds	Top of L1	16 yrs	L1 = 100 percent

Transient (1953–68) simulation

INTEC	1.271	CPP 3	L1,2	1953–68	L1 = 82 percent; L2 = 18 percent
	.013	Disposal pit	Top of L1	1953–68	L1 = 100 percent
RTC	.132	TRA DISP	L1,2,4	1953–68	L1 = 63 percent; L2 = 32 percent; L4 = 5 percent
	.692	Warm-waste ponds	Top of L1	1953-68	L1 = 100 percent

Table 24. Summary of tritium disposal, wastewater disposal, and groundwater withdrawals at the Idaho Nuclear Technology and Engineering Center and the Reactor Technology Complex used to simulate the growth of tritium plumes in the eastern Snake River Plain aquifer under steady-state conditions from 1980 and transient flow conditions from 1953 to 1968, Idaho National Laboratory, Idaho.—Continued

[Locations of facilities, disposal wells, production wells, disposal pit, and warm-waste ponds are shown in figure 41 Tritium disposal, wastewater discharge, and groundwater withdrawal estimates are given in appendix A **Abbreviations**: INTEC, Idaho Nuclear Technology and Engineering Center; RTC, Reactor Technology Complex; Ci, Curies; Ci/yr, Curies per year; L, model layer; ft, foot; bls; below land surface; ft³/s, cubic foot per second]

Groundwater withdrawals

Historical (1953–68)

Facility	Annual withdrawal rates (ft³/s)		Withdrawal location	Withdrawal depths (ft-bls) or (model layer)	Calendar or number of years	
	Range	Average				
INTEC	0.44–0.92	0.67	CPP 1	460–577	1953–68	
	0.64–0.93	.74	CPP 2	458–600	1953–68	
RTC	0.55–2.31	1.38	TRA 1	480–580	1952–68	
	0–0.97	.18	TRA 2	558–601	1952–63	
	0.49–1.72	1.27	TRA 3	470–592	1957–68	
			TRA 4	900–965	1963–68	

Simulated

Steady-state (1966–80 average) simulation

INTEC	0.883	CPP 1	L1,2	16 yrs	L1 = 72 percent; L2 = 28 percent
	.883	CPP 2	L1,2	16 yrs	L1 = 43 percent; L2 = 57 percent
	.060	CPP 4	L1,2,3	16 yrs	L1 = 40 percent; L2 = 40 percent; L3 = 20 percent
RTC	.223	TRA 1	L1,2	16 yrs	L1 = 78 percent; L2 = 22 percent
	.024	TRA 3	L1,2	16 yrs	L1 = 67 percent; L2 = 33 percent
	2.504	TRA 4	L4	16 yrs	L4 = 100 percent

Transient (1953–68) simulation

INTEC	0.730	CPP 1	L1,2	1953–1968	L1 = 72 percent; L2 = 28 percent
	.757	CPP 2	L1,2	1953–1968	L1 = 43 percent; L2 = 57 percent
RTC	1.026	TRA 1	L1,2	1953–1968	L1 = 78 percent; L2 = 22 percent
	.576	TRA 2	L1,2	1953–1968	L1 = 8 percent; L2 = 92 percent
	.982	TRA 3	L1,2	1953–1968	L1 = 67 percent; L2 = 33 percent
	.334	TRA 4	L4	1953–1968	L4 = 100 percent

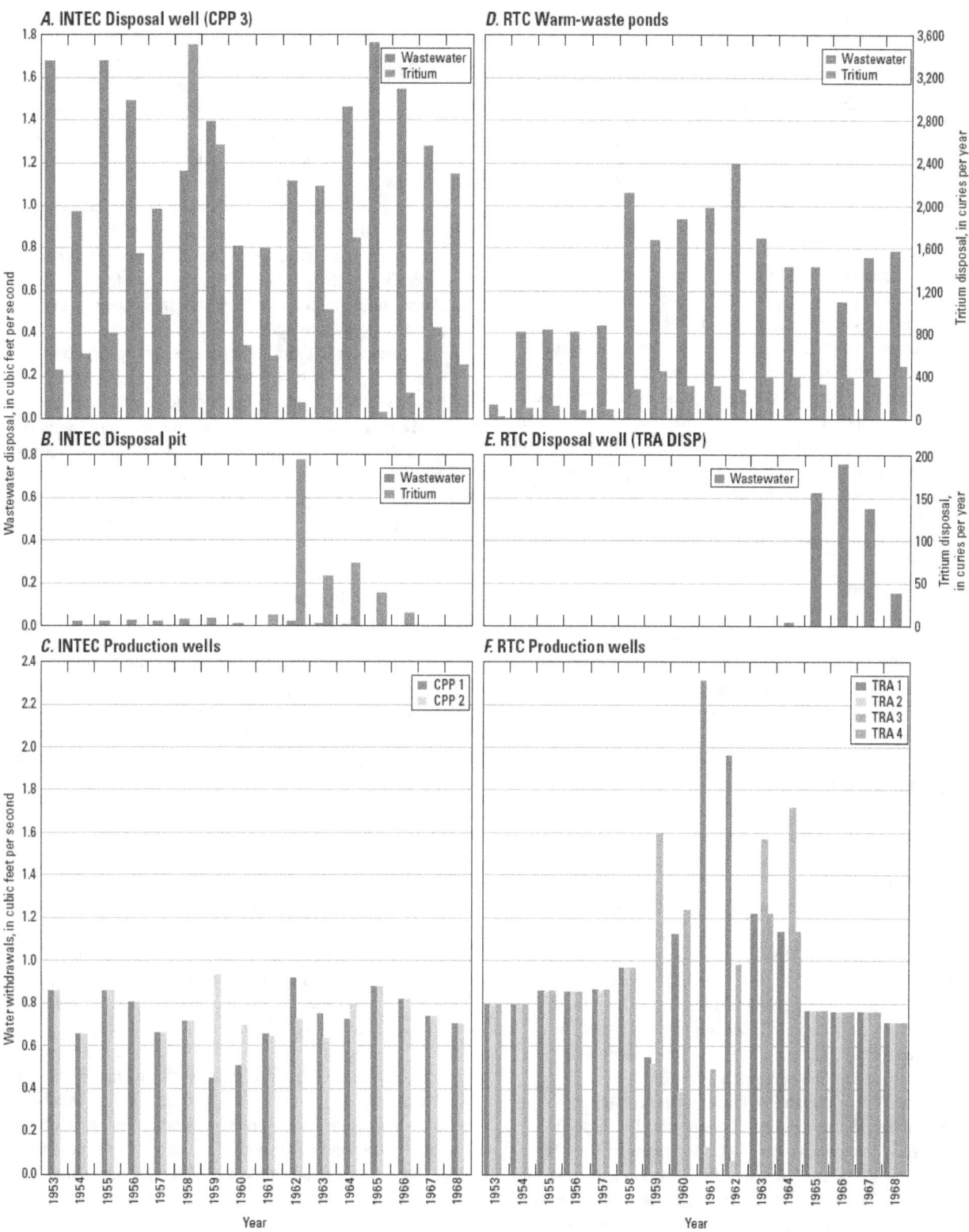

Figure 42. Annual amount of tritium discharged in wastewater, annual wastewater disposal, and annual groundwater withdrawals, Idaho Nuclear Technology and Engineering Center and Reactor Technology Complex, 1953–1968, Idaho National Laboratory, Idaho. The location and other information for each well are available in tables A1 and I1 and figures B6 and B7.

Comparison of Steady-State and Transient Wastewater Discharges and Groundwater Withdrawals at INTEC and RTC

In the steady-state simulation, wastewater discharges and groundwater withdrawals are based on 1966–80 average values. In the transient simulation wastewater discharges and groundwater withdrawals are based on semi-annual averages from 1953–68. These inflows and outflows are constant for each four-month transient stress period within a simulation year (fig. 42; table I1).

From January 1953 through February 1984, most of the low-level radioactive and chemical wastewater generated at the INTEC was discharged to the aquifer through the INTEC disposal well (CPP 3). During the early operational history of this well, 1953–66, much smaller quantities of low-level radioactive and chemical wastewater also were discharged to the INTEC disposal pit. The INTEC disposal well was taken out of routine service in 1984, used intermittently until 1986, permanently decommissioned in 1989, and replaced with two wastewater infiltration ponds about 0.5 mi south of the INTEC disposal well.

In the steady-state simulation, wastewater discharged to the INTEC disposal well averaged 1.694 ft^3/s and was simulated as inflow to the model layer 1 (82 percent) and 2 (18 percent), corresponding to the location of the open intervals in the disposal well (table 24). In the transient simulation, the rate of wastewater discharged from the INTEC disposal well averaged 1.271 ft^3/s and was allocated to the model layer 1 and 2 cells in the same proportions as in the steady-state simulation.

In the steady-state simulation, no wastewater was discharged to the INTEC disposal pit. In the transient simulation, wastewater discharged to the INTEC disposal pit averaged 0.013 ft^3/s and was simulated as recharge to the top of the model layer 1 cell corresponding to the location of the disposal pit.

Groundwater withdrawals from the INTEC production wells, about 0.4 mi north of the INTEC disposal well, averaged 1.826 ft^3/s (CPP 1, 2, and 4) in the steady-state simulation and 1.487 ft^3/s (CPP 1 and 2) in the transient simulation. In the transient simulation, water from these production wells was withdrawn from model layers 1 (72 percent) and 2 (28 percent) for CPP 1 and model layers 1 (43 percent) and 2 (57 percent) for CPP 2.

From 1952 through 1964, wastewater containing radioactive and non-radioactive waste was discharged to the warm-waste ponds at the RTC. Beginning in 1964, all non-radioactive wastewater was discharged to a 6- to 18–in. diameter, 1,267-ft deep well (TRA DISP) at inflow depths of 512-697, 935-1,005, and 1,045-1,070 ft below land surface while radioactive wastewater continued to be discharged to the warm-waste ponds through 1982. In 1982, the deep disposal well was replaced with two cold-waste infiltration ponds.

In the steady-state simulation, wastewater discharged to the RTC disposal well averaged 2.008 ft^3/s and was simulated as inflow to the model layer 1 (63 percent), 2 (32 percent), and 4 (5 percent) cells corresponding to the location of the open intervals in the RTC disposal well. In the transient simulation, wastewater discharge averaged 0.132 ft^3/s and was allocated to the model layer 1, 2, and 4 cells in the same proportions as in the steady-state simulation.

In the steady-state simulation, the rate of wastewater discharged to the warm-waste ponds averaged 0.241 ft^3/s and was simulated as recharge to the top of the model layer 1 cell corresponding to the location of the warm-waste ponds. In the transient simulation, the rate of wastewater discharged to the warm-waste ponds averaged 0.692 ft^3/s.

Groundwater withdrawals from the RTC production wells, about 0.3 mi north of the TRA disposal well, averaged 2.751 ft^3/s (TRA 1, 3, and 4) in the steady-state simulation and 2.918 ft^3/s (TRA 1, 2, 3, and 4) in the transient simulation. In the transient simulation water was withdrawn from model layers 1 (78 percent) and 2 (22 percent) for TRA 1, layers 1 (8 percent) and 2 (92 percent) for TRA 2, layers 1 (67 percent) and 2 (33 percent) for TRA 3, and layer 4 (100 percent) for TRA 4.

Comparison of Steady-State and Transient Streamflow Infiltration

In the steady-state simulation, streamflow infiltration is based on 1966–80 average values that are constant for each stream reach and for every 12-month stress period (table C1). In the transient simulation streamflow infiltration is based on 4-month averages from 1953–68 that generally vary for each 4-month transient stress period (fig. 43; table C1). Differences in the timing, location, and amount of streamflow infiltration represent the major differences in hydrologic conditions between the 1980 steady-state and the 1953–68 transient particle-tracking simulations.

In the 16-year steady-state simulation, the annual mean discharge from 1966 through 1980 into the model area at Big Lost River near Arco (13132500; figs. 1 and 10) was 119 ft^3/s. The annual mean diversions to the spreading areas during this period was 43 ft^3/s, the steady-state average based on discharge records at INL diversion at head near Arco (13132513; figs. 1 and 10). Prior to 1965, all streamflow infiltration from the Big Lost River was along the channel and at the sinks and playas near the west-central part of the INL. From 1966 through 1980, spreading-area diversions accounted for about 36 percent of the annual mean discharge into the model area. In the steady-state simulation, no distinction is made between a dry-climate and a wet-climate cycle.

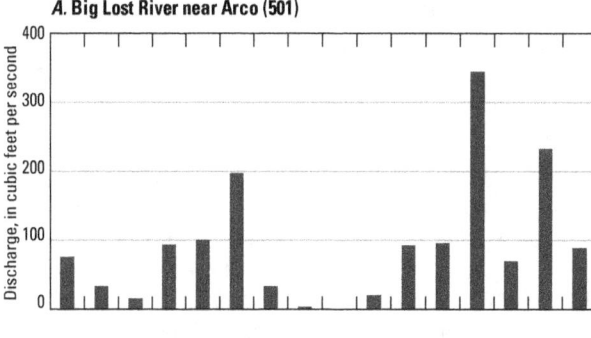

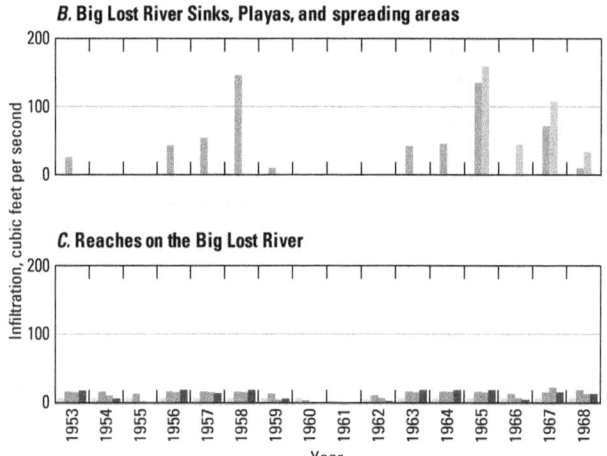

EXPLANATION

- Sinks and Playas (608–610)
- Spreading areas (602–605)
- Model Boundary to Big Lost River near Arco (600)
- Big Lost River near Arco to INL Diversion (601)
- INL Diversion to Lincoln Boulevard (606)
- Lincoln Boulevard to Big Lost River Sinks (607)

Map numbers in parenthesis refer to identifiers used to locate streamflow-gaging stations and stream reaches described in appendix table A2, and located on maps in appendix figures B1, B2, with data provided in appendix table C1.

Figure 43. Annual mean discharge at (*A*) streamflow-gaging station Big Lost River near Arco, and streamflow infiltration at (*B*) Big Lost River Sinks, Playas, and spreading areas, and from (*C*) model boundary to the Big Lost River Sinks, 1953 through 1968, Idaho National Laboratory and vicinity, Idaho.

From 1953 through 1968, annual mean discharge into the model area at Big Lost River near Arco was 93 ft³/s and ranged from 0 to 343 ft³/s (fig. 43*A*; table C1). Peak annual mean discharges, substantially greater than the 16 year average for this period and greater than the 1966–80 steady-state average, 119 ft³/s, occurred in three intermittent years and ranged from 196 ft³/s in 1958 to 344 ft³/s in 1965 (fig. 43*A*). The period from 1953 through 1968 included one 2-year (1954–55) and one 4-year (1959–62) dry-climate cycle and one 3-year (1956–58) and one 6-year (1963–68) wet-climate cycle. The first diversions of streamflow from the Big Lost River to the spreading areas were in 1965 followed by diversions in each of the next 3 years. From 1965 through 1968, diversions to the spreading areas ranged from 35 to 159 ft³/s and averaged 87 ft³/s representing a total diversion of 1.1×10^{10} ft³ (251,940 acre-ft). From 1953 through 1968, streamflow infiltration from mean-annual spreading-area diversions averaged 22 ft³/s, or about 24 percent of the total streamflow into the model area at Big Lost River near Arco. On an annual basis, the streamflow infiltration rate at the spreading areas in the steady-state simulation is about twice as much as in the transient simulation; however, in the transient simulation this recharge is allocated over a 4-year period resulting in much higher recharge fluxes temporally and spatially.

Description and Comparison of Tritium and Particle Plumes

The ³H plumes are presented as a mapping of concentration isopleths that define the 2-D spatial distribution of ³H in the aquifer downgradient of the INTEC and the RTC in May 1968. Groundwater velocities are based on the inferred position of the 2,000 pCi/L isopleth that was used (or assumed) to define the leading edge of the ³H plume in 1968.

For discussion, the simulated particle plumes are subdivided into segments that are based on the release location, release layer, and hydrogeologic zone or zones through which the particles traveled. The geographical position of each particle is plotted at the end of every 1-year simulation period and at every location where a particle crosses a cell face. This plotting approach is designed to track the entire history of each particle's movement from time of release to the end of the 16-year simulation period. Simulated particle velocities are based on the position of the leading edge of each plume segment at various times during the simulation period.

Tritium Plumes Downgradient of INTEC and RTC in 1968 and 1985

The areal extents of the 1968 ^{3}H plumes downgradient of the INTEC and RTC, as depicted by Robertson and others (1974), are shown in figure 41. Robertson's depiction of the plume boundaries is based on ^{3}H concentrations greater than 2,000 pCi/L, the analytical detection limit[3] in use at the time. The location of the 2,000 pCi/L isopleth in 1968 is based on very sparse well coverage and ^{3}H non-detections that are a considerable distance away from the inferred 2,000 pCi/L isopleth, particularly along the southern[4] (practically nonexistent) and much of the western and eastern boundaries of the mapped plumes (fig. 41). Additionally, locally elevated ^{3}H concentrations in the aquifer in 1968, derived from rapid focused recharge in the spreading areas, the Big Lost River channel, sinks, and playas in 1965 and 1967 may have been as high as 5,800 pCi/L[4], implying that measured concentrations in areas affected by streamflow infiltration may include a substantial component of ^{3}H, albeit diluted, from sources other than the INTEC and RTC, with uncertain implications for the mapped distribution of ^{3}H that is assumed to be derived exclusively from wastewater disposal at the INTEC and RTC. In areas not affected by rapid focused recharge, background ^{3}H concentrations in 1968 would have ranged from about 160 to 320 pCi/L[4]. In this case, detections would not have included a substantial component of ^{3}H from non-facility sources; however, the analytical detection limit, 6 to 12 times greater than regional background, would not have been sufficiently sensitive to detect the absence (or near absence) of facility-derived ^{3}H regardless of well coverage. The foregoing indicates that considerable uncertainty is associated with the location of the dashed boundary used to define the areal extent and assumed leading edge of the ^{3}H plume in 1968 (fig. 41). Limitations of the analytical detection limit and their effect on interpretations of the extent of ^{3}H in the aquifer were expressed by Robertson and others (1974, p. 170) who noted,

The detection limit improved from 5 pCi/mL [5,000 pCi/L] in 1963 to 2 pCi/mL [2,000 pCi/L] in 1966. This may lead to the mistaken impression that the outer fringe of the waste actually expanded to the extent indicated by the outer lines from 1963 to 1966.

Also shown in figure 41 are the areal extents of the 1985 ^{3}H plumes and the location of the 2,000 pCi/L isopleth as depicted by Pittman and others (1988). Pittman's depiction of the plume boundaries is based on ^{3}H concentrations greater than an analytical detection limit of 500 pCi/L and much improved, although still less than ideal, well coverage over that available in 1968. Pittman's depiction of the leading edge includes ^{3}H detections and non-detections in much closer proximity to the mapped 500 pCi/L isopleth than were available to map the 1968 2,000 pCi/L isopleth. In 1985, background concentrations of ^{3}H in the aquifer in areas away from rapid focused recharge would have been about 165–200 pCi/L for water representing recharge between 1945 and 1980—water greater than 5 years old and less than 40 years old. In areas close to inputs of ^{3}H from rapid focused recharge, concentrations would have been about 60–160 pCi/L based on a 1 year residence-time model for water in the Big Lost River that recharged the aquifer in 1982, 1983, and 1984[2] (Busenberg and others, 2001, fig. 16). The areal influence of these recharge events in the early 1980s on groundwater levels, and presumably their effect on ^{3}H concentrations, was quite extensive as shown in the water-level rise map for July 1981 to July 1985 (fig. 12).

From the foregoing, it can be inferred that the assumed leading edge of the 1985 ^{3}H plume probably was closer to its true position[5] than the assumed leading edge of the 1968 ^{3}H plume. The leading edge of the 1985 ^{3}H plume would not have been as affected by limitations of well coverage, sensitivity of the analytical detection limit, or by ambiguity over the effect of regional background and locally elevated levels of ^{3}H in the aquifer from non-facility sources.

---

[3] In 1968 and 1985, a tritium measurement was reported as a true detection if the measurement was greater than or equal to the analytical detection limit and greater than 2 times the counting uncertainty (95 percent probability that the true value is within ±2 sigmas) For measurements close to the analytical detection limit and (or) with counting uncertainties that approach the limiting value, this reporting criteria implies that there are 50 chances in 1,000 that the reported value is a false positive

[4] The 1968 2,000 pCi/L isopleth south of USGS 83 (fig 41) is based on an analysis of a thief sample of groundwater from this borehole in May 1968 Pumped samples from this borehole have never produced tritium concentrations greater than the reporting level despite the fact that boreholes south of USGS 83, along the southern boundary of the INL for example, have had positive detections in later years (Bartholomay and others, 2000) USGS 83 penetrates about 250 ft of the aquifer and is open from about 15 ft below the water table to total depth Sampling in 1995 indicated that strontium and chlorofluorocarbon -11, -12, and -113 were anomalously low for this well given its location within the contaminant plume area These anomalously low values and the lack of repeatable tritium detections above the analytical detection limit suggest that water pumped from this borehole does not come from the shallower parts of the aquifer where facility-derived tritium would normally be detected (Busenberg and others, 2001, figs 5, 7, 8, 9 and p 87)

[5] Tritium concentrations greater than local background levels, were first detected at USGS 103 (800 pCi/L) in July 1983, at USGS 105 (from less than the reporting level to 3,400 pCi/L with positive detections in 3 of 9 samples), and at USGS 108 (from 830 to 3,400 pCi/L with positive detections in 11 of 11 samples) in October and November 1983 using a thief sampler Repeat sampling of USGS 103 in October and November 1983 did not result in a reportable detection in 10 of 10 samples Monitoring of these boreholes began in September 1980 In late November and early December 1983, the thief sampling system was replaced with dedicated submersible pumps and measured tritium concentrations declined in boreholes with positive detections in October and November 1983, indicating nonuniform distribution of tritium in the aquifer and mixing of tritium over the 150–200 ft sampled intervals in these boreholes (Mann and Cecil, 1990, p 27)

A comparison of the concentration gradients (fig. 44) for the 1968 and 1985 plumes along lines of section tracing downgradient concentrations towards the leading edge of the 1968 and 1985 plumes suggests that the inferred location of the 2,000 pCi/L isopleth in 1968 may have substantially underestimated the downgradient extent of facility-derived ^{3}H in the aquifer. This underestimation was due to a combination of inadequate well coverage, unknown effects of mixing of ^{3}H from various sources, and constraints of the analytical detection limit as previously described. The shape of the 1985 concentration gradient indicates a gradually varying and smooth transition leading into the 500 pCi/L isopleth that was used to define the leading edge of the 1985 plume, whereas the 1968 concentration gradient terminates abruptly at the assumed leading edge of the plume at a concentration that is 6 to 12 times greater than the estimated regional background level (160–320 pCi/L). Calculated travel distances, in a constant velocity flow field, using ^{3}H/^{3}He-derived velocity estimates from the INTEC to wells USGS 103 (4.3 ft/d), USGS 105 (6.2 ft/d), and USGS 109 (7.2 ft/d) (fig. 39; table 21) and travel times of 15.4 years and 32.8 years indicate the theoretical position of the leading edge of the ^{3}H plume in May 1968 and in October 1985 (fig. 44) for ^{3}H releases at the INTEC disposal well (CPP 3) beginning in January 1953 (fig. 42; table I1). The difference in the shape of the 1968 and 1985 concentration gradients suggests that the downgradient extent of facility-derived ^{3}H in 1968, attributable to advective flow, may have been several thousand to perhaps as much as 10,000 ft south of the inferred 2,000 pCi/L isopleth before facility-derived ^{3}H would have become indistinguishable from regional background levels, 160 to 320 pCi/L. This rendering, however, does not take into account the attenuating effects of radioactive decay (1.3 and 2.7 half-lives) for the initial releases of ^{3}H that presumably would mark the leading edge of the 1968 and 1985 plumes, respectively. This rendering also does not account for the effects of locally elevated ^{3}H concentrations, a largely unknown complication that is presumed to have progressively less influence with increasing distance from areas of rapid focused recharge. A 10,000 ft difference represents an uncertainty of as much as 38 percent in groundwater velocity estimates between the INTEC and the assumed leading edge of the ^{3}H plume in 1968.

For several reasons, meaningful comparisons of the areal extent, shape, and dimensional features of the simulated particle plumes are limited to a comparison of the ^{3}H plume that originated at the INTEC. The areal extent of the ^{3}H plume downgradient of the RTC was much less extensive than that of the 1968 ^{3}H plume downgradient of the INTEC (fig. 41). The evolution of the RTC plume also was quite different from that of the INTEC plume. At the RTC all ^{3}H was disposed of in the warm-waste ponds, about 450 ft above the water table.

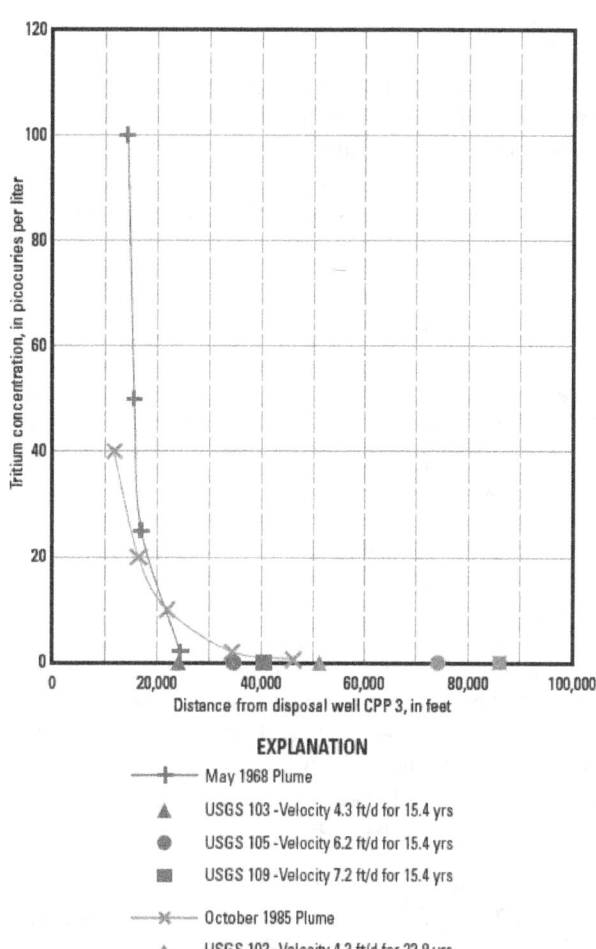

EXPLANATION

—+—	May 1968 Plume
▲	USGS 103 - Velocity 4.3 ft/d for 15.4 yrs
●	USGS 105 - Velocity 6.2 ft/d for 15.4 yrs
■	USGS 109 - Velocity 7.2 ft/d for 15.4 yrs
—✕—	October 1985 Plume
▲	USGS 103 - Velocity 4.3 ft/d for 32.8 yrs
●	USGS 105 - Velocity 6.2 ft/d for 32.8 yrs
■	USGS 109 - Velocity 7.2 ft/d for 32.8 yrs

Figure 44. Tritium concentration gradients defining the leading edge of the tritium plume downgradient of the Idaho Nuclear Technology and Engineering Center in 1968 (2,000 pCi/L isopleth) and in 1985 (500 pCi/L isopleth), Idaho National Laboratory, Idaho. Theoretical positions of the leading edge based on tritium/helium-3-derived velocity estimates from the INTEC to wells USGS 103, USGS 105, and USGS 109, and travel times of 15.4 years and 32.8 years. Traces of concentration gradients are shown in figure 41. Velocity estimates are based on the tritium and helium-3 ages of the "young fraction" of groundwater (Busenberg and others, 2001).

Extensive perched water horizons were (and are) common beneath the RTC and the INTEC (Robertson and others, 1974, p. 95-101, 125-132) derived from Big Lost River infiltration, lateral flow in the unsaturated zone, and percolation of wastewater at the RTC infiltration ponds. As a result, the timing and location of ^{3}H releases to the aquifer at the RTC from 1952 through 1968 are poorly defined. Furthermore, the first systematic mapping of ^{3}H concentrations in the aquifer was in 1961 (Mann and Cecil, 1990, p. 17; Fromm and others, 1994, p. 224). The analytical detection limit in use in 1961 was 5,000 pCi/L. Maximum background concentrations of ^{3}H from infiltration in the nearby Big Lost River would have been less than about 600 pCi/L[2] (Busenberg and others, 2001, p. 25, figs. 15, 16) from streamflow infiltration in 1958 (fig. 43) (assuming no dilution with water in the aquifer). Concentrations of ^{3}H above the analytical detection limit were reported after the first sampling round in all but one of the wells (USGS 84) downgradient of the RTC. The single downgradient well without a reportable detection likely contained facility-derived ^{3}H that was less than the 5,000 pCi/L analytical detection limit in 1961. Because of the sampling history of wells downgradient of the RTC and poor definition of the timing and release locations of ^{3}H to the aquifer, useful first-arrival travel-time estimates and travel distances are not available to evaluate the results of the RTC particle-plume simulation.

Based on the depiction of Robertson and others (1974), the 1968 ^{3}H plume downgradient of the INTEC was exceptionally wide (6 mi) relative to its length (5 mi), with a ratio of 1.2 (fig. 41). The maximum linear velocity along the regional direction of groundwater flow that transported ^{3}H to the leading edge of the plume, as defined by the 2,000 pCi/L isopleth in May 1968, was 4.7 ft/d, but may have been as much as 6.5 ft/d using the extended definition of the position of the leading edge in 1968 (10,000 ft south of the 2,000 pCi/L isopleth) previously described (table 25).

The north to south orientation and wide triangular shape of the 1968 plume may have resulted from variable streamflow infiltration from the Big Lost River. From 1953 through 1964, streamflow in the Big Lost River was average to less than average and no diversions to the spreading areas took place (fig. 43A). During this period, ^{3}H probably moved downgradient along the regional northeast to southwest flow direction. Above average streamflow in 1965 and 1967 and diversions to the spreading areas in 1965 resulted in a rise in the water table of 2–10 ft over a large area centered beneath the spreading areas, the Big Lost River channel, sinks, and playas and extending as much as 6 mi outward from these areas of rapid focused recharge (Robertson, 1974, fig. 9). This water-table rise would have changed the direction of the hydraulic gradient and caused shallow groundwater (and ^{3}H) between the Big Lost River and the RTC to move slower and in a more westerly direction, and shallow groundwater between the river and the INTEC to move faster and in a more

easterly direction as illustrated in the wet-cycle and dry-cycle particle-tracking simulations (particles K, N and O; fig. 40; table 21).

Simulated Particle Plumes Downgradient of INTEC and RTC

Particle plumes generated using the steady-state and transient flow models are shown in figure 45. The most noticeable differences between the steady-state and transient plumes are the (1) larger widths of transient plume segments, particularly downgradient of the boundary between hydrogeologic zones 11 and 1, compared to the equivalent steady-state plume segments; (2) slightly shorter travel distances of particles forming transient plume segments INTEC-D and RTC-G compared to the equivalent steady-state plume segments; and (3) westerly movement of particles from the INTEC disposal well (CPP 3) toward the RTC in the transient simulation that was not in the steady-state simulation. The larger widths of the transient plume segments and the westerly movement of particles from the INTEC towards the RTC were caused by variable wastewater disposal, groundwater withdrawal, and streamflow infiltration, which indicates that large-scale advective dispersion is greater under transient flow conditions than under steady-state flow conditions as illustrated in the steady-state and transient particle-tracking simulations (particles K, L, M, and N; fig. 38; table 20).

Plume Segment INTEC-A

Plume segment INTEC-A (fig. 45B) is composed of particles, released to model layer 1 at the INTEC disposal well and the INTEC disposal pit, 0.30 mi south of the disposal well, that traveled exclusively in hydrogeologic zone 11 (fig. 45B). The maximum distance the particles traveled downgradient in hydrogeologic zone 11 was 3.4 mi in 13.7 years, which corresponds to an average linear groundwater velocity of 3.6 ft/d. This velocity is about 23 percent slower than the velocity represented by the leading edge of the ^{3}H plume in 1968 and about 45 percent slower than the velocity represented by the extended definition of the leading edge (fig. 41; table 25).

The wide distribution of particles and the north to south orientation of the plume segment were caused by (1) simulated radial flow in all directions from the cell where the particles originated, (2) transient effects of streamflow infiltration from 1953 through 1968 (fig. 43), and (3) a gradient direction partly controlled by the large contrast in HC between hydrogeologic zones 1 and 11 (fig. 30). Radial flow away from the cell where the particles originated produced a wide distribution of particles. This wide distribution reflects the effects of the large fluxes of wastewater discharge at the INTEC disposal well. These fluxes represent more than 90 percent of the total flux passing through the model layer 1 cell at the INTEC disposal well.

Table 25. Groundwater and particle velocities based on the assumed position and extended definition of the leading edge of the tritium plume of 1968, transient particle-plume and steady-state particle-pathline simulations, peak and first arrivals of tritium, tritium/helium-3 ages of the "young fraction" of groundwater, and first and peak arrivals of chlorine-36 at wells downgradient of the Idaho Nuclear Technology and Engineering Center and the Reactor Technology Complex, Idaho National Laboratory and vicinity, Idaho.

[Tritium plume and downgradient wells for 1968 are shown in figure 41, simulated transient particle plume in figure 45B, and simulated steady-state pathline in figure 47. Distances measured from INTEC disposal well (CPP 3) or RTC warm-waste ponds. ^{3}H/^{3}He ages, Busenberg and others (2001); ^{36}Cl arrivals, Cecil and others (2000). **Abbreviations: INTEC, Idaho Nuclear Technology and Engineering Center; RTC, Reactor Technology Complex;** ^{3}H, tritium; ^{3}He, helium-3; ^{36}Cl, chlorine-36; vel, velocity; mi, mile; yrs, years; ft/d, foot per day; pCi/L, picocurie per liter]

Criteria	Leading edge of ^{3}H plume from May 1968 Distance (mi)	Time (yr)	Velocity (ft/d)	Transient particle-plume simulation Plume segment	Distance (mi)	Time (yrs)	Velocity (ft/d)	Steady-state particle-pathline simulation Well	Distance (mi)	Time (yrs)	Velocity (ft/d)	^{3}H peak arrival or first arrival Well	Distance (mi)	Time (yrs)	Velocity (ft/d)	^{3}H/^{3}He ages Well	Velocity (ft/d)	^{36}Cl arrivals First Well	Velocity (ft/d)	Peak Velocity (ft/d)
2,000 pCi/L isopleth [1]	5.0	15.4	4.7	INTEC-A	3.4	13.7	3.6	USGS 36	0.95	1.8	7.6	USGS 36	0.95	1.0	[2]14	USGS 103	4.3			
Extended definition [3]	6.9	15.4	6.5	INTEC-B	2.4	4.7	7.4	CFA 2	3.05	8.0	5.5	CFA 2	3.05	8.7	[4]5.1	USGS 105	6.2			
				B-C	1.6	4.3	5.4	USGS 90	7.57	10.0	10.9	USGS 90	7.57	22.3	4.9					
				INTEC-C	3.8	9.1	6.0	USGS 108	8.33	10.5	11.5	USGS 103	7.95	30.6	3.8	USGS 109	6.2–7.2			
				C-D	19.7	6.9	41	USGS 105	8.86	10.5	12.2	USGS 108	8.33	30.8	3.9	USGS 124	7.5			
				A-E	10.4	2.3	65					USGS 105	8.86	30.8	4.2	USGS 125	10.2			
				A-F	14.1	3.3	62									USGS 11	12.5	USGS 11	[5]15.9	8.4
				INTEC-D	23.3	16.0	211									USGS 14	9.2	USGS 14	[5]11.9	8.1
				RTC-G	30.1	15.9	274													
				INTEC-USGS 11	14.9	11.0	196													
				RTC-USGS 11	15.1	14.5	151													
				INTEC-USGS 36 [6]	9.5	13.7	1													
				INTEC-USGS 36 [7]	9.5	1.5	9.2													

[1] Robertson (1974).

[2] Time series correlation of peak ^{3}H concentrations at USGS 36 with monthly releases of ^{3}H at CPP 3 - this report.

[3] Based on a comparison of the shape of the ^{3}H concentration gradients used to define the leading edge of the ^{3}H plumes of 1968 and 1985 - this report.

[4] Positive detection occurred with first sampling event following well construction. Reliability of first arrival time is questionable.

[5] USGS 11 ^{36}Cl concentration greater than 1.5 times background; USGS 14 ^{36}Cl concentration greater than 1.8 times background.

[6] Model Layer 1.

[7] Model Layer 2.

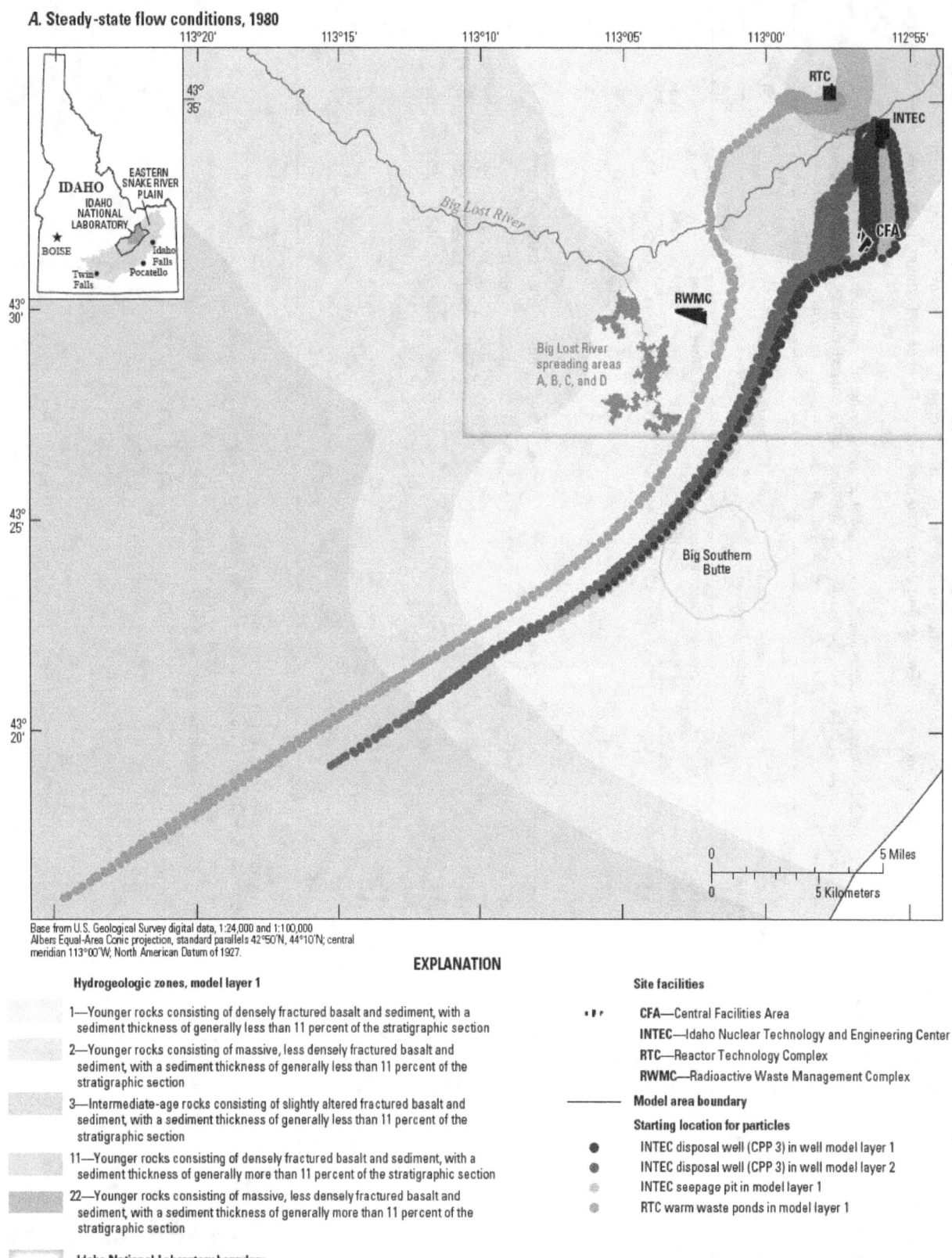

A. Steady-state flow conditions, 1980

Base from U.S. Geological Survey digital data, 1:24,000 and 1:100,000
Albers Equal-Area Conic projection, standard parallels 42°50'N, 44°10'N; central
meridian 113°00'W; North American Datum of 1927.

EXPLANATION

Hydrogeologic zones, model layer 1

1—Younger rocks consisting of densely fractured basalt and sediment, with a sediment thickness of generally less than 11 percent of the stratigraphic section

2—Younger rocks consisting of massive, less densely fractured basalt and sediment, with a sediment thickness of generally less than 11 percent of the stratigraphic section

3—Intermediate-age rocks consisting of slightly altered fractured basalt and sediment, with a sediment thickness of generally less than 11 percent of the stratigraphic section

11—Younger rocks consisting of densely fractured basalt and sediment, with a sediment thickness of generally more than 11 percent of the stratigraphic section

22—Younger rocks consisting of massive, less densely fractured basalt and sediment, with a sediment thickness of generally more than 11 percent of the stratigraphic section

Idaho National Laboratory boundary

Site facilities

CFA—Central Facilities Area
INTEC—Idaho Nuclear Technology and Engineering Center
RTC—Reactor Technology Complex
RWMC—Radioactive Waste Management Complex

Model area boundary

Starting location for particles

INTEC disposal well (CPP 3) in well model layer 1
INTEC disposal well (CPP 3) in well model layer 2
INTEC seepage pit in model layer 1
RTC warm waste ponds in model layer 1

Figure 45.   Particle plumes simulating the advective transport of tritium from the Idaho Nuclear Technology and Engineering Center and the Reactor Technology Complex for 16 years under (*A*) steady-state flow conditions in 1980 and (*B*) transient flow conditions from 1953 through 1968, Idaho National Laboratory and vicinity, Idaho.

B. Transient flow conditions, 1953–68

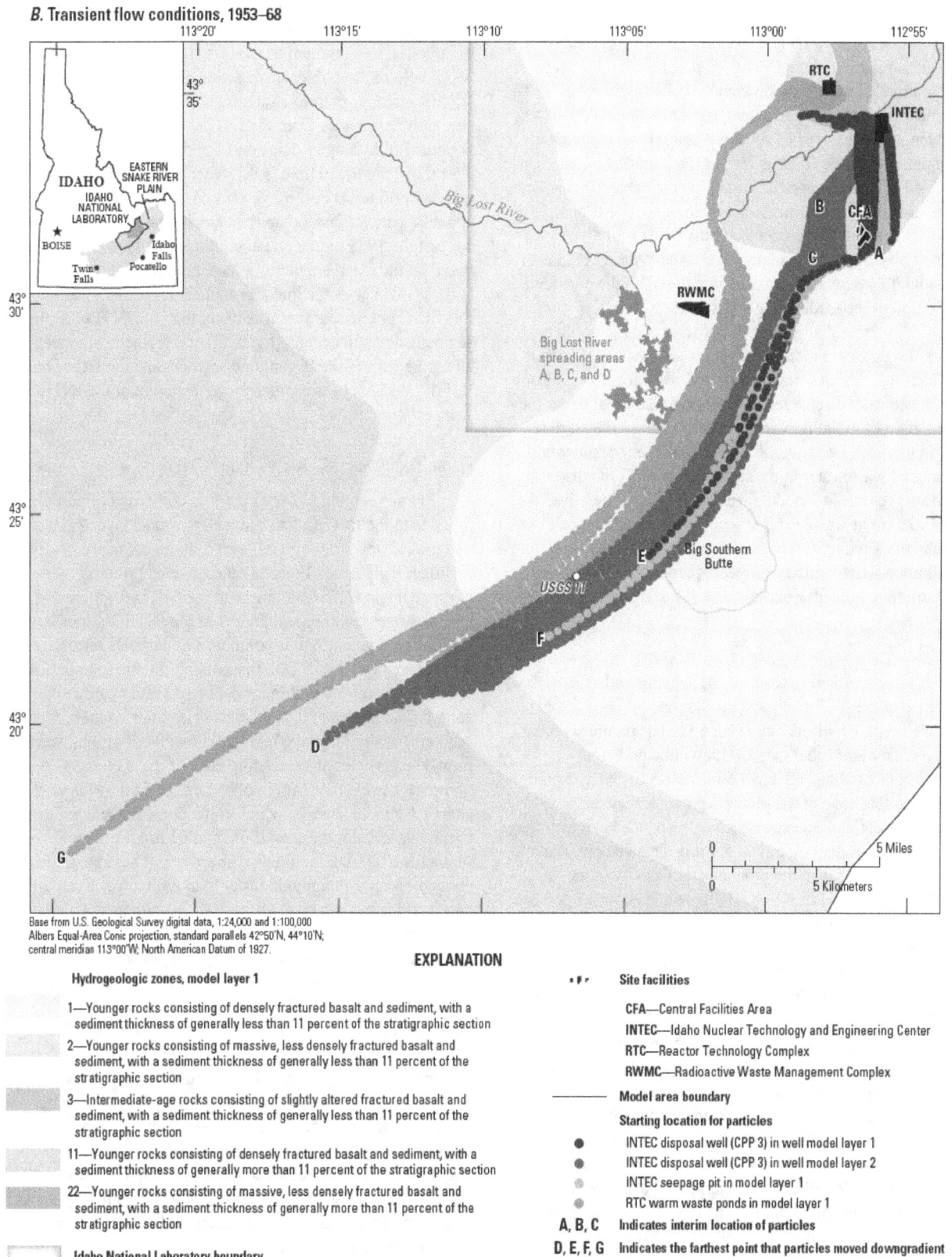

Base from U.S. Geological Survey digital data, 1:24,000 and 1:100,000
Albers Equal-Area Conic projection, standard parallels 42°50'N, 44°10'N;
central meridian 113°00'W; North American Datum of 1927.

EXPLANATION

Hydrogeologic zones, model layer 1

1—Younger rocks consisting of densely fractured basalt and sediment, with a sediment thickness of generally less than 11 percent of the stratigraphic section

2—Younger rocks consisting of massive, less densely fractured basalt and sediment, with a sediment thickness of generally less than 11 percent of the stratigraphic section

3—Intermediate-age rocks consisting of slightly altered fractured basalt and sediment, with a sediment thickness of generally less than 11 percent of the stratigraphic section

11—Younger rocks consisting of densely fractured basalt and sediment, with a sediment thickness of generally more than 11 percent of the stratigraphic section

22—Younger rocks consisting of massive, less densely fractured basalt and sediment, with a sediment thickness of generally more than 11 percent of the stratigraphic section

Idaho National Laboratory boundary

Site facilities

CFA—Central Facilities Area
INTEC—Idaho Nuclear Technology and Engineering Center
RTC—Reactor Technology Complex
RWMC—Radioactive Waste Management Complex

Model area boundary

Starting location for particles

INTEC disposal well (CPP 3) in well model layer 1
INTEC disposal well (CPP 3) in well model layer 2
INTEC seepage pit in model layer 1
RTC warm waste ponds in model layer 1

A, B, C   Indicates interim location of particles

D, E, F, G   Indicates the farthest point that particles moved downgradient

Figure 45.—Continued.

Plume Segment INTEC-B

Plume segment INTEC-B (fig. 45B) is composed of particles, released at the INTEC disposal well to model layer 2 that traveled exclusively within hydrogeologic zone 22. The plume segment had a narrow trapezoidal shape, with width and length dimensions of about 0.8 and 2.8 mi, respectively. The particles at the leading edge of the plume segment traveled 2.4 mi downgradient in 4.7 years, which corresponds to an average linear groundwater velocity of 7.4 ft/d. This velocity is about 57 percent faster than the groundwater velocity based on the leading edge of the 1968 ^{3}H plume and about 14 percent faster than that based on the extended definition of the leading edge (fig. 41; table 25).

The narrow distribution of particles and the northeast to southwest orientation of the plume segment were a result of releasing particles in model layer 2 to hydrogeologic zone 22. Wastewater discharge in model layer 2 did not cause the simulation of radial flow in all directions from the particle origination cell because only about 18 percent of the total simulated wastewater injected through the INTEC disposal well was released to model layer 2 (table 24). Therefore, the contribution of wastewater to total flow through the cell is small compared to background flows. Additionally, streamflow infiltration was simulated as recharge to model layer 1 so it had minimal influence on flow directions in model layer 2.

Plume Segment B-C

Plume segment B-C (fig. 45B) is composed of particles, released at the INTEC disposal well to model layer 2 that traveled within hydrogeologic zone 11. The plume segment had a wide trapezoidal shape, with width and length dimensions of about 1.3 and 1.5 mi, respectively. The particles at the leading edge of the plume segment traveled 1.6 mi downgradient in 4.3 years, which corresponds to an average linear groundwater velocity of 5.4 ft/d. This velocity is about 15 percent faster than the groundwater velocity based on the leading edge of the 1968 ^{3}H plume and about 17 percent slower than that based on the extended definition of the leading edge (fig. 41; table 25).

The directional deviation of the B-C plume segment and the relatively wider distribution of particles defined by the B-C plume segment compared to the INTEC-B segment resulted from groundwater flow refraction at the boundary of hydrogeologic zone 22, a medium with a large HC, and hydrogeologic zone 11, a medium with a smaller HC. Flow refraction across a boundary between media with different HC obeys a tangent law (Freeze and Cherry, 1979, p. 172) and is dependent on the HC ratio between the two media and the contact angle of flowlines intersecting at the boundary. In moving from a medium with a large HC to a medium with a smaller HC, flowlines will refract toward a line drawn perpendicular to the boundary. Refraction results in a change in the direction of flow and, in the case of flow from a medium of large HC to a medium of smaller HC, a widening of the width between flowlines intersecting at the boundary.

Combining plume segments INTEC-B and B-C, the straight line distance from the INTEC disposal well to the boundary of hydrogeologic zone 1 and zone 11 (INTEC-C) is 3.8 mi and the travel time is 9.1 years resulting in a composite velocity of 6.0 ft/d (table 25). This composite velocity is about 28 percent faster than the velocity based on the leading edge of the 1968 plume and about 8 percent slower than that based on the extended definition of the leading edge (fig. 41; table 25). The average linear groundwater velocity for the INTEC-C plume segment was faster than the INTEC-A plume segment, 3.6 ft/d, because the hydraulic gradient was steeper along the path of the B-C plume segment and the HC of the INTEC-B segment was much larger than that of the INTEC-A segment.

Plume Segments C-D, A-E, A-F, and INTEC-D

Plume segment C-D (fig. 45B) is composed of particles, released at the INTEC disposal well to model layer 2 that moved upward to model layer 1 at the boundary of hydrogeologic zones 1 and 11, and traveled primarily within hydrogeologic zone 1. Plume segment A-E is composed of particles released in model layer 1 at the INTEC disposal well, and plume segment A-F is composed of particles released in model layer 1 at the INTEC disposal pit. Plume segments A-E and A-F are entirely within model layer 1 and hydrogeologic zone 1. The northeast to southwest orientation of these plume segments is similar to the regional direction of groundwater flow. The particles at the leading edge of these plume segments traveled 19.7 mi in 6.9 years, 10.4 mi in 2.3 years, and 14.1 mi in 3.3 years, which correspond to average linear groundwater velocities of 41, 65, and 62 ft/d for plume segments C-D, A-E, and A-F, respectively. These velocities are about 1 order of magnitude faster than groundwater velocities based on the leading edge of the 1968 ^{3}H plume and on the extended definition of the leading edge (fig. 41; table 25).

Plume segments C-D, A-E, and A-F are long and narrow. The widths of these plumes are initially controlled by groundwater flow refraction across the boundary of hydrogeologic zone 1 and zone 11. At this boundary, flow is from a medium with a small HC to a medium with a higher HC and flowlines intersecting this boundary refract away from a line drawn perpendicular to the boundary. Flow refraction in this case results in a narrowing of the width between flowlines. Farther downgradient the widths of these plume segments gradually increase in response to the dispersive effects of streamflow infiltration under transient flow conditions as was described in the steady-state/transient and wet-cycle/dry-cycle particle-tracking simulations.

From the INTEC disposal well to the end of plume segment C-D (INTEC-D) particles released in model layer 2 traveled a distance of 23.3 mi in 16.0 years, which corresponds to an average linear groundwater velocity of 21.1 ft/d. From the INTEC disposal well to USGS 11 the particles traveled a distance of 14.9 mi in 11.0 years, which corresponds to an average linear groundwater velocity of 19.6 ft/d, about 23 percent faster than the velocity estimate for the assumed first arrival of [36]Cl at USGS 11, 16 ft/d (fig. 41; table 25).

Simulated velocities for particles released at the INTEC disposal well and disposal pit indicate that the velocity of particles composing plume segment INTEC-D (21 ft/d) represents the composite velocity from flow through three velocity zones: (1) a slow-velocity zone (1.0–9.2 ft/d) represented by hydrogeologic zones 11 and 22, with a gradient of about 3 ft/mi, a HC of 227–4,780 ft/d, and an effective porosity of 0.07–0.15; (2) a fast-velocity zone (62–65 ft/d) represented by hydrogeologic zone 1, with a gradient of about 3 ft/mi, a HC of 11,700 ft/d, and an effective porosity of 0.07; and (3) a medium-velocity zone (20–22 ft/d) represented by hydrogeologic zones 2 and 3, with a gradient of about 20 ft/mi, a HC of 384–435 ft/d, and an effective porosity of 0.03–0.14 (tables 14, 19, and 24). VANI for all three zones is 14,800 (table 14).

Plume Segments INTEC-A, INTEC-B, and B-C Combined

Particle movement within the upper 200 ft of the aquifer is represented by the assimilation of plume segments INTEC-A, INTEC-B, and B-C into a single composite plume (fig. 45B). This combined plume represents particles released at the INTEC to model layers 1 and 2 that traveled through hydrogeologic zones 11 and 22. The shape of the combined plume was elliptical, with a width (2.8 mi) to length (4.0 mi) ratio of 0.70 (fig. 45B). The combined plume, although larger, most closely resembles the shape and aspect ratio of the ^{3}H plume defined by the 1968 25,000 pCi/L isopleth (fig. 41). The 25,000 pCi/L isopleth encloses a plume area with a width (2.0 mi) to length (3.2 mi) ratio of 0.62. The long axis of the combined plume is oriented north-northeast to south-southwest, an orientation that is slightly different from the north to south orientation of the ^{3}H plume represented by the 25,000 pCi/L isopleth.

The distance downgradient from the INTEC that particle plumes were able to reasonably reproduce the 1968 ^{3}H plume extended only to the boundary of hydrogeologic zones 1 and 11 (fig. 45B). This boundary encompasses the entire area represented by the 1968 25,000 pCi/L isopleth (fig. 41). Particle plumes simulated beyond this boundary were long and narrow, and did not reasonably reproduce the shape, dimensions, or position of the ^{3}H plume as depicted by Robertson and others (1974); however as noted previously not enough data are available to characterize the true areal extent and shape of the 1968 ^{3}H plume. The long, narrow

shape of the simulated particle plumes downgradient of the hydrogeologic zone 1-zone 11 boundary was caused by the large HC of zone 1 and groundwater flow refraction at the boundary of zone 1 and zone 11.

Plume Segment RTC-G

The plume segment RTC-G is composed of particles released at the RTC warm-waste ponds to model layer 1 that traveled within hydrogeologic zones 22, 1, 2, and 3 (fig. 45B). Initially the particles traveled west and south along the edge of the hydrogeologic zone 1-zone 11 boundary, and then traveled northeast to southwest along the regional direction of groundwater flow. The resulting plume segment was long and narrow, with length and width dimensions of about 30.1 and 2.0 mi (fig. 45B), respectively. The orientation of the plume segment varied geographically, but the overall orientation, northeast to southwest, was similar to the regional direction of groundwater flow. The particles at the leading edge of the plume segment traveled 30.1 mi downgradient in 15.9 years, which corresponds to an average linear groundwater velocity of 27.4 ft/d.

The long length of plume segment RTC-G was caused by the particles traveling mostly through hydrogeologic zone 1, which has a large HC. This large HC produced an average linear groundwater velocity of 27.4 ft/d, which was 5.8 times faster than the velocity based on the leading edge of the 1968 ^{3}H plume and 4.2 times faster than the velocity based on the extended definition of the leading edge.

Simulated particles released at the RTC initially traveled west and south to areas outside of the 1968 ^{3}H plume and to areas where INL-derived ^{3}H has not been historically observed. Few monitoring wells are southwest of the RTC, resulting in poor definition of where ^{3}H released from the RTC may reside in the aquifer. However, sampling of monitoring well Middle 2051 (table A1; figs. B1, B2), a well equipped with five packer-isolated water-sampling ports about 3 mi southwest of the RTC, in September 2005 and May-June 2006, found ^{3}H concentrations ranging from 428 to 745 pCi/L at depths of 179, 257, 521, and 571 ft below the water table (table 26). These concentrations and the RTC-G plume segment indicate that ^{3}H released at the RTC may reside in the aquifer southwest of the RTC.

Simulated Particle Velocities Compared to Velocities Based on Peak and First Arrivals of Tritium

To avoid some ambiguity and uncertainty associated with the assumed position of the leading edge of the 1968 ^{3}H plume and the extended definition of its position, model-derived average linear groundwater velocities also were compared to the peak and first arrival times of ^{3}H at five

Table 26.   Tritium concentrations in water collected from well Middle 2051, Idaho National Laboratory, Idaho.

[Analytical results and uncertainties—for example 68 ± 116—in picocuries per liter Analytical results are reported to 1s
Concentrations that equal or exceed the reporting level of 3s are shown in **bold** Data are from Renee Bowser, CWI, written commun ,
2007, and http://nwis waterdata usgs gov/nwis/, accessed March 6, 2007 **Abbreviations**: CWI, CH2M-WG Idaho LLC; USGS, U S
Geological Survey; pCi/L, picocuries per liter; ft, foot]

Approximate depth below water table (ft)	Corresponding flow model layer	Tritium concentration (pCi/L)		
		September 2005 (CWI)	September 2005 (USGS)	May–June 2006 (CWI)
33	1	68 ± 116	-60 ± 100	7.7 ± 112
179	2	**632 ± 136**	**680 ± 130**	**449 ± 123**
257	3	**745 ± 141**	**670 ± 130**	**[1]571 ± 127**
521	5	123 ± 116	240 ± 110	**428 ± 125**
571	5	172 ± 120	-70 ± 100	**587 ± 123**

[1] Mean of sample + replicate

wells downgradient of the INTEC disposal well (CPP 3) that have had at least one ^{3}H detection exceeding the analytical detection limit. Travel time estimates are based on (1) a time-series correlation of semi-annual ^{3}H concentrations at USGS 36 with semi-annual ^{3}H releases at the INTEC disposal well (fig. 46) herein referred to as peak arrival-time estimates; (2) the assumed first arrivals of ^{3}H at CFA 2, USGS 90, USGS 105, and USGS 108 (fig. 47); and (3) a steady-state simulation of the time required for a particle originating at the INTEC disposal well to arrive at or near these five monitoring wells (fig. 47). Monitoring wells USGS 36, CFA 2, USGS 90, USGS 105, and USGS 108 extend about 90, 210, 50, 130, and 150 ft below the water table, and are about 1.06, 2.89, 7.10, 8.97, and 8.56 mi downgradient of the INTEC disposal well, respectively (fig. 46).

The travel time between the INTEC disposal well and USGS 36 was estimated by correlating the monthly activity of ^{3}H discharged at the disposal well with the semi-annual concentrations of ^{3}H in water from USGS 36 (fig. 46). A 352 day travel time for ^{3}H traveling between the INTEC disposal well and USGS 36 produces the optimal time-lagged correlation for this analysis (Pearson correlation coefficient [r] = 0.72, samples [n] = 44). The 352-day travel time correlates with the ^{3}H concentrations at USGS 36 from January 1966 to November 1988 with the monthly activity of ^{3}H discharged at the INTEC from January 1965 to December 1987. The resulting average linear groundwater velocity was about 14 ft/d. Uncertainty in the correlation resulted from using semi-annual ^{3}H concentrations at USGS 36 and monthly estimates of ^{3}H discharged at the INTEC disposal well and assuming that ^{3}H concentrations at USGS 36 were derived solely from the INTEC disposal well.

The assumed first arrival times of INTEC-derived ^{3}H at CFA 2, USGS 90, USGS 105, and USGS 108 correspond to the first detections of ^{3}H in groundwater from these wells.

These arrival times are assumed because (1) ^{3}H may have reached these wells at earlier times at concentrations below the analytical detection limit and (2) ^{3}H was detected in the first water samples collected from CFA 2 at relatively high concentrations, suggesting that ^{3}H likely reached CFA 2 much earlier than the time of its first detection. The routine analytical detection limit for ^{3}H in use at the time of these first-arrival detections was 2,000 pCi/L for the CFA 2 and USGS 90 detections, and 500 pCi/L for the USGS 105 and 108 detections (Mann and Cecil, 1990, p. 20 and 22).

The assumed first arrival times at CFA 2 (August 1961; 14,000 pCi/L), USGS 90 (April 1975; 1,200 pCi/L), USGS 105 (October/November 1983; 3,400 pCi/L), and USGS 108 (October/November 1983; 830 to 3,400 pCi/L) (Mann and Cecil, 1990, p. 22, 25, and 29) were 8.7, 22.3, and 30.8 years, respectively, after ^{3}H was initially discharged at the INTEC disposal well in January 1953. These assumed first arrival times resulted in average linear groundwater velocities of 3.9 to 5.1 ft/d (table 25).

Travel times of ^{3}H and average linear velocities were simulated by releasing a single particle, under steady-state flow conditions, at the center of the model layer 2 cell in hydrogeologic zone 22 at the INTEC disposal well. Travel time estimates based on this simulation scenario represent minimum travel times and maximum velocities because the HC of hydrogeologic zone 22 is more than one order of magnitude larger than the HC of hydrogeologic zone 11 (table 14) in model layer 1. Average linear particle velocities simulated under steady-state flow conditions are faster than velocities under transient flow conditions because particle movement is not affected by temporal changes in the local hydraulic gradient and pathline divergence (figs. 38, 40, 45; tables 20, 21). The resulting particle path, plotted as a pathline, is shown in figure 47.

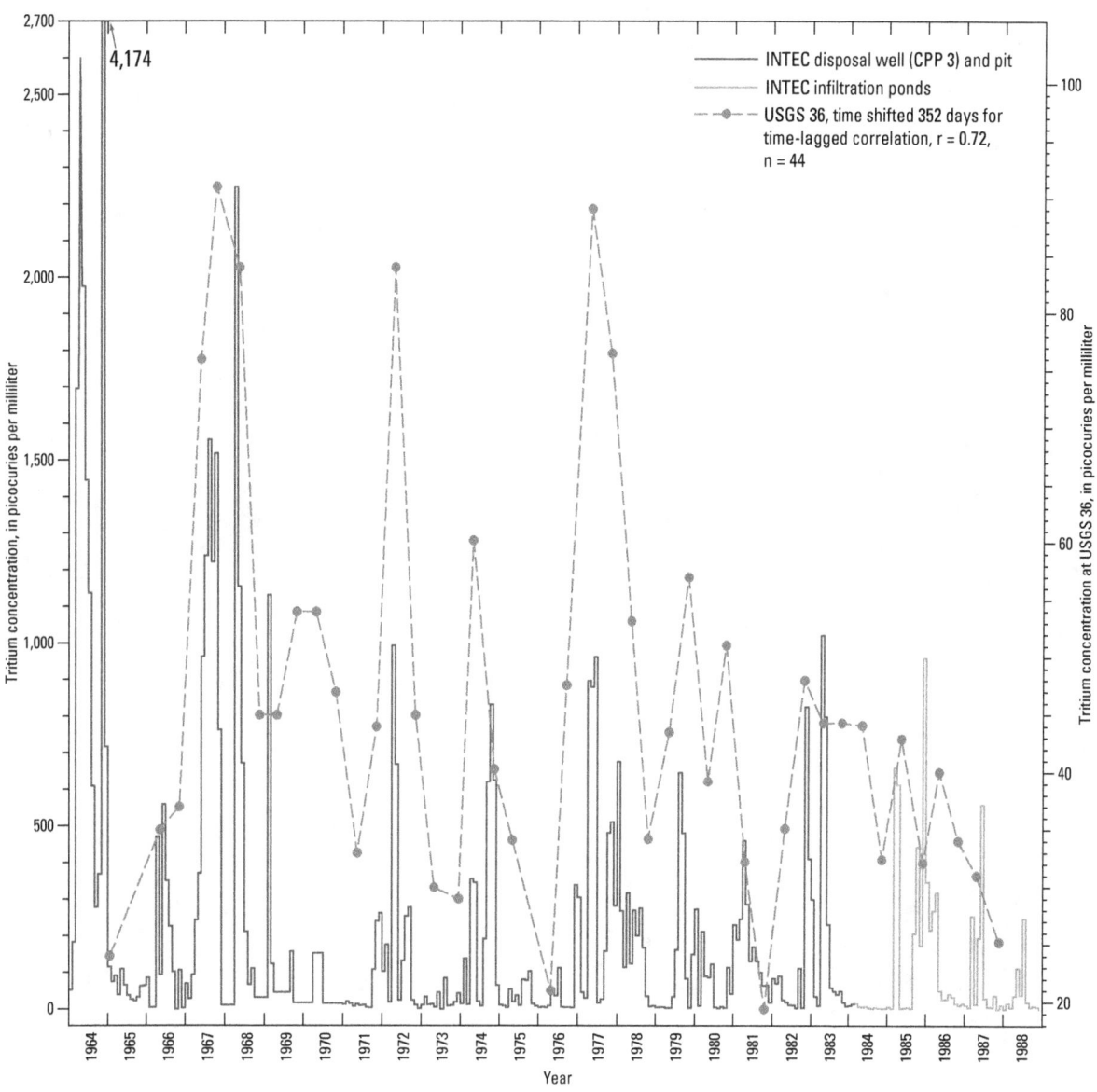

Figure 46. Concentrations of tritium in wastewater discharged at the Idaho Nuclear Technology and Engineering Center from 1964 through 1988 and concentrations of tritium in water from monitoring well USGS 36 from 1966 through 1988, Idaho National Laboratory, Idaho.

The simulated travel times of a particle released at the location of the INTEC disposal well to reach USGS 36, CFA 2, USGS 90, USGS 105, and USGS 108 (fig. 47) were 1.8, 8.0, 10.0, 10.5, and 10.5 years resulting in average linear velocities of about 7.6, 5.5, 10.9, 12.2, and 11.5 ft/d, respectively (table 25). For wells outside the particle pathline (for example, CFA 2, USGS 90, USGS 105, and USGS 108) travel times were estimated using the time required for a particle to reach a location within closest proximity to a well. Because wells

USGS 36 and USGS 90 penetrate only model layer 1, using travel times to these wells for particles released in model-layer 2 may not be valid. Simulation of the particle travel time between the disposal well and USGS 36, for a particle released in model layer 1, resulted in a travel time of 3.6 years and a velocity of 4.6 ft/d, or one-half the velocity of the particle released in model layer 2. However, the particle traveled south and not southwest towards USGS 36. For particles released in model layers 1 and 2, the range of travel times between the

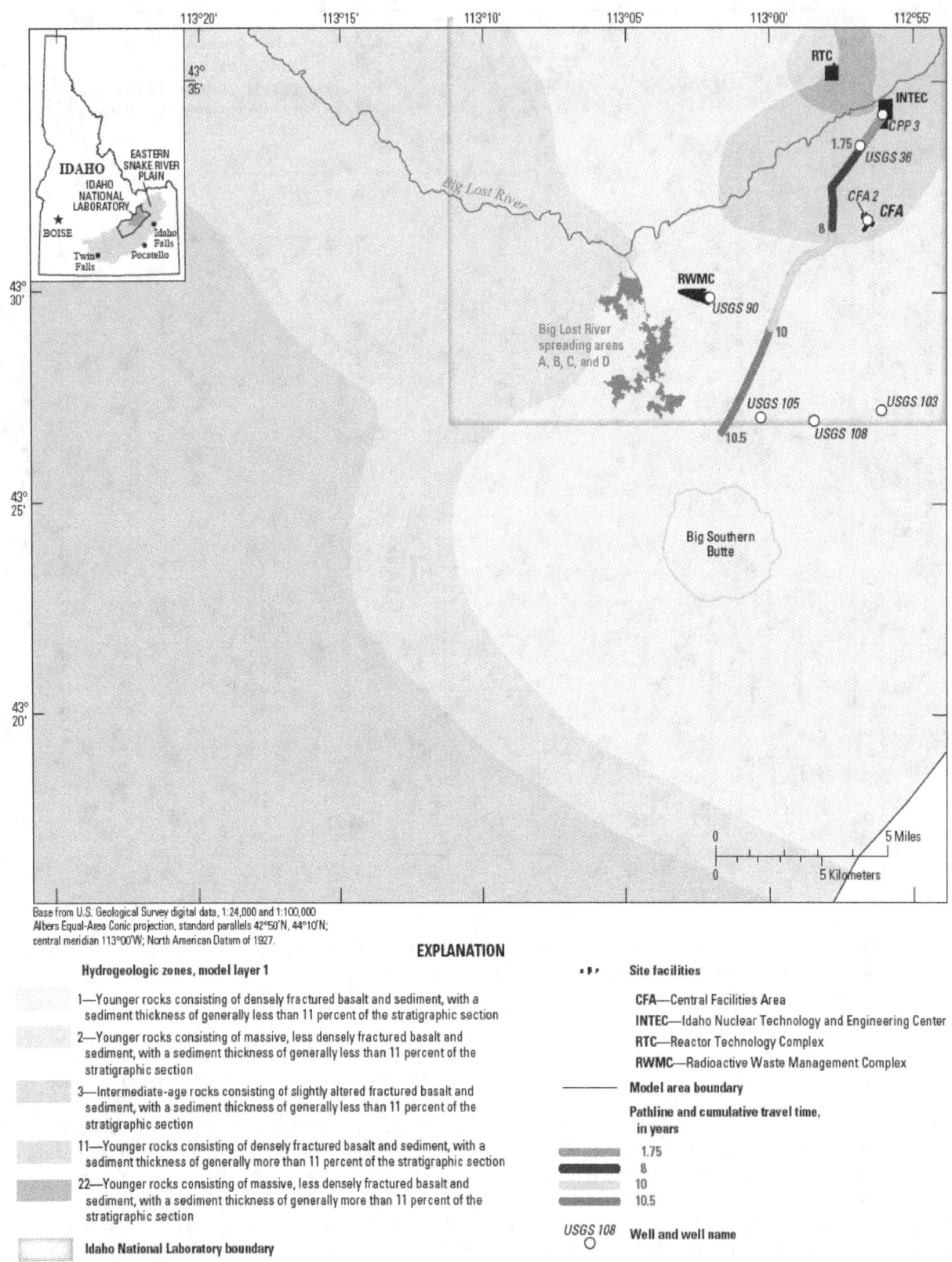

Figure 47.   Pathline and cumulative travel times for a particle simulating steady-state groundwater flow from the Idaho Nuclear Technology and Engineering Center disposal well (CPP 3) to the southern boundary of the Idaho National Laboratory used to evaluate model-derived travel times to selected wells, Idaho National Laboratory and vicinity, Idaho.

disposal well and USGS 36 was 1.8–3.6 years, resulting in a range of average linear velocities of 4.6–9.2 ft/d. The travel time between the disposal well and USGS 90 for the particle released in model layer 2 is valid because the particle moved upward to model layer 1 before reaching the vicinity of USGS 90. The particle paths intersected or nearly intersected USGS 36 (layer 2 particle) and USGS 105, but did not intersect CFA 2, USGS 90, or USGS 108 (fig. 47). The large differences in calculated (16 to 33 ft/d) and simulated (4.6–9.2 ft/d) average linear groundwater velocities between the INTEC disposal well and USGS 36 may result from the injection of wastewater into the aquifer at the disposal well. The influence of wastewater injection on contaminant velocities was observed by Jones (1963, p. 228) who noted that wastewater discharge increased contaminant velocities by a factor of 2 to 4 between the contaminant injection point and a monitoring well approximately 1,000 ft downgradient. The rate of wastewater discharged from the disposal well during the correlation period of 1966–83 was 1.8 ft^3/s, or about 28 percent more than the 1.4 ft^3/s simulated with the steady-state flow model.

Comparison of ^{3}H velocities and simulated average linear particle velocities between the disposal well and CFA 2, USGS 90, USGS 105, and USGS 108 provided an indication of the quality of the model. The velocities for ^{3}H and the particle between the disposal well and CFA 2 were similar (table 25), which indicates that the model may be fairly representative. In contrast, the simulated velocity of the particle between the INTEC disposal well and wells USGS 90 (11 ft/d), USGS 105 (12 ft/d), and USGS 108 (11 ft/d) was about 2 to 3 times faster than the velocity calculated from the assumed first arrival of ^{3}H at these wells (table 25).

Summary and Conclusions

Three-dimensional (3-D) steady-state and transient models of groundwater flow and advective transport in the eastern Snake River Plain (ESRP) aquifer were developed by the U.S. Geological Survey (USGS) in cooperation with the U.S. Department of Energy (DOE) to better understand the aquifer system and the movement of wastes in the aquifer. A 50-plus year history of waste disposal associated with nuclear-reactor research and nuclear-fuel reprocessing at the Idaho National Laboratory (INL) has resulted in measurable concentrations of contaminants in the ESRP aquifer beneath the INL. The model area, in the west-central part of the ESRP, is 1,940 mi^2, extends 35 mi from northwest to southeast and 75 mi from northeast to southwest, and includes most of the INL (890 mi^2). Model simulation results can be used in numerical simulations to evaluate the movement of contaminants in the ESRP aquifer at the INL.

Saturated flow in the ESRP aquifer was simulated using the USGS modular, finite-difference groundwater flow model MODFLOW–2000. Steady-state flow was simulated to represent conditions of 1980 with average streamflow

infiltration from 1966–80 for the Big Lost River, the major variable inflow to the system. The transient flow model simulates groundwater flow between 1980 and 1995, a period that included a 5-year wet cycle (1982–86) followed by an 8-year dry cycle (1987–94). The years 1983–84 and 1987–94 produced the maximum streamflow and the longest dry period, respectively, in the Big Lost River over the past 60 years. Particle-tracking computations using the USGS particle-tracking program MODPATH and various graphics programs were used to simulate advective transport with the flow models and produce 2-D maps of 3-D flow in the aquifer and to evaluate how (1) simulated groundwater flow paths and travel times vary between the steady-state and transient flow models, (2) wet- and dry-climate cycles affect groundwater flow paths and travel times, and (3) model-derived groundwater flow directions and velocities compare to independently derived estimates in that part of the aquifer most affected by contamination.

The fractured basalts, interflow zones, rhyolitic rock units, and interbedded sediments of the ESRP aquifer are represented as porous media and are grouped into four primary hydrogeologic units. In areas where sediment constitutes more than 11 percent of the stratigraphic section, hydrogeologic units are further subdivided to distinguish sediment-rich areas from sediment-poor areas. The four primary hydrogeologic units were subdivided into eight hydrogeologic zones based on sediment abundance. A ninth hydrogeologic zone represents rhyolitic rocks. Model parameters estimated for the steady-state calibration represented horizontal hydraulic conductivity (HC) for seven of nine hydrogeologic zones and a global value of the vertical anisotropy (VANI)—the ratio of horizontal to vertical hydraulic conductivity. In this representation of the aquifer, the small-scale heterogeneities and anisotropies of individual basalt flows, basalt flow groups, rhyolites, and interbedded sediments are not preserved, and the resulting HC of the hydrogeologic zones reflects the aggregate lithology, thickness, and number of basalt flows, interflow zones, and sedimentary interbeds in each hydrogeologic zone.

Three physical and three artificial boundaries define the model area. The physical boundaries are the water table, the northwest mountain-front, and the base of the aquifer. The artificial boundaries are the northeast regional-underflow, the southeast flowline, and the southwest regional-underflow. Inflow to the aquifer is across the water table, northwest mountain-front, and northeast regional-underflow boundaries; outflow is to wells and the southwest regional-underflow boundaries. The base of the aquifer and the southeast flowline boundary are treated as no-flow boundaries. In the steady-state model, flow across inflow and outflow boundaries is represented as temporally constant with spatially uniform and nonuniform flow distributions. Streamflow infiltration from the Big Lost River channel, spreading areas, sinks, and playas represent the largest transient stress. The volume of groundwater flow in the aquifer increases progressively in a direction downgradient of the northeast boundary. Most increased flow is the result of the addition of underflow from the tributary-valleys of the Big Lost River, Little Lost River, and Birch Creek and streamflow infiltration across the water-table boundary, primarily from the Big Lost River. Together these additions account for about 40 percent of the outflow across the southwest boundary; the remaining outflow is from regional aquifer underflow across the northeast boundary. Specified flows into or out of the active model grid define the conditions on all boundaries except the southwest (outflow) boundary, which is simulated with head-dependent flow. Time was discretized in the transient model as 1 year of steady-state conditions (1980) followed by 15 years with 4-month stress periods to represent seasonal cycles of streamflow infiltration, industrial withdrawal and returns, and irrigation withdrawal and infiltration.

Calibration of the models used parameter-estimation techniques incorporated in MODFLOW-2000 that uses the nonlinear least-squares regression method. Calibration data consisted of 201 head observations for steady state. Transient calibration data consisted of 328 head and 8,171 head-difference observations. In the calibration process, models were evaluated by visual and statistical comparisons of estimated parameter values to expected values, the fit of simulated to observed heads or head differences, and the statistical reliability of parameter estimations. The same types of statistical measures, graphs, and maps were used to evaluate the alternative models with changes in system geometry, boundary conditions, and parameters.

Estimates of HC ranged from 227 to 11,700 ft/d, and the estimated value of VANI was 14,800. All estimates of HC were nearly within 2 orders of magnitude of the maximum expected value in a range that exceeds 6 orders of magnitude. The estimated values are more consistent with the greater expected values of HC parameters derived from large-scale aquifer tests than with lesser values derived from small-scale aquifer tests or laboratory tests on core samples. Vertical anisotropy of the model domain was greater than the highest value of a few aquifer tests of short aquifer intervals, but only slightly greater than an independent estimate based on the contrasts in hydraulic conductivity of basalts and interbedded sediment. Parameter reliability, as measured by the confidence interval size and its relation to the expected range of values, was least for two HC parameters and VANI, which were least constrained by observations.

The steady-state model reasonably simulated the observed water-table altitude, orientation, and gradients. Simulation of observed vertical flow directions and gradients was best for greater vertical gradients. Simulation of heads near the water table was better than for heads representing multiple open intervals below the water table or from deeper parts of the aquifer, although weighted residuals were distributed evenly about zero. The collection of additional head data at multiple levels in the aquifer may improve the estimates of aquifer properties and the simulations of heads at depth and vertical flow. Recent geostatistical modeling of sediment abundance in the upper 300 ft of the aquifer in the study area suggests a better spatial resolution for delineating

relative sediment abundance within the model layers and aquifer parameter zones. This information may provide a parameterization with more spatial resolution of HC within zones of abundant sediment.

Recalibration and analysis of the steady-state model to changes in aspects of the conceptual model and implementation of the numerical model revealed that little improvement in the model could come from the tested alternate conceptualizations of sediment content, aquifer thickness, streamflow infiltration, or changes in vertical head distribution on the downgradient boundary. Of the tested alternative estimates of flow to or from the aquifer, only a 20 percent decrease in the largest flow, the northeast boundary underflow, resulted in a recalibrated parameter value just outside the confidence interval of the base-case calibrated value. Features of the flow system, such as the dominance of horizontal flow or downward flow downgradient of the INL, were not affected. Other recalibrations did not improve calibration or result in significant changes to estimated parameter values.

Model parameters estimated for the transient calibration represented specific yield (SY) for five of the seven hydrogeologic zones present at the water table. Estimates of SY ranged from 0.029 to 0.116. All estimates of SY and their confidence intervals were within the ranges of values expected for the parameters and the range of porosity of basalt. Confidence intervals for two of the parameters were 2 to 3 times larger than confidence intervals of other parameters. The confidence intervals probably were larger because fewer observations constrained the estimated values and because of the increased distance of the corresponding hydrogeologic zones from the Big Lost River, the locus of change in water levels. Simulation of transient conditions accurately represented changes in the flow system resulting from episodic infiltration from the Big Lost River and facilitated understanding and visualization of the relative importance of differences in infiltration flux due to (1) extended periods of flow and drought, (2) differences in infiltration rates, and (3) changes in diversions. Simulation of transient conditions could not reproduce annual fluctuations of water levels in the northeast quarter of the model where seasonal irrigation pumpage occurs. The quality of the simulated water-level change is good, however, in the other parts of the model where more information is available and that coincide with the area of greatest interest near observed contamination of the aquifer. Recalibration and analysis of the transient model to changes in boundary conditions revealed little improvement in the simulation of alternate conceptualizations of annual fluctuations of inflow in the northeast quarter of the model area.

As described in the conceptual model, the numerical models simulate flow that is (1) dominantly horizontal through interflow zones in basalt and vertical anisotropy resulting from contrasts in hydraulic conductivity of various types of basalt and the interbedded sediments, (2) temporally variable due to streamflow infiltration from the Big Lost River, and (3) moving downward downgradient of the INL.

Single-particle tracking simulations were used to evaluate the effect of transient stresses on groundwater flowpaths and velocities. Ten-year steady-state particle pathlines were compared to 10-year transient pathlines that included the last 4 years of a wet-climate cycle from 1983 through 1986 and the first 6 years of a dry-climate cycle from 1987 through 1992. Particle tracking also was used to compare the effects of streamflow infiltration from the Big Lost River channel, spreading areas, sinks, and playas during a wet-climate cycle to the effects of no recharge from these sources during a dry-climate cycle.

Collectively, these particle-tracking simulations indicate that average linear groundwater velocities, based on uncalibrated estimates of porosity, and flow paths are influenced by two primary factors: (1) the dynamic character of the water table and (2) the large contrasts in the hydraulic properties of the media, primarily hydraulic conductivity. The simulated growth and decay of large groundwater mounds as much as 34 ft above the steady-state water table beneath the Big Lost River spreading areas, sinks, and playas, and to a lesser extent beneath the Big Lost River channel lead to nonuniform changes in the altitude of the water table throughout the model area. These changes affect the orientation and magnitude of water-table gradients and groundwater flow directions and velocities to a greater or lesser degree depending on the magnitude, duration, and proximity of the transient stress. In areas that are in close proximity to streamflow infiltration, simulation results indicate that pathline divergence, caused by climate-induced temporal changes in the local hydraulic gradient, can account for some of the observed dispersion of contaminants in the aquifer near the Idaho Nuclear Technology and Engineering Center (INTEC) and the Reactor Technology Complex (RTC) and perhaps most observed dispersion several miles downgradient of these facilities.

Multiple-particle tracking simulations were used to simulate the growth of tritium (^{3}H) plumes at the INTEC and RTC under steady-state conditions for 16 years and under transient conditions from 1953 to 1968. In the transient simulation, initial conditions were defined by the steady-state model and transient stresses were based on annual, and when available, semi-annual estimates of groundwater withdrawals, wastewater disposal, and streamflow infiltration from 1953 through 1968 allocated over 4-month simulation periods.

The 2-D shape, dimensions, and areal extent of these particle plumes were compared to maps of the ^{3}H plumes in the aquifer in 1968 that originated from ^{3}H releases at the INTEC and RTC beginning in 1952. Model-derived average linear groundwater velocities were compared to velocity estimates based on (1) the position of the assumed leading edge of the ^{3}H plume in 1968, (2) an extended definition of the leading edge based on a comparison of the shape of the ^{3}H concentration gradients used to define the position of the leading edge of the ^{3}H plumes in 1968 and 1985, and (3) the assumed first arrival of ^{3}H and chlorine-36 (^{36}Cl) and the tritium/helium-3 ages of the young fraction of groundwater in wells downgradient of the INTEC and the RTC.

Results of these multiple-particle simulations indicate that the velocity of particles defining the leading edge of the simulated plume originating at the INTEC represents the composite velocity of flow through three velocity zones: (1) a slow-velocity zone, 1.0–9.2 ft/d, represented by hydrogeologic zones 11 and 22, zones of abundant sediment, with a hydraulic gradient of about 3 ft/mi, a HC of 227–4,780 ft/d, and an effective porosity of 0.07–0.15 ft/d; (2) a fast-velocity zone, 62–65 ft/d, represented by hydrogeologic zone 1, with a gradient of about 3 ft/mi, a HC of 11,700 ft/d, and an effective porosity of 0.07; and (3) a medium-velocity zone, 20–22 ft/d, represented by hydrogeologic zones 2 and 3, with a gradient of about 20 ft/mi, a HC of 384–435 ft/d, and an effective porosity of 0.03–0.14. Model-derived velocities compare favorably with independently derived velocity estimates within areas of abundant sediment, hydrogeologic zones 11 and 22. Except for the assumed first-arrival velocity of ^{36}Cl at USGS 11 and USGS 14, model-derived velocities were much faster than independent velocity estimates such as those based on the assumed first arrival of ^{3}H or the tritium/helium-3 ages of the young fraction of groundwater at wells downgradient of the boundary of hydrogeologic zones with abundant sediment. The reliability of velocity estimates based on the assumed first arrival time of ^{3}H, however, is questionable because of sampling frequency and limitations in the analytical detection limit used to distinguish background ^{3}H concentrations from facility-derived ^{3}H concentrations. The simulated composite velocity of 20 ft/d for particles released at the INTEC disposal well (CPP 3) and measured at USGS 11, a distance of 14.9 mi, compares favorably with an independent velocity estimate of 16 ft/d for the assumed first-arrival of ^{36}Cl at USGS 11.

The distance downgradient of the INTEC that simulated particle plumes were able to reproduce the shape and dimensions reasonably of the 1968 ^{3}H plume extended to the boundary of areas of abundant sediment, hydrogeologic zones 22 and 11, a distance of about 4 mi. This boundary encompasses the entire area represented by the 1968 25,000 picocuries/liter (pCi/L) isopleth. The shape of the simulated particle plume was elliptical with a width (2.8 mi) to length (4.0 mi) ratio of 0.70, a ratio that approximates the 0.62 width (2.0 mi) to length (3.2 mi) ratio of the 1968 plume defined by the 25,000 pCi/L isopleth. Particle plumes simulated beyond this boundary were narrow and long, and did not reasonably reproduce the shape, dimensions, or position of the leading edge of the ^{3}H plume as depicted in earlier reports; however, as noted in an assessment of the 1968 plume, few data were available to characterize its true areal extent and shape.

Although the numerical models described in this report are intended to provide a large-scale representation of a complex flow system that includes the integrated effects of many small-scale features and interactions, several apparent shortcomings are worth noting:

1. The optimized HC for hydrogeologic zone 22 in the area of abundant sediment is larger than that of zone 2 in the area of little or no sediment. This result is inconsistent with current conceptual interpretations of the hydrogeologic properties of hydrogeologic units affected by the presence of abundant sediment.

2. Simulated vertical head gradients in the first 600–500 ft at two wells, USGS 103 (near the southern boundary of the INL), and USGS 132 (south of the RWMC) are 0.1 (downward) as compared to approximately 0.001 upward or downward when compared to recent multi-level piezometric heads. The difference may reflect error in the specified positions of hydrogeologic-zone boundaries or insufficient depth-dependent head data to calibrate parameter values.

3. The single optimized value of VANI is an order of magnitude larger than local measured values, but is similar to other estimated values. The local variation of VANI may be present but not supported by the distribution of head at depth within the aquifer in the calibration data set.

4. The simulated vertical movement of contaminants in the south of the INL as indicated by particle tracking may not be as deep as suggested by recent detection of ^{3}H in multi-level piezometers near the south boundary. These ^{3}H detections, if confirmed, are a few hundred feet deeper in the aquifer than indicated by the INTEC and RTC particle-tracking simulations presented in this report. This is a tentative conclusion that may be related to causes similar to the larger simulated vertical head gradients.

5. Model-derived average linear groundwater velocities within hydrogeologic zone 1 are generally faster than independently derived estimates; however, the reliability of these independent estimates also is subject to considerable uncertainty.

These apparent shortcomings indicate a need for further study and refinement of the existing models. Several of these shortcomings, particularly (2), (3), and (4) suggest that additional observations of head at depth would support estimation of additional parameters and improve calibration. Recalibration of the existing models using recently acquired vertical head data and their error estimates from eight instrumented wells in the southwestern part of the INL, the part of the aquifer most affected by contamination, may be able to substantiate and resolve some of the noted shortcomings. These wells are instrumented to measure pressure heads and temperature, and sample aquifer water in isolated intervals down to depths of about 300 (model layer 3) to 800 ft (model layer 5) below the water table. Repeated measurements since 2007 indicate that vertical head gradients are generally stable as to direction and fluctuate within a relatively small range. Additional improvement of model calibration and finer resolution of the aquifer parameter values in the area of abundant sediment also may be possible by incorporating geostatistical estimates of sediment abundance.

The numerical models described in this report provide a large-scale integrated framework within which to incorporate smaller-scale contaminant transport models. Hydrogeologic zones used to represent the aquifer in the present models are assumed to be adequately represented as homogeneous, anisotropic porous media; small-scale heterogeneity within individual hydrogeologic zones is not represented. Because of this limitation, the particle plume simulations do not incorporate the effects of small-scale tortuosity and likely underestimate the full extent of dispersion. Modeling the dispersive effects of smaller-scale heterogeneities and anisotropies within this larger-scale model is possible using local grid refinement that includes refinement of the hydrogeologic framework in areas affected by contamination and where data are sufficient for refinement.

Acknowledgments

The authors are grateful for many helpful suggestions and comments made by Claire Tiedeman and Dave Pollock as part of the USGS internal program reviews for the study. Their guidance and encouragement, are appreciated.

References Cited

Ackerman, D.J., 1991, Transmissivity of the Snake River Plain aquifer at the Idaho National Engineering Laboratory, eastern Snake River Plain: U.S. Geological Survey Water-Resources Investigations Report 91-4058 (DOE/ID-22097), 35 p.

Ackerman, D.J., 1995, Analysis of steady-state flow and advective transport in the eastern Snake River Plain aquifer system, Idaho: U.S. Geological Survey Water-Resources Investigations Report 94-4257 (DOE/ID-22120), 25 p.

Ackerman, D.J., Rattray, G.W., Rousseau, J.P., Davis, L.C., and Orr, B.R., 2006, A conceptual model of ground-water flow in the eastern Snake River Plain aquifer at the Idaho National Laboratory and vicinity with implications for contaminant transport: U.S. Geological Survey Scientific Investigations Report 2006-5122, (DOE/ID-22198), 62 p., available online at http://pubs.water.usgs.gov/sir2006-5122/.

Anderson, S.R. and Liszewski, M.J., 1997, Stratigraphy of the unsaturated zone and the Snake River Plain Aquifer at and near the Idaho National Engineering Laboratory, Idaho: U.S. Geological Survey Water-Resources Investigations Report 97-4183 (DOE/ID-22142), 65 p.

Anderson, S.R., Ackerman, D.J., Liszewski, M.J., and Freiburger, R.M., 1996, Stratigraphic data for wells at and near the Idaho National Engineering Laboratory, Idaho: U.S. Geological Survey Open-File Report 96-248 (DOE/ID-22127), 27 p. and 1 diskette.

Anderson, S.R., Liszewski, M.J., and Ackerman, D.J., 1996, Thickness of surficial sediment at and near the Idaho National Engineering Laboratory, Idaho: U.S. Geological Survey Open-File Report 96-330 (DOE/ID-22128), 16 p.

Anderson, S.R., Liszewski, M.J., and Cecil, L.D., 1997, Geologic ages and accumulation rates of basalt-flow groups and sedimentary interbeds in selected wells at the Idaho National Engineering Laboratory, Idaho: U.S. Geological Survey Water-Resources Investigations Report 97-4010 (DOE/ID-22134), 39 p.

Anderson, S.R., Kuntz, M.A., and Davis, L.C., 1999, Geologic controls of hydraulic conductivity in the Snake river Plain aquifer at and near the Idaho National Engineering and Environmental Laboratory, Idaho: U.S. Geological Survey Water-Resources Investigations Report 99-4033 (DOE/ID-22155), 38 p.

Anderson, M.P., and Woessner, W.W., 1992, Applied Groundwater Modeling: San Diego, Calif., Academic Press, 3814 p.

Barraclough, J.T., Lewis, B.D., and Jensen, R.G., 1981, Hydrologic conditions at the Idaho National Engineering Laboratory, Idaho, emphasis; 1974–1978: U.S. Geological Survey Open-File Report 81-526, (IDO-22060), 116 p.

Barraclough, J.T., Robertson, J.B., Janzer, V.J., and Saindon, L.G., 1976, Hydrology of the solid waste burial ground, as related to the potential migration of radionuclides, Idaho National Engineering Laboratory: U.S. Geological Survey Open-File Report 76-471, (IDO-22056), 183 p.

Barraclough, J.T., Teasdale, W.E., and Jensen, R.G., 1967, Hydrology of the National Reactor Testing Station, Idaho, 1965: U.S. Geological Survey Open-File Report, (IDO-22048), 107 p.

Barraclough, J.T., Teasdale, W.E., Robertson, J.B., and Jensen, R.G., 1967, Hydrology of the National Reactor Testing Station, Idaho, 1966: U.S. Geological Survey Open-File Report, (IDO-22049), 95 p.

Bartholomay, R.C., Orr, B.R., Liszewski, M.J., and Jensen, R.G., 1995, Hydrologic conditions and distribution of selected radiochemical and chemical constituents in water, Snake River Plain aquifer, Idaho National Engineering Laboratory, Idaho, 1989 through 1991: U.S. Geological Survey Water-Resources Investigations Report, 95-4175 (DOE/ID-22123), 47 p.

Bartholomay, R.C., Tucker, B.J., Ackerman, D.J., and Liszewski, M.J., 1997, Hydrologic conditions and distribution of selected radiochemical and chemical constituents in water, Snake River Plain aquifer, Idaho National Engineering Laboratory, Idaho, 1992 through 1995: U.S. Geological Survey Water-Resources Investigations Report, 97-4086 (DOE/ID-22137), 57 p.

Bartholomay, R.C., Tucker, B.J., Davis, L.C., and Greene, M.R., 2000, Hydrologic conditions and distribution of selected constituents in water, Snake River Plain aquifer, Idaho National Engineering and Environmental Laboratory, Idaho, 1996 through 1998: U.S. Geological Survey Water-Resources Investigations Report 00-4192 (DOE/ID-22167), 52 p.

Beasley, T.M., 1995, Inventory of site-derived ^{36}Cl in the Snake River Plain aquifer, Idaho National Engineering Laboratory, Idaho: Department of Energy Report EML-567, 51 p.

Becker, B.H., Burgess, J.D., Holdren, K.J., Jorgenson, D.K., Magnuson, S.O., and Sondrup, A.J., 1998, Interim risk assessment and contaminant screening for the Waste Area Group 7 remedial investigation: Lockheed Martin Idaho Technologies Company, DOE/ID-10569, Draft rev. 1.

Bennett, C.M., 1990, Streamflow losses and ground-water level changes along the Big Lost River at the Idaho National Engineering Laboratory, Idaho: U.S. Geological Survey Water-Resources Investigations Report 90-4067 (DOE/ID-22091), 49 p.

Brennan, T.S., O'Dell, I., and Lehmann, A.K., 2005, Water resources data, Idaho, water year 2004, Volume 1. Surface water records for Great Basin and Snake River above King Hill: U.S. Geological Survey Water-Data Report ID-04-1, 249 p., available at http://pubs.usgs.gov/wdr/2004/wdr-id-04-1/.

Bukowski, J.M., Bullock, H., and Neher, E.R., 1998, Site conceptual model—1997 activities, data analysis, and interpretation for Test Area North, Operable Unit 1-07B: Lockheed Martin Technologies Company Report INEEL/EXT-98- 00575 Rev. 0, [variously paged].

Busenberg, E., Plummer, L.N., and Bartholomay, R.C., 2001, Estimated age and source of the young fraction of ground water at the Idaho National Engineering and Environmental Laboratory: U.S. Geological Survey Water-Resources Investigations Report 01-4265, DOE/ID-22177, 144 p.

Busenberg, E., Weeks, E.P., Plummer, L.N., and Bartholomay, R.C., 1993, Age dating ground water by use of chlorofluorocarbons (CCl$_3$F and CCl$_2$F$_2$), and distribution of chlorofluorocarbons in the unsaturated zone, Snake River Plain Aquifer, Idaho National Engineering Laboratory, Idaho: U.S. Geological Survey Water-Resources Investigations Report 93-4054, DOE/ID-22107, 47 p.

Cecil, L.D., Pittman, J.R., Beasley, T.M., Michel, R.L., Kubik, P.W., Sharma, P., Fehn, U., and Gove, H.E., 1992, Water infiltration rates in the unsaturated zone at the Idaho National Engineering Laboratory estimated from chlorine-36 and tritium profiles, and neutron logging, in Kharaka, Y.K., and Maest, A.S., eds., Water-Rock Interaction–Proceedings. Volume1, Low temperature environments: Rotterdam, A.A. Balkema, p. 709-715.

Cecil, L.D., Welhan, J.A., Green, J.R., Frape, S.K., and Sudicky, E.R., 2000, Use of chlorine-36 to determine regional scale aquifer dispersivity, eastern Snake River Plain aquifer, Idaho/USA: Nuclear Instruments and Methods in Physics Research B 172, p. 679-687.

Clawson, K.L., Start, G.E., and Ricks, N.R., 1989, Climatography of the Idaho National Engineering Laboratory (2d ed.): National Oceanic and Atmospheric Administration, DOE/ID-12118, 155 p.

Crosthwaite, E.G., Thomas, C.A., and Dyer, K.L., 1970, Water Resources in the Big Lost River Basin, south-central Idaho: U.S. Geological Survey Open-File Report, 109 p.

Domenico, P. A., 1972, Concepts and models in groundwater hydrology, New York, McGraw-Hill, 405 p.

Duffy, J.D., and Harrison, J.H., 1987, The statistical structure and filter characteristics of tritium fluctuations in fractured basalt, Water Resources Research, v. 23, no. 5, p. 894-902.

Frederick, D.B., and Johnson, G.S., 1996, Estimation of hydraulic properties and development of a layered conceptual model for the Snake River Plain aquifer at the Idaho National Engineering Laboratory, Idaho: Idaho Water Resources Research Institute, Research Technical Completion Report, 100 p.

Freeze, R.A., and Cherry, J.A., 1979, Groundwater: Upper Saddle River, N.J., Prentice-Hall Inc., 604 p.

French, D.L., Tallman, R.E., and Taylor, K.A., 1999, Radioactive waste information for 1998 and record-to-date: Lockheed Martin Idaho Technologies, DOE/ID-10054(98), 134 p.

Fromm, J., Welhan, J., McCurry, M., Hackett, W., 1994, Idaho Chemical Processing Plant (ICCP) injection well: operations history and hydrochemical inventory of the waste stream: Proceedings 30th symposium, Engineering Geology and Geotechnical Engineering, p. 221-237.

Garabedian, S.P., 1986, Application of a parameter-estimation technique to modeling the regional aquifer underlying the eastern Snake River Plain, Idaho: U.S. Geological Survey Water-Supply Paper 2278, 60 p.

Garabedian, S.P., 1992, Hydrology and digital simulation of the regional aquifer system, eastern Snake River Plain, Idaho: U.S. Geological Survey Professional Paper 1408-F, 102 p., 10 pl.

Gelhar, L.W., Welty, C., and Rehfeldt, K.R., 1992, A critical review of data on field-scale dispersion in aquifers: Water Resources Research, v. 28, no. 7, p. 1955-1974.

Geslin, J.K., Gianniny, G.L., Link, P.K., and Riesterer, J.W., 1997, Subsurface sedimentary facies and Pleistocene stratigraphy of the northeastern Idaho National Engineering Laboratory–Controls on hydrogeology, in Sharma, S., and Hardcastle, J.H., eds., Symposium on Engineering Geology and Geotechnical Engineering, 32nd, Boise, Idaho, 1997 [Proceedings], p.15-28.

Gianniny, G.L., Geslin, J.K., Riesterer, J.W., Link, P.K., and Thackray, G.D., 1997, Quaternary surficial sediments near Test Area North (TAN) northeastern Snake River Plain–an actualistic guide to aquifer characterization, in Sharma, S., and Hardcastle, J.H., eds., Symposium on Engineering Geology and Geotechnical Engineering, 32nd, Boise, Idaho,1997 [Proceedings], p. 29-44.

Goode, D.J., and L.F. Konikow, 1990a, Apparent dispersion in transient groundwater flow: Water Resources Research, v. 26, no. 10, p. 2339-2351.

Goode, D.J., and L.F. Konikow, 1990b, Reevaluation of large-scale dispersivities for a waste chloride plume–Effects of transient flow, in Kovar, Karel, ed., ModelCARE 90: Calibration and Reliability in Groundwater Modeling (Proceedings of the conference held in The Hague, September 1990), International Association of Hydrological Sciences, Publication no. 195, p. 417-426.

Goodell, S.A., 1988, Water use on the Snake River Plain, Idaho and eastern Oregon: U.S. Geological Survey Professional Paper 1408-E, 51 p.

Harbaugh, A.W., Banta, E.R., Hill, M.C., and McDonald, M.G., 2000, MODFLOW-2000, the U.S. Geological Survey modular ground-water model–User guide to modularization concepts and the Ground-Water Flow Process: U.S. Geological Survey Open-File Report 00-92, 121 p.

Hill, M.C. 1990, Preconditioned Conjugate Gradient 2 (PCG2), A computer program for solving ground-water flow equations: U.S. Geological Survey Open-File Report 98-4048, 43 p.

Hill, M.C., 1994, Five computer programs for testing weighted residuals and calculating linear confidence and prediction intervals on results from the ground-water parameter-estimation computer program MODFLOWP: U.S. Geological Survey Open-File Report 93-481, 81 p.

Hill, M.C., 1998, Methods and guidelines for effective model calibration: U.S. Geological Survey Water-Resources Investigations Report 98-4005, 90 p.

Hill, M.C., Banta, E.R., Harbaugh, A.W., and Anderman, E.R., 2000, MODFLOW-2000, the U.S. Geological Survey modular ground-water model–User guide to the Observation, Sensitivity, and Parameter-Estimation Processes and three post-processing programs: U.S. Geological Survey Open-File Report 00-184, 209 p.

Hill, M.C., and Tiedeman, C.R., 2007, Effective groundwater model calibration–With analysis of data, sensitivities, predictions, and uncertainty: Hoboken, N.J., John Wiley & Sons, Inc., 455 p.

Hughes, S.S., Smith, R.P., Hackett, W.R., and Anderson, S.R., 1999, Mafic volcanism and environmental geology of the eastern Snake River Plain, Idaho, in Hughes, S.S., and Thackray, G.D., eds., Guidebook to the geology of eastern Idaho: Pocatello, Idaho, Idaho Museum of Natural History, p. 143-168.

Jones, P.H., 1963, The velocity of ground-water flow in basalt aquifers of the Snake River Plain, Idaho: International Association of Scientific Hydrology, no. 64, p. 224-234.

Kjelstrom, L.C., 1986, Flow characteristics of the Snake River and water budget for the Snake River Plain, Idaho and eastern Oregon: U.S. Geological Survey Hydrologic Investigations Atlas HA-680, scale 1:1,000,000, 2 sheets.

Kjelstrom, L.C., 1995, Streamflow gains and losses in the Snake River and ground-water budgets for the Snake River Plain, Idaho and eastern Oregon: U.S. Geological Survey Professional Paper 1408-C, 47 p.

Knutson, C.F., McCormick, K.A., Crocker, J.C., Glenn, M.A., and Fishel, M.L., 1992, 3D RWMC vadose zone modeling (including FY-89 to FY-90 basalt characterization results): EG&G Idaho, Inc., Report EGG-ERD-10246 [variously paged].

Kuntz, M.A., and Dalrymple, G.B., 1979, Geology, geochronology, and potential volcanic hazards in the Lava Ridge-Hell's Half Acre area, eastern Snake River Plain, Idaho: U.S. Geological Survey Open-File Report 79-1657, 70 p.

Kuntz, M.A., Skipp, Betty, Lanphere, M.A., Scott, W.E., Pierce, K.L., Dalrymple, G.B., Champion, D.E., Embree, G.F., Page, W.R., Morgan, L.A., Smith, R.P., Hackett, W.R., and Rodgers, D.W., 1994, Geologic map of the Idaho National Engineering Laboratory and adjoining areas, eastern Idaho: U.S. Geological Survey Miscellaneous Investigations Map I-2330, scale 1:100,000.

Leake, S.A., and Lilly, M.R., 1997, Documentation of a computer program (FHB1) for assignment of transient specified-flow and specified-head boundaries in applications of the modular finite-difference ground-water flow model (MODFLOW): U.S. Geological Survey Open-File Report 97-571, 50 p.

Lindholm, G.F., 1996, Summary of the Snake River Plain Regional Aquifer-System Analysis in Idaho and eastern Oregon: U.S. Geological Survey Professional Paper 1408-A, 59 p.

Lindholm, G.F., Garabedian, S.P., Newton, G.D., and Whitehead, R.L., 1988, Configuration of the water table and depth to water, spring 1980, water-level fluctuations, and water movement in the Snake River Plain regional aquifer system, Idaho and eastern Oregon: U.S. Geological Survey Hydrologic Atlas HA-703, scale 1:500,000.

Magnuson, S.O., and Sondrup, A.J., 1998, Development, calibration, and predictive results of a simulator for subsurface pathway fate and transport of aqueous- and gaseous-phase contaminants in the Subsurface Disposal Area at the Idaho National Engineering and Environmental Laboratory: D.O.E. contractor report no. INEEL/EXT-97-00609, 239 p.

Mann, L.J., 1986, Hydraulic properties of rock units and chemical quality of water for INEL-1, a 10,365-foot deep test hole drilled at the Idaho National Engineering Laboratory, Idaho: U.S. Geological Survey Water-Resources Investigations Report 86-4020, IDO-22070, 23 p.

Mann, L.J., and Beasley, T.M., 1994, Iodine-129 in the Snake River Plain aquifer at and near the Idaho National Engineering Laboratory, Idaho, 1990–91: U.S. Geological Survey Water-Resources Investigations Report 94-4053, DOE/ID-22115, 27 p.

Mann, L.J., and Cecil, L.D., 1990, Tritium in ground water at the Idaho National Engineering Laboratory, Idaho: U.S. Geological Survey Water-Resources Investigations Report 90-4090, DOE/ID-22090, 35 p.

Mann, L.J., and Knobel, L.L., 1987, Purgeable organic compounds in ground water at the Idaho National Engineering Laboratory, Idaho: U.S. Geological Survey Open-File Report 87-766, DOE/ID-22074, 23 p.

Mazurek, J., McCurry, M., and Portner, R., 2004, Alteration characteristics and authigenic mineralogy of basalts in middle 1823, eastern Snake River Plain, SE Idaho: Geological Society of America, 2004 Abstracts with Programs, v. 36, no. 4, p. 86.

McCarthy, J.M., Arnett, R.C., Neupauer, R.M., Rohe, M.J., and Smith, C., 1995, Development of a regional groundwater flow model for the area of the Idaho National Engineering Laboratory, eastern Snake River Plain aquifer: Lockheed Idaho Technologies Company, INEL-95-0169, Revision 1, variously paged.

McCurry, M., Hackett, W.R., and Hayden, K., 1999, Cedar Butte and cogenetic Quaternary rhyolite domes of the eastern Snake River Plain, in Hughes, S.S., and Thackray, G.D., eds., Guidebook to the geology of eastern Idaho: Pocatello, Idaho, Idaho Museum of Natural History, p. 169-179.

McDonald, M.G., and Harbaugh, A.W., 1988, A modular three-dimensional finite-difference ground-water flow model: U.S. Geological Survey Techniques of Water-Resources Investigations, book 6, chap. A1, 586 p.

Morin, R.H., Barrash, Warren, Paillet, F.L., and Taylor, T.A., 1993, Geophysical logging studies in the Snake River Plain aquifer at the Idaho National Engineering Laboratory—wells 44, 45, and 46: U.S. Geological Survey Water-Resources Investigations Report 92-4184, 44 p.

Morris, D.A., Barraclough, J.T., Chase, G.H., Teasdale, W.E., and Jensen, R. G., 1965, Hydrology of subsurface waste disposal, National Reactor Testing Station, Idaho, annual progress report, 1964: U.S. Geological Survey Open-File Report, (IDO-22047), [variously paged].

Morris, D.A., Teasdale, W.E., Barraclough, J.T., Chase, G.H., Hogenson, G.M., Jensen, R. G., Ralston, D.A., and Shuter, Eugene, 1964, Hydrology of subsurface waste disposal, National reactor Testing Station, Idaho, annual progress report, 1963: U.S. Geological Survey Open-File Report, IDO-22046, [variously paged].

Morse, L.H., and McCurry, M., 1997, Possible correlations between basalt alteration and the effective base of the Snake River Plain aquifer at the Idaho National Engineering and Environmental Laboratory, in Sharma, S., and Hardcastle, J.H., eds., Symposium on Engineering Geology and Geotechnical Engineering, 32nd, Boise, Idaho, 1997 [Proceedings], p. 1-14.

Morse, L.H., and McCurry, M., 2002, Genesis of alteration of Quaternary basalt within a portion of the eastern Snake River Plain aquifer, in Link, P.K., and Mink, L.L., eds., Geology, hydrogeology, and environmental remediation: Idaho National Engineering and Environmental Laboratory, Eastern Snake River Plain, Idaho: Geological Society of America Special Paper 353, p. 213-224.

Mundorff, M.J., Broom, H.C., and Kilburn, Chabot, 1963, Reconnaissance of the hydrology of the Little Lost River basin, Idaho: U.S. Geological Survey Water-Supply Paper 1539-Q, 51 p.

Mundorff, M.J., Crosthwaite, E.G., and Kilburn, Chabot, 1964, Ground water for irrigation in the Snake River basin in Idaho: U.S. Geological Survey Water-Supply Paper 1654, 224 p.

Nace, R.L., and Barraclough, J.T., 1952, Ground-water recharge from the Big Lost River below Arco, Idaho: U.S. Geological Survey Open-File Report, IDO-22016, 31 p.

Nace, R.L., Voegeli, P.T., Jones, J.R., and Deutsch, Morris, 1975, Generalized geologic framework of the National Reactor Testing Station, Idaho: U.S. Geological Survey Professional Paper 725-B, 49 p.

Nimmo, J.R., Perkins, K.S., Rose, P.A., Rousseau, J.P., Orr, B.R., Twining, B.V., and Anderson, S.R., 2002, Kilometer-scale rapid transport of naphthalene sulfonate tracer in the unsaturated zone at the Idaho National Engineering and Environmental Laboratory: Vadose Zone Journal, v. 1, no. 1, p. 89-101.

Perkins, K.S., 2008, Laboratory-measured and property-transfer modeled saturated hydraulic conductivity of Snake River Plain aquifer sediments at the Idaho National Laboratory, Idaho: U.S. Geological Survey Scientific Investigations Report, 2008-5169, DOE/ID-22207, 14 p., available online at http://pubs.usgs.gov/sir/2008/5169/.

Pittman, J.R., Jensen, R.G., and Fischer, P.R., 1988, Hydrologic conditions at the Idaho National Engineering Laboratory, 1982 to 1985: U.S. Geological Survey Water-Resources Investigations Report 89-4008, DOE/ID-22078, 73 p.

Plummer, L.N., Rupert, M.G., Busenberg, E., and Schlosser, P., 2000, Age of irrigation water in groundwater from the Snake River Plain aquifer, South-Central Idaho: Ground Water, v. 38, p. 264-283.

Pollock, D.W., 1994, User's guide for MODPATH/MODPATH-PLOT, version 3–a particle tracking post-processing package for MODFLOW, the U.S. Geological Survey finite-difference ground-water flow model: U.S. Geological Survey Open-File Report 94-464, 245 p.

Robertson, J.B., 1974, Digital modeling of radioactive and chemical waste transport in the Snake River Plain aquifer at the National Reactor Testing Station, Idaho: U.S. Geological Survey Open-File Report, IDO-22054, 41 p.

Robertson, J.B., Schoen, Robert, and Barraclough, J.T., 1974, The influence of liquid waste disposal on the geochemistry of water at the National Reactor Testing Station, Idaho—1952–1970: U.S. Geological Survey Open-File Report, (IDO-22053), 231 p.

Schaffer-Perini., A.L., 1993, Test Area North groundwater remedial investigation/feasibility study contaminant fate and transport modeling results: Idaho Falls, Idaho, EWR WAG1-21, EG&G Idaho, Inc., Idaho National Engineering and Environmental Laboratory [variously paged].

Spear, D.B., and King, J.S., 1982, The geology of Big Southern Butte, Idaho, in Bonnichsen, B., and Breckenridge, R.M., eds., Cenozoic Geology of Idaho: Idaho Bureau of Mines and Geology Bulletin 26, p. 395-403

Spinazola, J.M., 1994, Geohydrology and simulation of flow and water levels in the aquifer system in the Mud Lake area of the eastern Snake River Plain, eastern Idaho: U.S. Geological Survey Water–Resources Investigations Report 93-4227, 78 p.

Stearns, H.T., Bryan, L.L., and Crandall, Lynn, 1939, Geology and water resources of the Mud Lake basin, Idaho, including the Island Park area: U.S. Geological Survey Water-Supply Paper 818, 125 p

Stearns, H.T., Crandall, Lynn, and Steward, W.G., 1938, Geology and ground-water resources of the Snake River Plain in southeastern Idaho: U.S. Geological Survey Water-Supply Paper 774, 268 p.

Stroup, C.N., Welhan, J.A., and Davis, L.C., 2008, Statistical stationarity of sediment interbed thicknesses in a basalt aquifer, Idaho National Laboratory, eastern Snake River Plain, Idaho: U.S. Geological Survey Scientific Investigations Report 2008–5167, DOE/ID-22204, 20 p., available online at http://pubs.usgs.gov/sir/2008/5167/.

U.S. Environmental Protection Agency, 1991 Sole source designation of the Eastern Snake River Plain Aquifer, Southern Idaho: Federal Register, v. 56, no. 194, p. 50634.

Welhan, J.A., Clemo, T.M., and Gégo, E.L., 2002, Stochastic simulation of aquifer heterogeneity in a layered basalt aquifer system, eastern Snake River Plain, Idaho, in Link, P.K., and Mink, L.L., eds., Geology, Hydrogeology, and Environmental Remediation–Idaho National Engineering and Environmental Laboratory, Eastern Snake River Plain, Idaho: Geological Society of America Special Paper 353, p. 225-247.

Welhan, J.A., Farabaugh, R.L., Merrick, M.J., and Anderson, S.R., 2007, Geostatistical modeling of sediment abundance in a heterogeneous basalt aquifer at the Idaho National Laboratory, Idaho: U.S. Geological Survey Scientific Investigations Report 2006-5316, DOE/ID-22201, 32 p., available online at http://pubs.usgs.gov/sir/2006/5316/.

Welhan, J.A., and Wylie, A.H., 1997, Stochastic modeling of hydraulic conductivity in the Snake River Plain aquifer–2. Evaluation of lithologic controls at the core and borehole scales, in Sharma, S., and Hardcastle, J.H., eds., Symposium on Engineering Geology and Geotechnical Engineering, 32nd, Boise, Idaho, 1997 [Proceedings], p. 93-107.

Whitehead, R.L., 1986, Geohydrologic framework of the Snake River Plain, Idaho and eastern Oregon: U.S. Geological Survey Hydrologic Investigations Atlas HA-681, scale 1:1,000,000, 3 sheets.

Whitehead, R.L., 1992, Geohydrologic framework of the Snake River Plain regional aquifer system, Idaho and eastern Oregon: U.S. Geological Survey Professional Paper 1408-B, 32 p.

Wood, T.R., and Norrell, G.T., 1996, Integrated large-scale aquifer pumping and infiltration tests, groundwater pathways, OU 7-06, summary report: INEL-96/0256, Rev.0 [variously paged].

Appendix A. Data for Wells, Boreholes, and Streams Used in the Construction and Calibration of Steady-State and Transient Models of Groundwater Flow, Idaho National Laboratory and Vicinity, Idaho

Table A1.   Data for wells and boreholes used in the construction and calibration of steady-state and transient models of groundwater flow, Idaho National Laboratory and vicinity, Idaho.

[**Map No.:** Identifier used to locate wells in appendix B figures and as a cross reference with data in other appendixes **Site identifier:** Unique numerical identifier used to access well data (http://waterdata.usgs.gov/nwis) **Local name:** Local well identifier used in this study **Stratigraphic data No.:** Cross reference to borehole locations used by Anderson, Ackerman, and others (1996) and Anderson, Liszewski, and Ackerman (1996) used to construct the hydrogeologic framework of the models **Abbreviation:** ft-bls, foot below land surface **Symbols:** –, data not available; X, entry in specified appendix tables]

Map No.	Site identifier	Local name	Depth of hole (ft-bls)	Depth of well (ft-bls)	Appendix figure No.	Stratigraphic data No.	Data in table			
							G1	H1	C1	F1
1	434819112380501	2ND OWSLEY	310	302	B2		X	X		
2	434556112575601	434556	350	350	B2		X	X		
3	434647112534101	434647	600	600	B2		X	X		
4	434650112545501	434650	816	816	B2		X	X		
5	434714112175801	434714	300	300	B2		X	X		
6	434726112244101	434726	–	–	B2		X	X		
7	434756112212101	434756	292	290	B2		X	X		
8	435026112253101	435026	270	270	B2		X	X		
9	435100112271601	435100	247	247	B2		X	X		
10	440109112391301	440109	380	380	B2		X	X		
11	432853113021701	A11A31	678	675	B1, B3	128				
12	433534112392901	ANL MW 11	677	654	B9		X	X		
13	433545112394102	ANL MW 13	668	665	B9		X	X		
14	433545112394101	ANL OBS A 001	1,910	1,910	B9		X	X		
15	–	ANL-IWP-M1	54	54	B9	129				
16	–	ANL-IWP-M2	80	80	B9	130				
17	–	ANL-IWP-M3	60	45	B9	131				
18	–	ANL-IWP-M4	68	30	B9	132				
19	–	ANL-IWP-M5	64	64	B9	133				
20	–	ANL-IWP-M6	423	406	B9	134				
21	435308112454101	ANP 5	396	396	B2		X	X		
22	435152112443101	ANP 6	305	305	B2, B4	135	X	X		
23	435522112444201	ANP 7	436	433	B2	136	X	X		
24	434952112411301	ANP 8	309	309	B4					X
25	434856112400001	ANP 9	322	322	B2	137	X	X		
26	434909112400401	ANP 10	681	681	B2	138	X	X		
27	435053112423201	ANP DISP 3	310	310	B4		X	X		
28	433107112492201	ARA 2	787	787	B2					X
29	433156112494401	ARA 3	1,340	1,340	B2					X
30	433509112384801	ARBOR TEST	790	790	B9	140	X	X		
31	433223112470201	AREA 2	877	877	B2	139	X	X		
32	435016112311201	Ashcraft	420	360	B2	142				
33	432638112484101	ATOMIC CITY WELL	–	–	B2					
34	433042112535101	BADGING FACILITY	–	–	B2	221				X
35	435003112313101	Barney North	660	660	B2	152				
36	434950112311201	Barney South	596	–	B2	153				

Table A1.   Data for wells and boreholes used in the construction and calibration of steady-state and transient models of groundwater flow, Idaho National Laboratory and vicinity, Idaho.—Continued

[**Map No.:** Identifier used to locate wells in appendix B figures and as a cross reference with data in other appendixes **Site identifier:** Unique numerical identifier used to access well data (http://waterdata.usgs.gov/nwis) **Local name:** Local well identifier used in this study **Stratigraphic data No.:** Cross reference to borehole locations used by Anderson, Ackerman, and others (1996) and Anderson, Liszewski, and Ackerman (1996) used to construct the hydrogeologic framework of the models **Abbreviation:** ft-bls, foot below land surface **Symbols**: –, data not available; X, entry in appendix tables]

Map No.	Site identifier	Local name	Depth of hole (ft-bls)	Depth of well (ft-bls)	Appendix figure No.	Stratigraphic data No.	G1	H1	C1	F1
37	–	BG-76-1	228	–	B3	143				
38	–	BG-76-2	253	–	B3	144				
39	–	BG-76-3	240	–	B3	145				
40	–	BG-76-4	215	–	B3	146				
41	–	BG-76-4A	254	–	B3	147				
42	–	BG-76-5	245	–	B3	148				
43	–	BG-76-6	244	–	B3	149				
44	–	BG-77-1	600	–	B3	150				
45	–	BG-77-2	87	–	B3	151				
46	433631113143702	Butte City #2	850	475	B1	154				
47	–	C-1	664	–	B3	155				
47	–	C-1A	1,805	–	B3	156				
48	434949112300101	Callaway	650	650	B2	172				
49	432618112555501	CERRO GRANDE	564	564	B2	173	X	X		
50	433204112562001	CFA 1	685	639	B8	157				X
51	433144112563501	CFA 2	681	681	B8	158	X	X		X
52	–	CFA 4	–	–	B2, B8	159				
53	433216112563201	CFA LF 2-8	526	495	B8	160				
54	433217112563401	CFA LF 2-9	676	497	B8	161				
55	433216112563301	CFA LF 2-10	816	716	B8	162	X	X		
56	433230112561701	CFA LF 2-11	511	499	B8	163	X	X		
57	433217112563601	CFA LF 2-12	517	490	B8	164				
58	433218112571001	CFA LF 3-8	526	510	B8	165				
59	433216112571001	CFA LF 3-9	517	500	B8	166	X	X		
60	433222112571901	CFA LF 3-10	530	501	B8	167				
61	433249112565501	CFA LF 3-11	532	492	B8	168				
62	435213112302001	Cope	784	784	B2	174				
63	432927112410101	COREHOLE 1	2,002	2,000	B2	175	X	X		
64	434558112444801	COREHOLE 2A	3,000	3,000	B2	176				
65	431519113112901	COX WELL	777	777	B1		X	X		
66	433433112560201	CPP 1	586	586	B7					X
67	433432112560801	CPP 2	605	605	B7	169				X
68	433413112560401	CPP 3	598	598	B7	170				X
69	433440112554401	CPP 4	700	700	B7	171	X	X		X
69	433440112554402	CPP 5	721	721	B7		X	X		
70	432128113092701	CROSS ROAD	796	796	B1					

Table A1.   Data for wells and boreholes used in the construction and calibration of steady-state and transient models of groundwater flow, Idaho National Laboratory and vicinity, Idaho.—Continued

[**Map No.:** Identifier used to locate wells in appendix B figures and as a cross reference with data in other appendixes **Site identifier:** Unique numerical identifier used to access well data (http://waterdata.usgs.gov/nwis) **Local name:** Local well identifier used in this study **Stratigraphic data No.:** Cross reference to borehole locations used by Anderson, Ackerman, and others (1996) and Anderson, Liszewski, and Ackerman (1996) used to construct the hydrogeologic framework of the models **Abbreviation:** ft-bls, foot below land surface **Symbols:** –, data not available; X, entry in specified appendix tables]

Map No.	Site identifier	Local name	Depth of hole (ft-bls)	Depth of well (ft-bls)	Appendix figure No.	Stratigraphic data No.	G1	H1	C1	F1
71	–	D-10	–	–	B3	177				
72	–	D-15	–	–	B3	178				
73	435817112365401	Dahle	232	232	B2	186				
74	434611112504301	DH 1B	400	400	B2	179	X	X		
75	434547112512801	DH 2A	430	425	B2	180	X	X		
76	–	DH3	–	–	B2	181				
77	–	DH-50	250	–	B2, B9	182				
78	–	DO-2	235	–	B3	183				
79	–	DO-6	126	–	B3	184				
79	–	DO-6A	50	–	B3	185				
80	433051113002601	EBR 1	1,075	1,075	B1, B2	187				X
81	433546112391601	EBR 2-1	747	745	B9					X
82	433544112391301	EBR II-2	753	753	B9					X
83	431831113312901	ELLSWORTH	1,305	1,305	B1		X	X		
84	433120112535101	EOCR	1,237	1,237	B2	189				
85	–	EOCR (Disp)	–	–	B2	190				
86	435120112432101	FET 1	339	330	B4					X
87	435119112431801	FET 2	462	455	B4					X
88	435124112433701	FET DISP 3	302	300	B4	191	X	X		X
89	433548112562301	FIRE STA 2	518	510	B2					X
90	432424113165301	FNGR BUTTE	1,056	1,056	B1		X	X		
91	434947112414301	GIN 1 (new)	373	364	B4			X		
91	434947112414301	GIN 1 (old)	373	373	B4	192	X	X		
92	434949112413401	GIN 2	402	381	B4	193	X	X		
93	434945112413101	GIN 3 (new)	386	378	B4			X		
93	434945112413101	GIN 3 (old)	386	386	B4	194	X	X		
94	434949112413601	GIN 4	306	300	B4	195	X	X		
95	434953112413301	GIN 5 (new)	430	285	B4			X		
95	434953112413301	GIN 5 (old)	430	285	B4	196	X	X		
96	–	GIN #6	–	–	B4	197				
97	–	GIN #7	–	–	B4	198				
98	–	GIN #8	–	–	B4	199				
99	–	GIN #9	–	–	B4	200				
100	–	GIN #10	–	–	B4	201				
101	–	GIN #11	–	–	B4	202				

Table A1. Data for wells and boreholes used in the construction and calibration of steady-state and transient models of groundwater flow, Idaho National Laboratory and vicinity, Idaho.—Continued

[**Map No.:** Identifier used to locate wells in appendix B figures and as a cross reference with data in other appendixes **Site identifier:** Unique numerical identifier used to access well data (http://waterdata usgs gov/nwis) **Local name:** Local well identifier used in this study **Stratigraphic data No.:** Cross reference to borehole locations used by Anderson, Ackerman, and others (1996) and Anderson, Liszewski, and Ackerman (1996) used to construct the hydrogeologic framework of the models **Abbreviation:** ft-bls, foot below land surface **Symbols**: –, data not available; X, entry in specified appendix tables]

Map No.	Site identifier	Local name	Depth of hole (ft-bls)	Depth of well (ft-bls)	Appendix figure No.	Stratigraphic data No.	Data in table			
							G1	H1	C1	F1
102	–	GIN #12	–	–	B4	203				
103	–	GIN #13	–	–	B4	204				
104	–	GIN #14	–	–	B4	205				
105	–	GIN #15	–	–	B4	206				
106	–	GIN #16	–	–	B4	207				
107	–	GIN #17	–	–	B4	208				
108	–	GIN #18	–	–	B4	209				
109	–	GIN #19	–	–	B4	210				
110	–	GIN #20	–	–	B4	211				
111	433218112191603	HIGHWAY 1A	1,302	1,147	B2	212				
111	433218112191602	HIGHWAY 1B	1,038	982	B2					
111	433218112191601	HIGHWAY 1C	883	800	B2					
112	433307112300001	HIGHWAY 2	786	786	B2	214				
113	433256113002501	HIGHWAY 3	750	750	B1, B2	215				
114	431439113071401	Houghland	775	775	B1		X	X		
115	435153112420501	IET 1 DISP	329	324	B2, B4	216	X	X		
116	433717112563501	INEL 1	10,365	10,333	B2	217				
117	432533112504901	LEO ROGERS 1	720	720	B2	278				
118	434946112412401	LPTF DISP	314	314	B4	219				X
119	433418112581701	MIDDLE 1823	1,653	720	B6					
120	433217113004901	MIDDLE 2051	1,179	1,175	B1, B2					
121	434700112530401	ML 11	601	601	B2		X	X		
122	434558112585301	ML 12	540	540	B1, B2		X	X		
123	434723112552701	ML 13	650	650	B2		X	X		
124	433520112572601	MTR TEST	588	588	B6	220	X	X		
125	432956113041401	NA 89-1	238	232	B1	222				
126	433056113045101	NA 89-2	235	230	B1	223				
127	432918113031701	NA 89-3	184	180	B3	224				
128	435038112453401	NO NAME 1	550	550	B2	320	X	X		
129	433449112523101	NPR TEST	600	600	B2	225	X	X		
130	433451112523201	NPR WO-2	5,000	5,000	B2	226				
131	433859112545401	NRF 1	535	535	B5					X
132	433854112545401	NRF 2	529	528	B5					X
133	433858112545501	NRF 3	546	546	B5					X
134	433853112545901	NRF 4	–	–	B5	228				X

Table A1. Data for wells and boreholes used in the construction and calibration of steady-state and transient models of groundwater flow, Idaho National Laboratory and vicinity, Idaho.—Continued

[**Map No.:** Identifier used to locate wells in appendix B figures and as a cross reference with data in other appendixes **Site identifier:** Unique numerical identifier used to access well data (http://waterdata.usgs.gov/nwis) **Local name:** Local well identifier used in this study **Stratigraphic data No.:** Cross reference to borehole locations used by Anderson, Ackerman, and others (1996) and Anderson, Liszewski, and Ackerman (1996) used to construct the hydrogeologic framework of the models **Abbreviation:** ft-bls, foot below land surface **Symbols:** –, data not available; X, entry in specified appendix tables]

Map No.	Site identifier	Local name	Depth of hole (ft-bls)	Depth of well (ft-bls)	Appendix figure No.	Stratigraphic data No.	G1	H1	C1	F1
135	433844112550201	NRF #5	–	–	B5	279				
136	433910112550101	NRF 6	417	417	B5	229	X	X		
137	–	NRF #6P	500	494	B5	230				
138	433920112543601	NRF 7	430	415	B5	231	X	X		
139	–	NRF #7P	500	–	B5	232				
140	433843112550901	NRF 8	425	423	B5		X	X		
141	433840112550201	NRF 9	425	422	B5		X	X		
142	433841112545201	NRF 10	450	427	B5		X	X		
143	433847112544201	NRF 11	425	417	B5		X	X		
144	433855112543201	NRF 12	425	421	B5		X	X		
145	433928112545401	NRF 13	425	425	B5		X	X		
146	–	NRF 89-04	248	–	B5	233				
147	–	NRF 89-05	242	–	B5	234				
148	–	OW-1	1,000	1,000	B3	235				
149	–	OW-2	1,000	1,000	B3	236				
150	435416112460401	P&W 1	432	432	B2	241	X	X		
151	435419112453101	P&W 2	386	386	B2	242	X	X		
152	435443112435801	P&W 3	406	406	B2	243	X	X		
153	–	PBF (CW)	–	–	B2	238				
154	–	PBF (WW)	–	–	B2	239				
155	434941112454201	PSTF TEST	322	319	B2	240	X	X		
156	433349112560701	PW-1	119	119	B7	244				
157	433344112555601	PW-2	131	131	B7	245				
158	433351112555701	PW 3	123	123	B7	246				
159	433348112554901	PW 4	150	150	B7	247				
160	433348112555701	PW 5	131	129	B7	248				
161	433353112562201	PW 6	135	125	B7	249				
162	433446112574602	PW 7	237	237	B6	250				
163	433456112572001	PW 8	178	170	B6	251				
164	433500112575401	PW 9	201	200	B6	252				
165	433512112573701	PW 10	150	128	B6	253				
166	433505112572201	PW 11	169	129	B6	254				
167	433510112574901	PW 12	142	128	B6	255				
168	433505112574101	PW 13	149	88	B6	256				
169	433518112573401	PW 14	136	123	B6	257				

Table A1. Data for wells and boreholes used in the construction and calibration of steady-state and transient models of groundwater flow, Idaho National Laboratory and vicinity, Idaho.—Continued

[**Map No.:** Identifier used to locate wells in appendix B figures and as a cross reference with data in other appendixes **Site identifier:** Unique numerical identifier used to access well data (http://waterdata.usgs.gov/nwis) **Local name:** Local well identifier used in this study **Stratigraphic data No.:** Cross reference to borehole locations used by Anderson, Ackerman, and others (1996) and Anderson, Liszewski, and Ackerman (1996) used to construct the hydrogeologic framework of the models **Abbreviation:** ft-bls, foot below land surface **Symbols:** –, data not available; X, entry in specified appendix tables]

Map No.	Site identifier	Local name	Depth of hole (ft-bls)	Depth of well (ft-bls)	Appendix figure No.	Stratigraphic data No.	G1	H1	C1	F1
170	432632113095901	QAB	1,115	1,115	B1	258	X	X		
171	434857112185801	R. Archer	260	260	B2	141				
172	433243112591101	RIFLE RANGE	–	–	B1, B2	277				
173	–	RWMC-78-1	82	–	B3	259				
174	–	RWMC-78-2	253	–	B3	260				
175	–	RWMC-78-3	248	–	B3	261				
176	–	RWMC-78-4	–	–	B3	262				
177	–	RWMC-78-5	250	–	B3	263				
178	–	RWMC-79-1	244	–	B3	264				
179	–	RWMC-79-2	223	–	B3	265				
180	–	RWMC-79-3	262	–	B3	266				
181	–	RWMC-88-02D	–	–	B3	268				
182	–	RWMC-88-1D	–	–	B3	267				
183	–	RWMC-89-01D	–	–	B3	269				
184	432956113030901	RWMC M1SA	678	638	B3	270	X	X		
185	433008113021801	RWMC M3S	660	633	B3	271	X	X		
186	432939113030101	RWMC M4D	838	828	B3	272	X	X		
187	432931113015001	RWMC M6S	697	668	B1, B3	273	X	X		
188	433023113014801	RWMC M7S	638	628	B3	274	X	X		
189	432949113024301	RWMC M10S	678	648	B3	275				
190	432919113031199	RWMC-PRO-A-064	848	848	B3					
191	433002113021701	RWMC PROD	685	685	B3	276				X
192	–	Sdd-1	–	–	B2	280				
193	–	Sdd-2	–	–	B2	281				
194	–	Sdd-3	–	–	B2	282				
195	434744112212202	Siddoway	930	715	B2	283				
196	432854113201002	SITE 1	1,053	1,053	B1	334	X	X		
197	431946113161401	SITE 2	1,041	1,041	B1	127	X	X		
198	433617112542001	SITE 4	495	495	B2	188				
199	433826112510701	SITE 6	523	523	B2	284	X	X		
200	433123112530101	SITE 9	1,131	1,057	B2	285	X	X		
201	434334112463101	SITE 14	717	717	B2	286	X	X		
202	433545112391501	SITE 16	758	758	B9	287	X	X		
203	434027112575701	SITE 17	600	600	B2	288	X	X		
204	433522112582101	SITE 19	865	860	B6	289	X	X		

Table A1.   Data for wells and boreholes used in the construction and calibration of steady-state and transient models of groundwater flow, Idaho National Laboratory and vicinity, Idaho.—Continued

[**Map No.:** Identifier used to locate wells in appendix B figures and as a cross reference with data in other appendixes **Site identifier:** Unique numerical identifier used to access well data (http://waterdata.usgs.gov/nwis) **Local name:** Local well identifier used in this study **Stratigraphic data No.:** Cross reference to borehole locations used by Anderson, Ackerman, and others (1996) and Anderson, Liszewski, and Ackerman (1996) used to construct the hydrogeologic framework of the models **Abbreviation:** ft-bls, foot below land surface **Symbols:** –, data not available; X, entry in specified appendix tables]

Map No.	Site identifier	Local name	Depth of hole (ft-bls)	Depth of well (ft-bls)	Appendix figure No.	Stratigraphic data No.	Data in table			
							G1	H1	C1	F1
205	433252112520301	SPERT 1	653	653	B2					X
206	433247112515201	SPERT 2	1,217	1,217	B2	237				X
207	435056112420001	TAN 1	365	355	B2, B4		X	X		X
208	435100112420701	TAN 2	346	335	B4		X	X		X
209	435104112420301	TAN 3	269	264	B4	290	X	X		
210	435055112421301	TAN 4	247	245	B4	291				
210	435055112421302	TAN 5	303	303	B4	292	X	X		
211	435039112412601	TAN 6	263	255	B4	293	X	X		
212	435038112412601	TAN 7	324	317	B4	294	X	X		
213	435034112421701	TAN 8	251	250	B4	295	X	X		
214	435053112423202	TAN 9	326	322	B4	296	X	X		
215	435050112423201	TAN #10	258	245	B4	297				
216	435051112423201	TAN 10 A	250	250	B4	298	X	X		
217	435050112423202	TAN 11	313	310	B4	299	X	X		
218	435050112423301	TAN 12	394	384	B4	300				
219	435040112423701	TAN 13	255	–	B4	301	X	X		
220	435040112423801	TAN 13A	244	236	B4	302	X	X		
221	435039112423701	TAN 14	404	396	B4	303	X	X		
222	435021112412701	TAN 15	255	252	B4	304	X	X		
223	435020112412701	TAN #16	325	299	B4	305				
224	435034112421601	TAN 17	351	340	B4	306	X	X		
225	435051112421401	TAN 18	519	516	B4	307	X	X		
226	435051112421501	TAN 19	453	416	B4	308	X	X		
227	435046112425001	TAN 20	400	372	B4	309	X	X		
228	435009112420001	TAN 21	520	451	B4	310	X	X		
229	435020112412702	TAN 22	513	–	B4	311	X	X		
230	435019112412701	TAN 22A	539	531	B4	312	X	X		
231	435020112412703	TAN 23	507	–	B4	313	X	X		
231	435020112412704	TAN 23A	467	455	B4	314	X	X		
232	434942112411101	TAN 24	481	–	B4	315	X	X		
232	434942112411001	TAN 24A	478	238	B4	316	X	X		
233	435058112423401	TAN CH 1	600	294	B4	321	X	X		
234	435033112421702	TAN CH 2B	1,114	1,090	B4	322	X	X		
235	435042112420901	TAN Drainage Disp.#1	325	315	B4	317				
236	435054112423201	TAN Drainage Disp.#2	262	252	B4	318				

Table A1. Data for wells and boreholes used in the construction and calibration of steady-state and transient models of groundwater flow, Idaho National Laboratory and vicinity, Idaho.—Continued

[**Map No.:** Identifier used to locate wells in appendix B figures and as a cross reference with data in other appendixes **Site identifier:** Unique numerical identifier used to access well data (http://waterdata.usgs.gov/nwis) **Local name:** Local well identifier used in this study **Stratigraphic data No.:** Cross reference to borehole locations used by Anderson, Ackerman, and others (1996) and Anderson, Liszewski, and Ackerman (1996) used to construct the hydrogeologic framework of the models **Abbreviation:** ft-bls, foot below land surface **Symbols**: –, data not available; X, entry in specified appendix tables]

Map No.	Site identifier	Local name	Depth of hole (ft-bls)	Depth of well (ft-bls)	Appendix figure No.	Stratigraphic data No.	G1	H1	C1	F1
237	435116112430301	TAN Drainage Disp.#3	302	300	B4	319				
238	433521112573801	TRA 1	600	600	B6					X
239	433521112574201	TRA 2	772	747	B6					
240	433522112573501	TRA 3	602	602	B6	323				X
241	433453112574901	TRA 5 PZ1	297	128	B6	325				
242	433446112574701	TRA 6	562	558	B6	326	X	X		
243	433449112575901	TRA 7	501	493	B6	327	X	X		
244	433431112580101	TRA 8	502	502	B6	328	X	X		
245	433506112572301	TRA DISP	1,275	1,267	B6	329	X	X		X
246	–	TW-1	238	–	B3	330				
247	432700112470801	USGS 1	636	636	B2	1	X	X		
248	433320112432301	USGS 2	704	699	B2	2	X	X		
249	433732112335401	USGS 3A	740	733	B2	3				
250	434657112282201	USGS 4	553	553	B2	4	X	X		
251	433543112493801	USGS 5	500	494	B2	5	X	X		
252	434031112453701	USGS 6	620	620	B2	6	X	X		
253	434915112443901	USGS 7	1,200	903	B2	7	X	X		
254	433121113115801	USGS 8	812	812	B1	8	X	X		
255	432740113044501	USGS 9	654	654	B1	9	X	X		
256	432336113064201	USGS 11	704	704	B1	10	X	X		
257	434126112550701	USGS 12	692	563	B2	11	X	X		
258	432731113143902	USGS 13	1,200	1,010	B1	12	X	X		
259	432019112563201	USGS 14	752	752	B2	13	X	X		
260	434234112551701	USGS 15	1,497	610	B2	14	X	X		
261	431333113001701	USGS 16	740	739	B1, B2	15				
262	433937112515401	USGS 17	498	498	B2	16	X	X		
263	434540112440901	USGS 18	329	329	B2	17	X	X		
264	434426112575701	USGS 19	405	399	B2	18	X	X		
265	433253112545901	USGS 20	676	658	B2	19	X	X		
266	434307112382601	USGS 21	406	363	B2	20	X	X		
267	433422113031701	USGS 22	657	657	B1	21	X	X		
268	434055112595901	USGS 23	467	458	B1, B2	22	X	X		
269	435053112420801	USGS 24	326	326	B4	23	X	X		
270	435339112444601	USGS 25	320	320	B2	24	X	X		
271	435212112394001	USGS 26	267	267	B2	25	X	X		

Table A1.   Data for wells and boreholes used in the construction and calibration of steady-state and transient models of groundwater flow, Idaho National Laboratory and vicinity, Idaho.—Continued

[**Map No.:** Identifier used to locate wells in appendix B figures and as a cross reference with data in other appendixes **Site identifier:** Unique numerical identifier used to access well data (http://waterdata usgs gov/nwis) **Local name:** Local well identifier used in this study **Stratigraphic data No.:** Cross reference to borehole locations used by Anderson, Ackerman, and others (1996) and Anderson, Liszewski, and Ackerman (1996) used to construct the hydrogeologic framework of the models **Abbreviation:** ft-bls, foot below land surface **Symbols:** –, data not available; X, entry in specified appendix tables]

Map No.	Site identifier	Local name	Depth of hole (ft-bls)	Depth of well (ft-bls)	Appendix figure No.	Stratigraphic data No.	Data in table			
							G1	H1	C1	F1
272	434851112321801	USGS 27	312	312	B2	26	X	X		
273	434600112360101	USGS 28	334	334	B2	27	X	X		
274	434407112285101	USGS 29	426	426	B2	28	X	X		
275	434601112315403	USGS 30A	1,007	725	B2	29	X	X		
275	434601112315402	USGS 30B	1,007	400	B2		X	X		
275	434601112315401	USGS 30C	1,007	300	B2		X	X		
276	434625112342101	USGS 31	428	428	B2	30	X	X		
277	434444112322101	USGS 32	392	392	B2	31	X	X		
278	434314112322901	USGS 33	516	516	B2	32				
279	433334112565501	USGS 34	700	700	B7	33	X	X		
280	433339112565801	USGS 35	579	579	B7	34	X	X		
281	433330112565201	USGS 36	567	567	B7	35	X	X		
282	433326112564801	USGS 37	573	572	B7	36	X	X		
283	433322112564301	USGS 38	729	612	B7	37	X	X		
284	433343112570001	USGS 39	572	494	B7	38	X	X		
285	433411112561101	USGS 40	679	483	B7	39	X	X		
286	433409112561301	USGS 41	674	666	B7	40	X	X		
287	433404112561301	USGS 42	678	678	B7	41	X	X		
288	433415112561501	USGS 43	676	676	B7	42	X	X		
289	433409112562101	USGS 44	650	650	B7	43	X	X		
290	433402112561801	USGS 45	651	651	B7	44	X	X		
291	433407112561501	USGS 46	651	651	B7	45	X	X		
292	433407112560301	USGS 47	651	651	B7	46	X	X		
293	433401112560301	USGS 48	750	750	B7	47	X	X		
294	433403112555401	USGS 49	656	458	B7	48	X	X		
295	433419112560201	USGS 50	405	395	B7	49				
296	433350112560601	USGS 51	659	647	B7	50	X	X		
297	433414112554201	USGS 52	650	602	B7	51	X	X		
298	433503112573401	USGS 53	90	90	B6	52				
299	433503112572801	USGS 54	91	91	B6	53				
300	433508112573001	USGS 55	81	81	B6	54				
301	433509112573501	USGS 56	80	80	B6	55				
302	433344112562601	USGS 57	732	582	B7	56	X	X		
303	433500112572502	USGS 58	503	503	B6	57	X	X		
304	433354112554701	USGS 59	657	587	B7	58	X	X		

Table A1. Data for wells and boreholes used in the construction and calibration of steady-state and transient models of groundwater flow, Idaho National Laboratory and vicinity, Idaho.—Continued

[**Map No.:** Identifier used to locate wells in <u>appendix B</u> figures and as a cross reference with data in other appendixes **Site identifier:** Unique numerical identifier used to access well data (<u>http://waterdata usgs gov/nwis</u>) **Local name:** Local well identifier used in this study **Stratigraphic data No.:** Cross reference to borehole locations used by Anderson, Ackerman, and others (1996) and Anderson, Liszewski, and Ackerman (1996) used to construct the hydrogeologic framework of the models **Abbreviation:** ft-bls, foot below land surface **Symbols**: –, data not available; X, entry in specified appendix tables]

Map No.	Site identifier	Local name	Depth of hole (ft-bls)	Depth of well (ft-bls)	Appendix figure No.	Stratigraphic data No.	G1	H1	C1	F1
305	433456112571901	USGS 60	117	117	B6	59				
306	433453112571601	USGS 61	123	123	B6	60				
307	433446112570701	USGS 62	165	165	B6, B7	61				
308	433455112574001	USGS 63	97	97	B6	62				
309	433513112571801	USGS 64	205	205	B6	63				
310	433447112574501	USGS 65	498	498	B6	64	X	X		
311	433436112564801	USGS 66	475	475	B6, B7	65				
312	433344112554101	USGS 67	694	694	B7	66	X	X		
313	433516112573901	USGS 68	128	128	B6	67				
314	433450112573001	USGS 69	115	115	B6	68				
315	433504112571001	USGS 70	100	100	B6	69				
316	433439112571501	USGS 71	184	171	B6	70				
317	433519112574601	USGS 72	200	200	B6	71				
318	433502112575401	USGS 73	127	127	B6	72				
319	433505112580501	USGS 74	192	192	B6	73				
320	433457112570001	USGS 75	212	212	B6	74				
321	433425112573201	USGS 76	718	718	B6	75	X	X		
322	433315112560301	USGS 77	610	586	B7	76	X	X		
323	433413112573501	USGS 78	204	204	B6	77				
324	433505112581901	USGS 79	702	702	B6	78	X	X		
325	433457112570002	USGS 80	204	204	B6	79				
326	433400112551001	USGS 81	104	104	B7	80				
327	433401112551001	USGS 82	700	693	B7	81	X	X		
328	433023112561501	USGS 83	752	752	B2	82	X	X		
329	433356112574201	USGS 84	505	505	B6	83	X	X		
330	433246112571201	USGS 85 (new)	637	614	B8			X		
330	433246112571201	USGS 85 (old)	637	637	B8	84	X	X		
331	432935113080001	USGS 86	691	691	B1	85	X	X		
332	433013113024201	USGS 87	673	673	B3	86	X	X		
333	432940113030201	USGS 88	663	663	B3	87	X			
334	433005113032801	USGS 89	650	637	B3	88	X	X		
335	432954113020501	USGS 90	626	626	B3	89	X	X		
336	–	USGS 91	255	–	B3	90				
337	433000113025301	USGS 92	247	214	B3	91				
338	–	USGS 93	246	–	B3	92				

Table A1.   Data for wells and boreholes used in the construction and calibration of steady-state and transient models of groundwater flow, Idaho National Laboratory and vicinity, Idaho.—Continued

[**Map No.:** Identifier used to locate wells in appendix B figures and as a cross reference with data in other appendixes **Site identifier:** Unique numerical identifier used to access well data (http://waterdata.usgs.gov/nwis) **Local name:** Local well identifier used in this study **Stratigraphic data No.:** Cross reference to borehole locations used by Anderson, Ackerman, and others (1996) and Anderson, Liszewski, and Ackerman (1996) used to construct the hydrogeologic framework of the models **Abbreviation:** ft-bls, foot below land surface **Symbols**: –, data not available; X, entry in specified appendix tables]

Map No.	Site identifier	Local name	Depth of hole (ft-bls)	Depth of well (ft-bls)	Appendix figure No.	Stratigraphic data No.	Data in table			
							G1	H1	C1	F1
338	–	USGS 93A	233	–	B3	93				
339	–	USGS 94	302	–	B3	94				
340	–	USGS 95	246	–	B3	95				
341	–	USGS 96	236	–	B3	96				
341	–	USGS 96A	120	–	B3	97				
342	–	USGS 96B	229	–	B3	98				
343	433807112551501	USGS 97	510	510	B2, B5	99	X	X		
344	433657112563601	USGS 98	508	508	B2	100	X	X		
345	433705112552101	USGS 99	450	440	B2	101	X	X		
346	433503112400701	USGS 100	750	750	B9	102	X	X		
347	433255112381801	USGS 101	865	842	B2	103	X	X		
348	433853112551601	USGS 102	445	445	B5	104	X	X		
349	432714112560701	USGS 103	1,307	1,297	B2	105	X	X		
350	432856112560801	USGS 104	700	700	B2	106	X	X		
351	432703113001801	USGS 105	800	800	B1, B2	107	X	X		
352	432959112593101	USGS 106	760	760	B1, B2	108	X	X		
353	432942112532801	USGS 107	690	690	B2	109	X	X		
354	432659112582601	USGS 108	760	760	B2	110	X	X		
355	432701113025601	USGS 109	800	800	B1	111	X	X		
356	432717112501501	USGS 110	780	780	B2	112	X	X		
356	432717112501502	USGS 110A	657	644	B2			X		
357	433331112560501	USGS 111	600	560	B7	113	X	X		
358	433314112563001	USGS 112	563	563	B7	114	X	X		
359	433314112561801	USGS 113	564	564	B7	115	X	X		
360	433318112555001	USGS 114	563	560	B7	116	X	X		
361	433320112554101	USGS 115	581	581	B7	117	X	X		
362	433331112553201	USGS 116	580	572	B7	118	X	X		
363	432955113025901	USGS 117	655	655	B3	119	X	X		
364	432947113023001	USGS 118	622	608	B3	120	X	X		
365	432945113023401	USGS 119	705	705	B3	121	X	X		
366	432919113031501	USGS 120	705	705	B3	122	X	X		
367	433450112560301	USGS 121	746	475	B7	123	X	X		
368	433353112555201	USGS 122	483	480	B7	124	X	X		
369	433352112561401	USGS 123	744	515	B7	125	X	X		
370	432307112583101	USGS 124	800	800	B2	126	X	X		

Table A1.   Data for wells and boreholes used in the construction and calibration of steady-state and transient models of groundwater flow, Idaho National Laboratory and vicinity, Idaho.—Continued

[**Map No.:** Identifier used to locate wells in appendix B figures and as a cross reference with data in other appendixes **Site identifier:** Unique numerical identifier used to access well data (http://waterdata usgs gov/nwis) **Local name:** Local well identifier used in this study **Stratigraphic data No.:** Cross reference to borehole locations used by Anderson, Ackerman, and others (1996) and Anderson, Liszewski, and Ackerman (1996) used to construct the hydrogeologic framework of the models **Abbreviation:** ft-bls, foot below land surface **Symbols**: –, data not available; X, entry in specified appendix tables]

Map No.	Site identifier	Local name	Depth of hole (ft-bls)	Depth of well (ft-bls)	Appendix figure No.	Stratigraphic data No.	Data in table			
							G1	H1	C1	F1
371	432602113052801	USGS 125	774	774	B1		X	X		
372	432906113025001	USGS 132	1,238	1,238	B3					
373	–	VZT-1	–	–	B3	331				
374	440813112532201	Wagoner Ranch	295	295	B2					
375	433445113202801	Weaver and Lowe	1,075	1,025	B1	335	X	X		
376	433716112563601	WS INEL 1	710	490	B2	218	X	X		
377	–	WWW#1	–	–	B3	332				
378	–	WWW#2	–	–	B3	333				
379	433116112534701	OMRE	943	943	B2					
380	433521112574201	TRA 4	970	965	B6	324				X

Table A2.   Data for streams used in the construction and calibration of steady-state and transient models of groundwater flow, Idaho National Laboratory and vicinity, Idaho.

[**Map No.:** Identifier used to locate streamflow-gaging stations and stream reaches on maps in appendix B figures and as a cross reference with appendix C **Site identifier:** Unique numerical identifier used to access streamflow-gaging station data (http://waterdata.usgs.gov/nwis) **Local name:** Local streamflow-gaging station or stream-reach identifier used in this study]

Map No.	Site identifier	Local name	Appendix figure No.
		Streamflow-gaging stations	
500	13127000	Big Lost River below Mackay Reservoir near Mackay	B1
501	12132500	Big Lost River near Arco	B1
502	13132513	INL Diversion at Head, near Arco	B1
503	13132515	INL Diversion at Outlet of Spreading Area A near Arco	B1
504	13132520	Big Lost River below INL diversion, near Arco	B1
505	13132535	Big Lost River at Lincoln Boulevard Bridge near Atomic City	B2, B6, B7
506	13132565	Big Lost River above Big Lost River Sinks near Howe	B2
		Stream reaches	
600		Big Lost River above Arco gage (13132500)	B1
601		Big Lost river between Arco gage (13132500) and INL Diversion	B1
602		Spreading Area A	B1
603		Spreading Area B	B1, B3
604		Spreading Area C	B1
605		Spreading Area D	B1
606		Big Lost River between INL diversion and Lincoln Boulevard gage (13132535)	B1, B2, B6, B7
607		Big Lost River between Lincoln Boulevard gage (13132535) and Big Lost River Sinks gage (13132565)	B2, B6, B7
608		Big Lost River Sinks	B2
609		Big Lost River Playas 1 and 2	B2
610		Big Lost River Playa 3	B2, B4
611		Little Lost River	B2
612		Birch Creek power diversion return	B2

Appendix B. Locations of Wells and Streams for Which Stratigraphic, Water-Level, Water-Use, and Discharge Data are Available, Idaho National Laboratory and Vicinity, Idaho

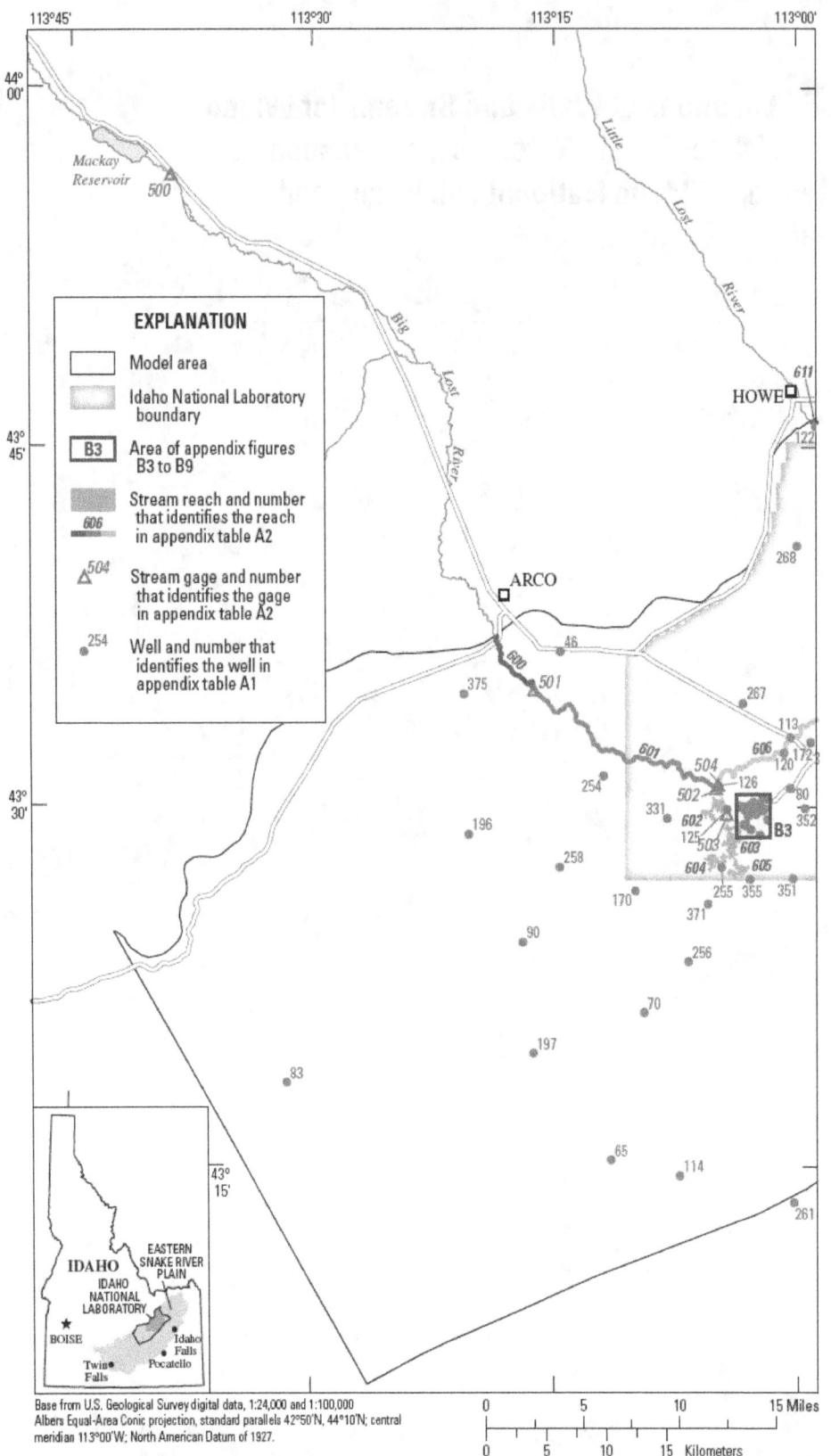

Figure B1. Locations of wells and streams for which stratigraphic, water-level, water-use, and discharge data are available, western half of study area, Idaho National Laboratory and vicinity, Idaho. Click on well, stream gage, or reach number to access information in tables A1 or A2.

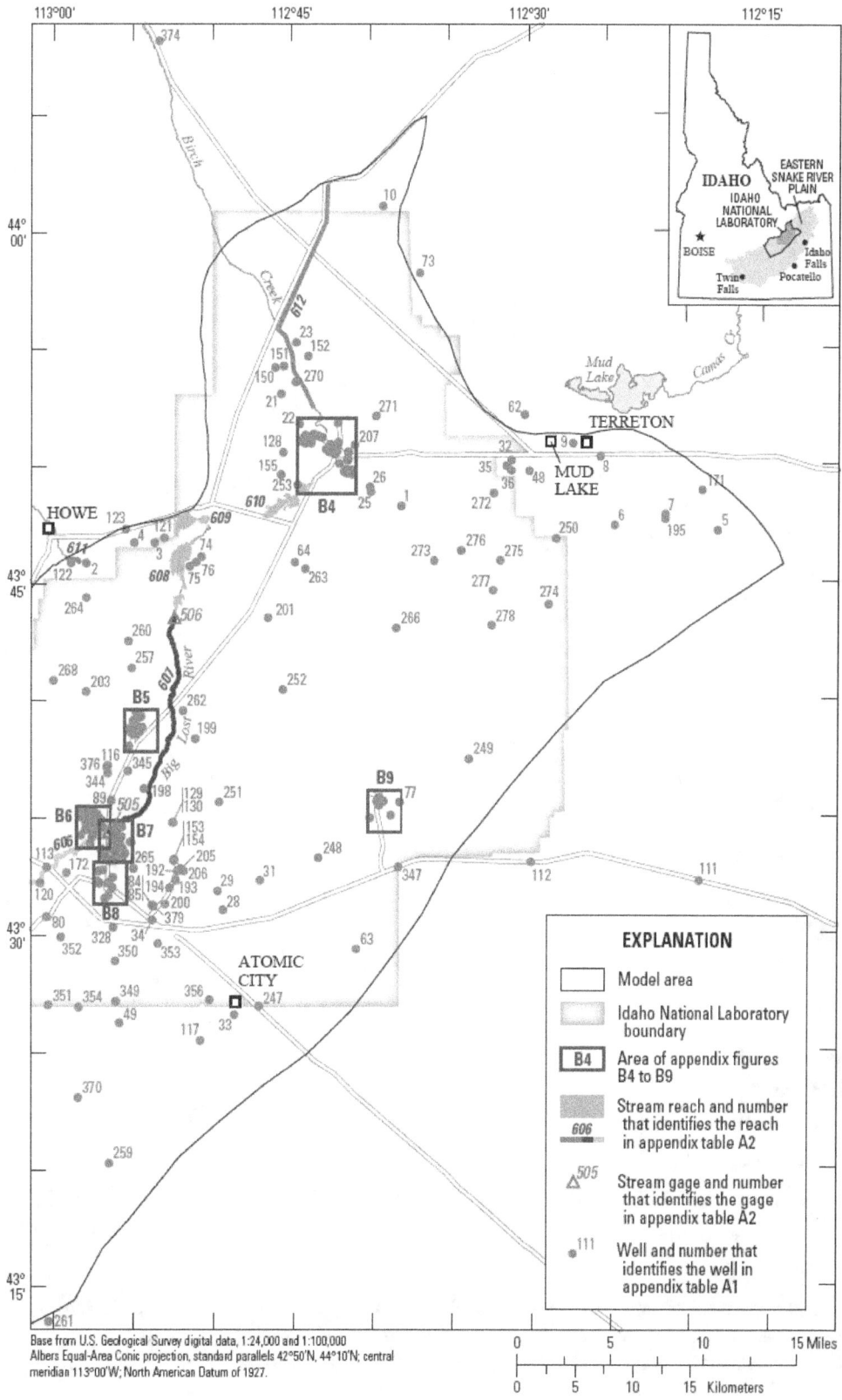

Figure B2.   Locations of wells and streams for which stratigraphic, water-level, water-use, and discharge data are available, eastern half of study area, Idaho National Laboratory and vicinity, Idaho. Click on well, stream gage, or reach number to access information in tables A1 or A2.

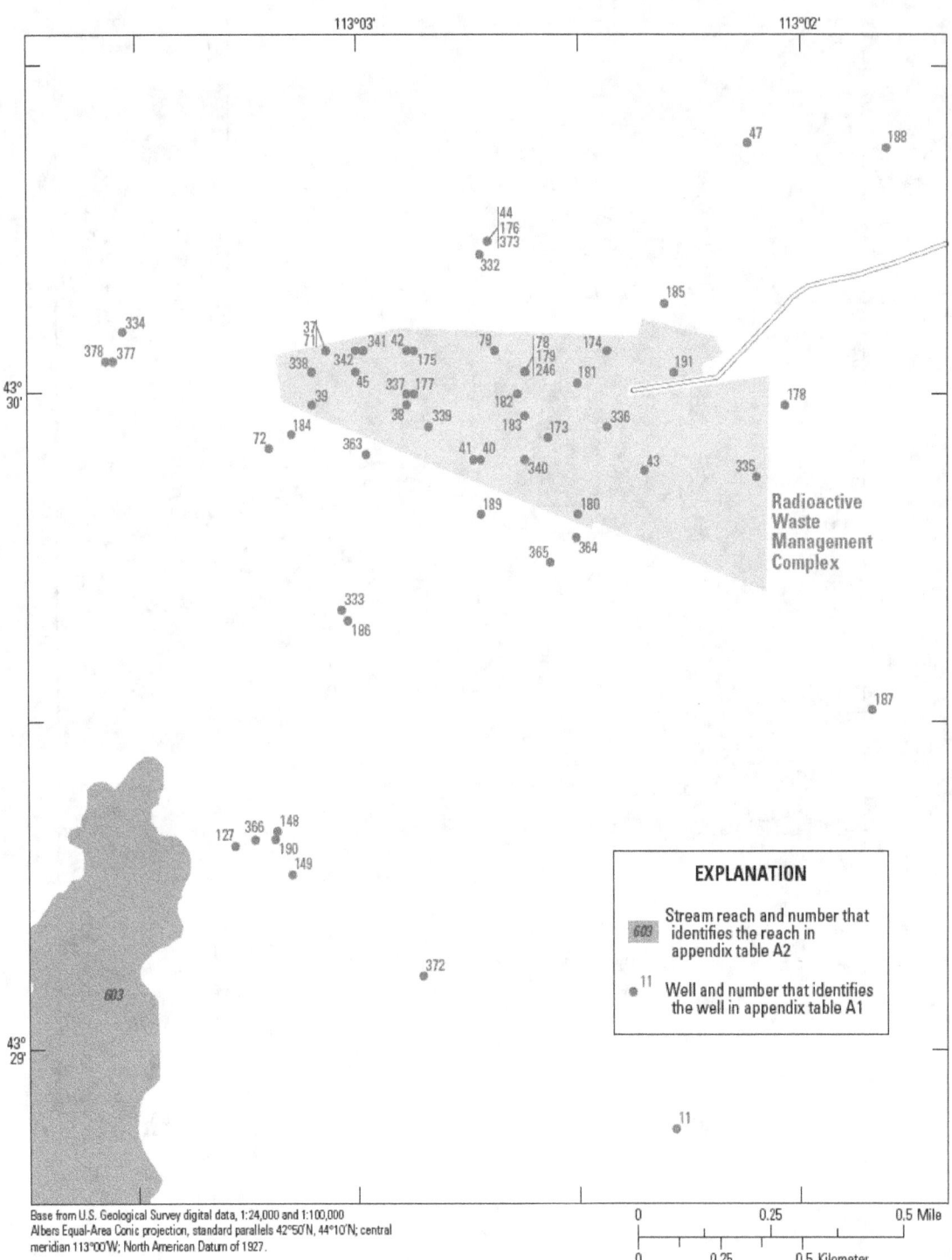

Figure B3. Locations of wells and streams for which stratigraphic, water-level, water-use, and discharge data are available at the Radioactive Waste Management Complex, Idaho National Laboratory, Idaho.
Click on well, stream gage, or reach number to access information in tables A1 or A2.

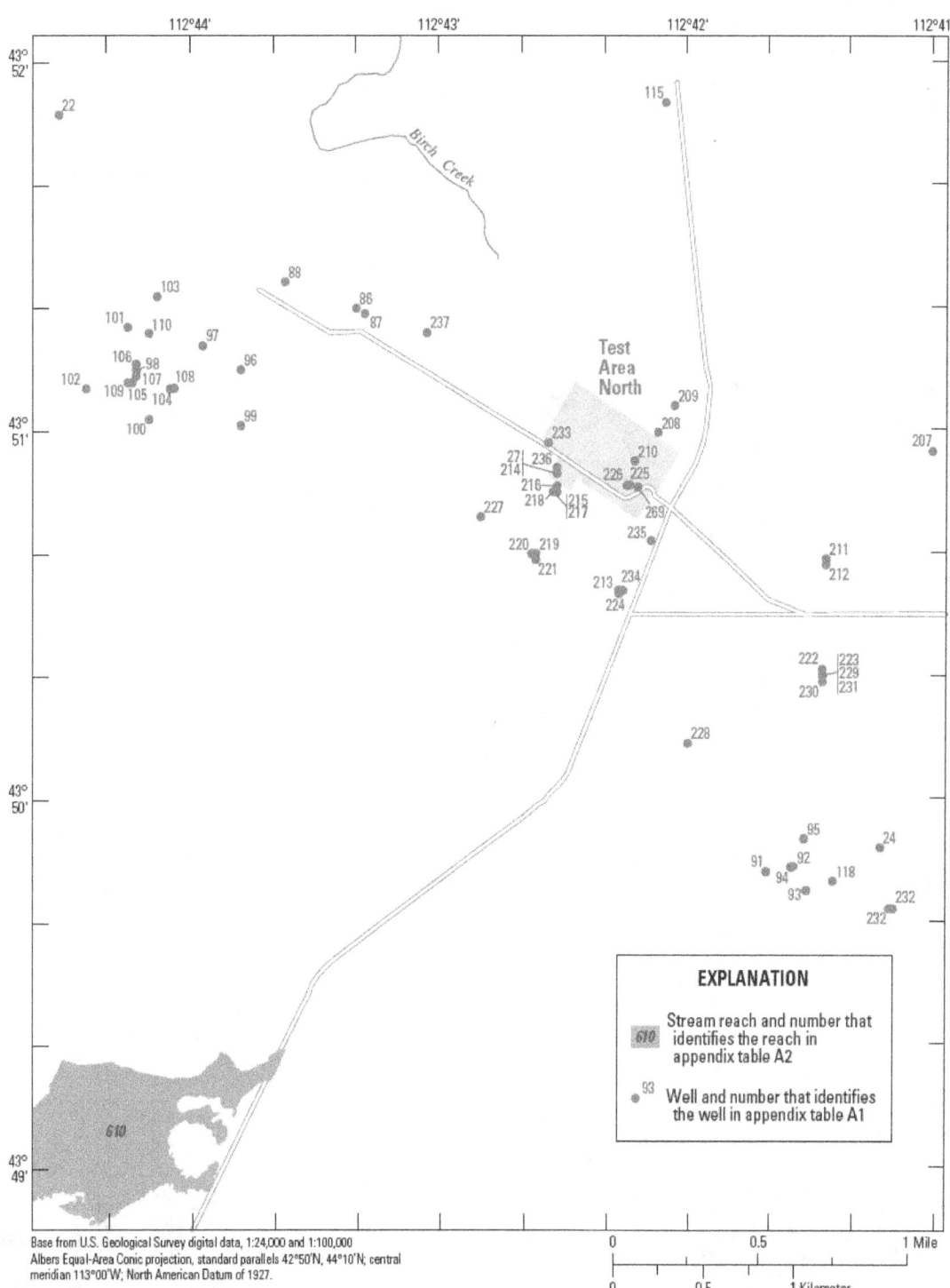

Figure B4. Locations of wells and streams for which stratigraphic, water-level, water-use, and discharge data are available at the Test Area North, Idaho National Laboratory, Idaho.

Click on well, stream gage, or reach number to access information in tables A1 or A2.

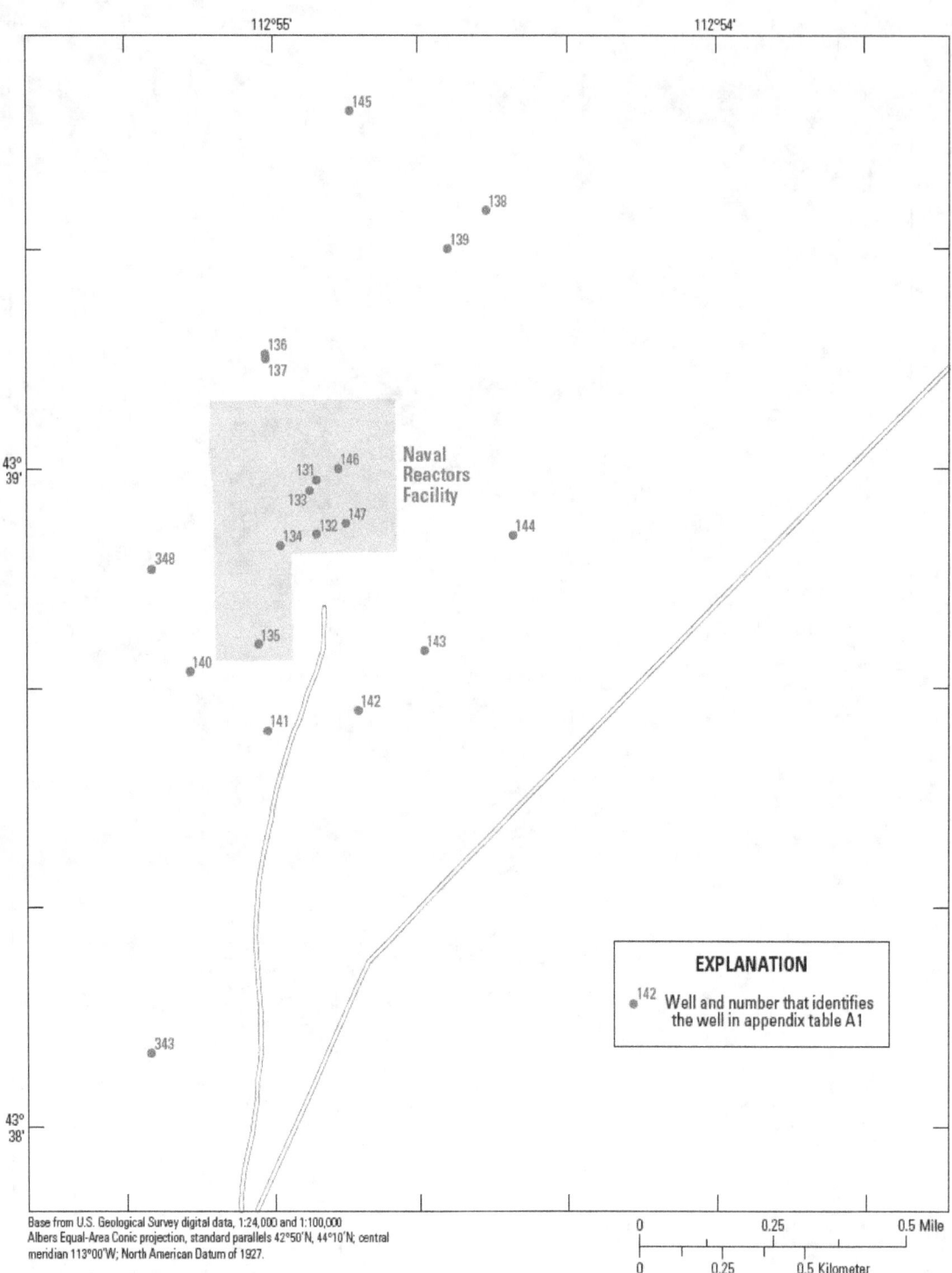

Figure B5.   Locations of wells and streams for which stratigraphic, water-level, water-use, and discharge data are available at the Naval Reactors Facility, Idaho National Laboratory, Idaho.
Click on well, stream gage, or reach number to access information in tables A1 or A2.

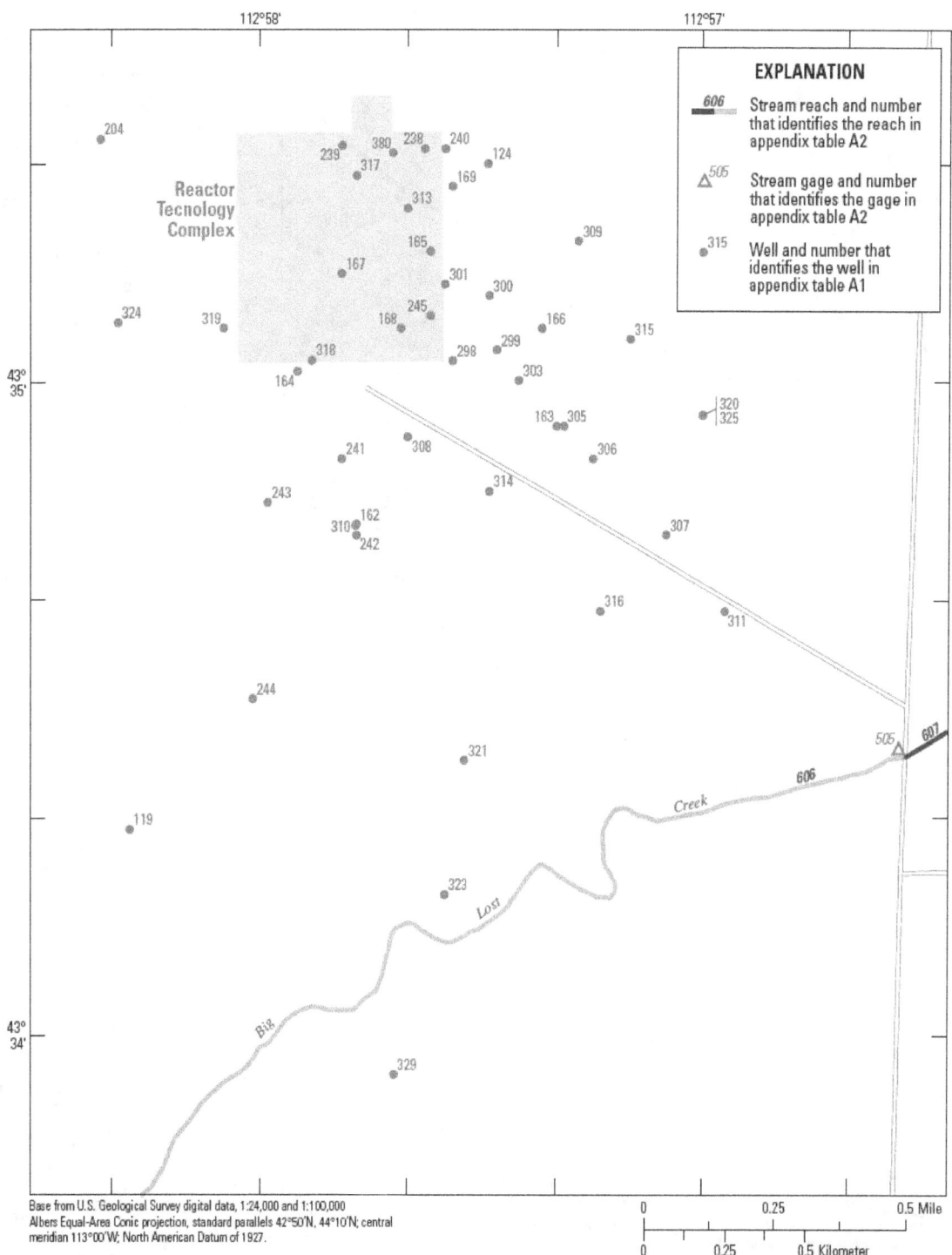

Figure B6. Locations of wells and streams for which stratigraphic, water-level, water-use, and discharge data are available at the Reactor Technology Complex, Idaho National Laboratory, Idaho.
Click on well, stream gage, or reach number to access information in tables A1 or A2.

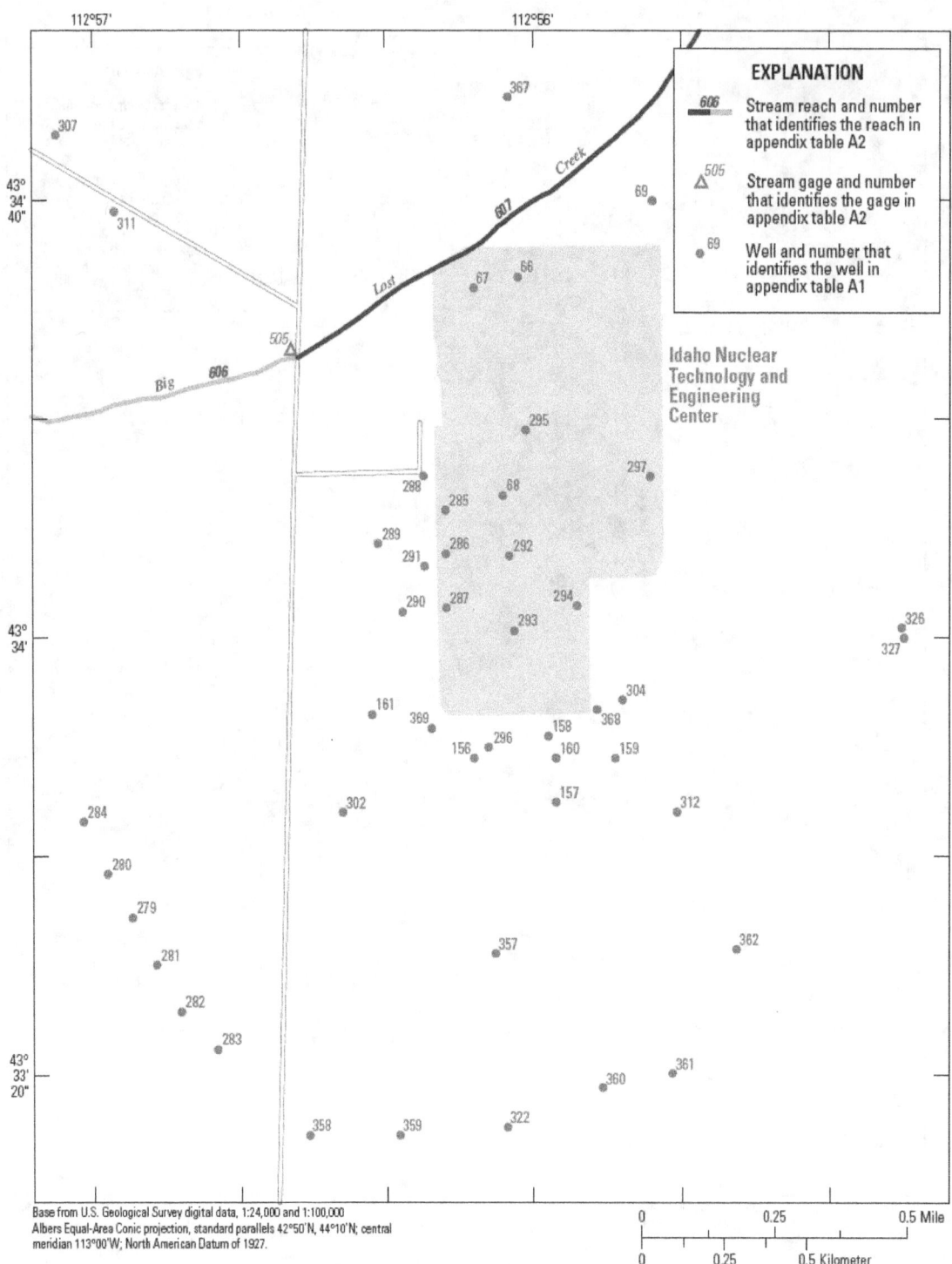

Figure B7.   Locations of wells and streams for which stratigraphic, water-level, water-use, and discharge data are available at the Idaho Nuclear Technology and Engineering Center, Idaho National Laboratory, Idaho.
Click on well, stream gage, or reach number to access information in tables A1 or A2.

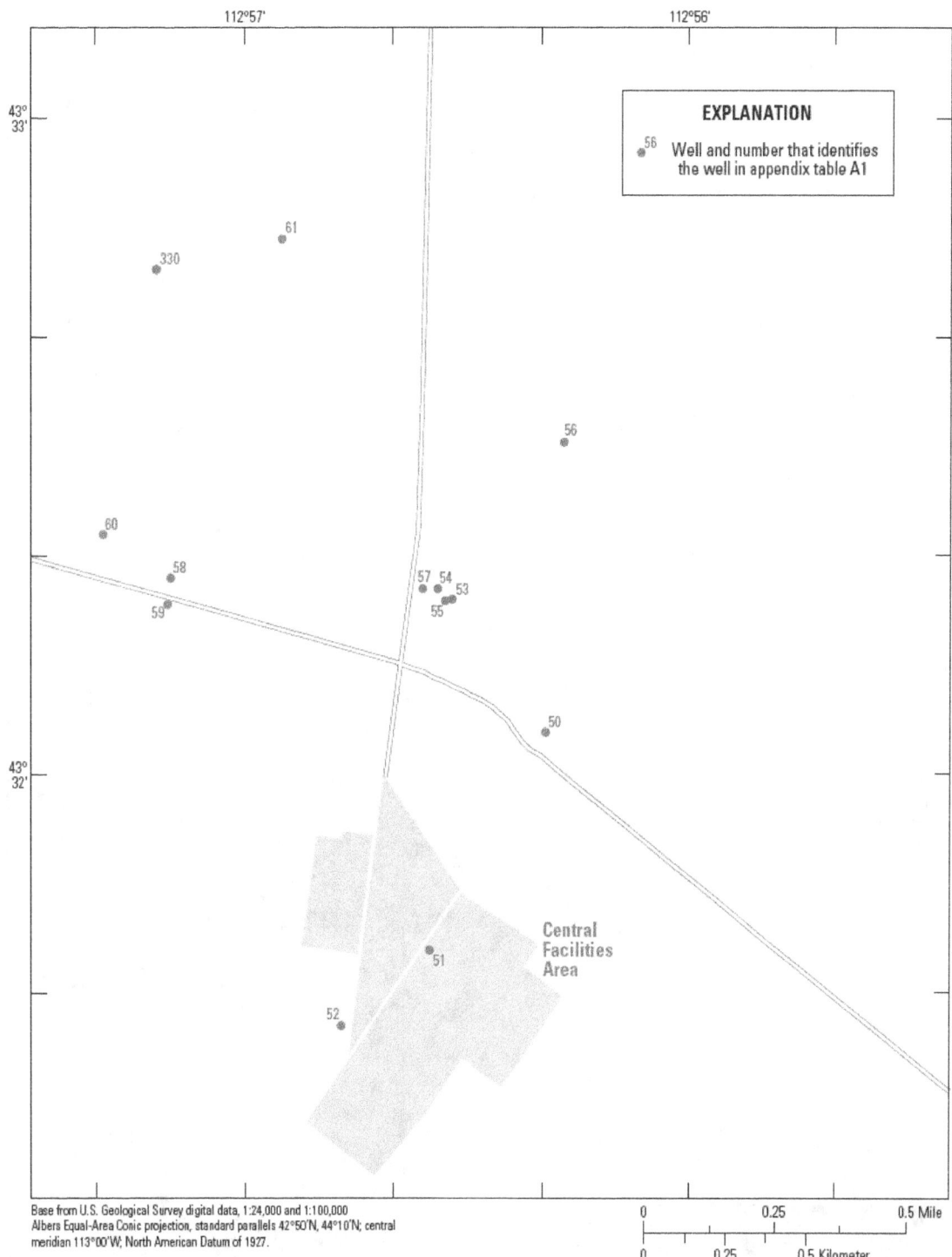

Figure B8. Locations of wells and streams for which stratigraphic, water-level, water-use, and discharge data are available at the Central Facilities Area, Idaho National Laboratory, Idaho.
Click on well, stream gage, or reach number to access information in tables A1 or A2.

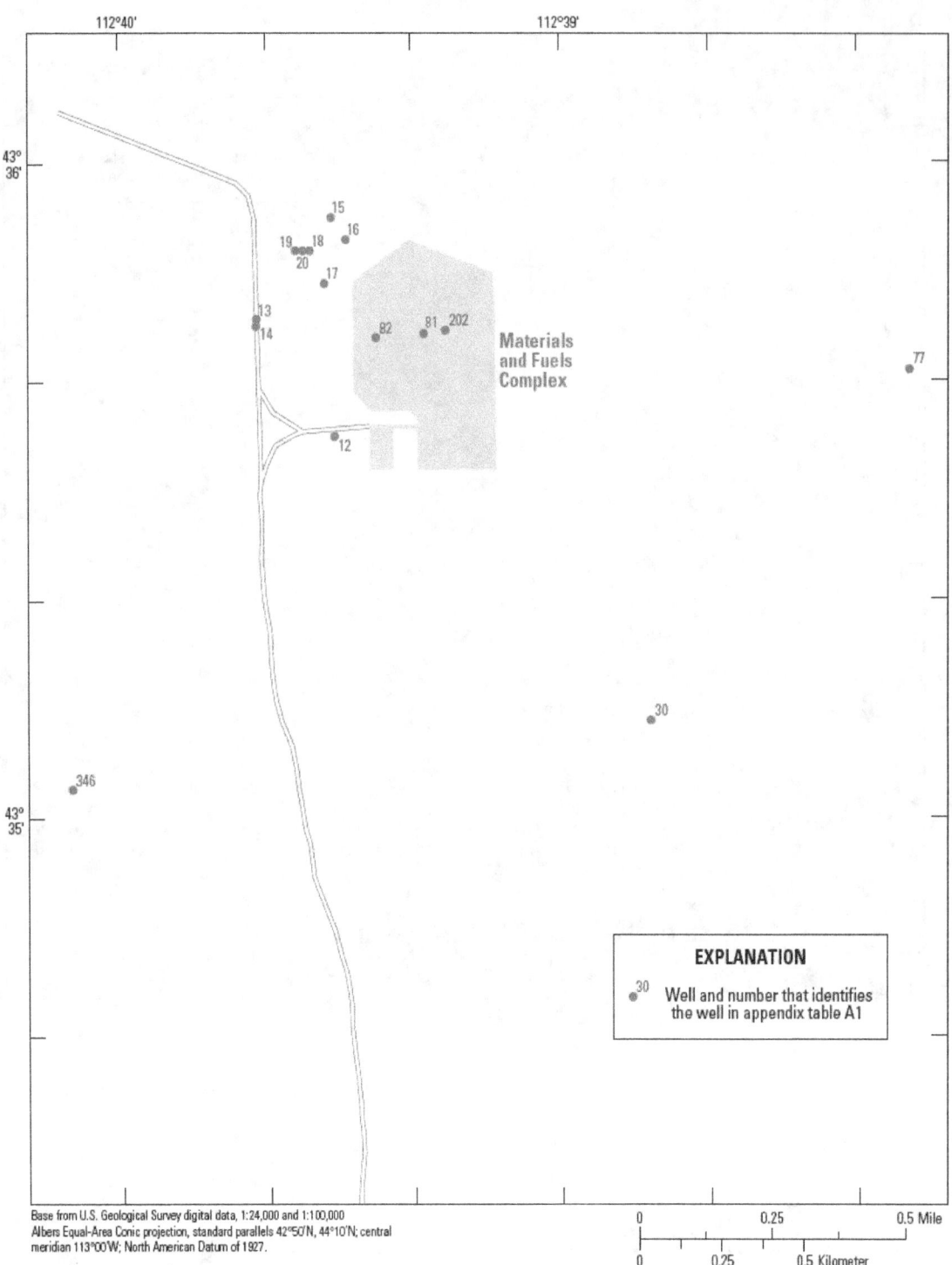

Figure B9.   Locations of wells and streams for which stratigraphic, water-level, water-use, and discharge data are available at the Materials and Fuels Complex, Idaho National Laboratory, Idaho.
Click on well, stream gage, or reach number to access information in tables A1 or A2.

Appendix C. Average Streamflow and Infiltration Used for Steady-State and Transient Models of Groundwater Flow, Idaho National Laboratory and Vicinity, Idaho

Table C1.   Average streamflow and infiltration for 1966–80, 1981–95, and 1953–68 used for steady-state and transient models of groundwater flow, Idaho National Laboratory and vicinity, Idaho.

[Values are in cubic feet per second. Map No. Identifier corresponding to a streamflow-gaging station or stream reach on maps located in appendix B figures and as a cross reference with data in other appendixes. Names of stations and reaches are given in table A2 and include the Big Lost River streamflow-gaging stations located near Arco (501), at the inlet to spreading area A (502), at the outlet of spreading area A (503), downstream of the diversion dam used to divert flow into spreading area A (504), at the Lincoln Boulevard bridge near the INTEC (505), and above the Big Lost River Sinks (506). Reaches are located between the model boundary and streamflow-gaging station 501 (600), between streamflow-gaging stations 501 and 504 (601), within spreading area A through D (602–605), between streamflow-gaging stations 504 and 505 (606), between streamflow-gaging station 505 and the Big Lost River Sinks (607), within the Big Lost River Sinks (608), and within Big Lost River Playas 1, 2 (609), and 3 (610). Annual transient stress periods are January through April, May through August, and September through December. **Abbreviations** I, infiltration; Q, discharge; INL, Idaho National Laboratory]

Year	Model boundary to INL diversion				INL spreading areas					INL diversion to Lincoln Blvd			Lincoln Blvd to Playa 3				
Map No.	600	501	601	502	602	503	603	604	605	504	606	505	607	506	608	609	610
	I	Q	Q	Q	I	Q	I	I	I	Q	I	Q	I	Q	I	I	I
				Average streamflow and infiltration recharge of 1966–80 for Big Lost River unadjusted for evaporation and used in steady state model													
1966		70.29	12.93	45.26	6.58	38.69	29.02	9.67	0.00	12.10	7.040	5.05	5.06	0.00	0.00	0.00	0.00
1967		232.55	15.82	108.22	13.42	94.79	71.09	23.7	.00	108.52	21.810	86.71	15.07	71.64	44.68	26.96	.00
1968		88.72	18.95	34.61	13.28	21.33	17.17	4.17	.00	35.15	12.860	22.29	13.46	8.84	8.84	.00	.00
1969		342.62	17.25	145.31	20.00	125.30	95.42	29.89	.00	180.90	19.460	161.44	24.23	137.21	58.08	23.59	55.54
1970		134.82	13.64	35.38	15.25	20.13	16.09	4.04	.00	85.68	13.750	71.93	17.05	54.87	40.04	14.84	.00
1971		256.21	20.54	65.90	20.00	45.90	36.1	9.79	.00	169.77	18.440	151.33	23.67	127.66	86.52	17.52	23.62
1972		96.47	17.29	35.64	13.28	22.37	17.15	5.21	.00	43.53	17.040	26.49	15.73	10.76	10.76	.00	.00
1973		54.81	20.12	23.43	8.53	14.89	11.17	3.72	.00	11.27	6.800	4.47	4.47	.00	.00	.00	.00
1974		157.80	17.06	44.72	20.00	24.72	19.41	5.31	.00	96.03	24.780	71.25	19.23	52.01	37.29	14.72	.00
1975		199.11	14.54	63.75	20.00	43.75	34.29	9.45	.00	120.82	21.010	99.81	19.94	79.87	48.42	31.45	.00
1976		89.00	13.13	51.62	13.28	38.34	28.76	9.59	.00	24.26	14.050	10.21	10.21	.00	.00	.00	.00
1977		16.66	8.49	4.51	4.51	.00	.00	.00	.00	3.66	3.660	.00	.00	.00	.00	.00	.00
1978		10.24	10.24	.00	.00	.00	.00	.00	.00	.00	.000	.00	.00	.00	.00	.00	.00
1979		11.42	10.79	.00	.00	.00	.00	.00	.00	.64	.640	.00	.00	.00	.00	.00	.00
1980		32.70	17.62	.04	.04	.00	.00	.00	.00	15.04	13.470	1.57	1.57	.00	.00	.00	.00
Simulated	6.56	119.53	15.23	43.88	12.05	30.99	23.78	7.21	0.00	60.47	12.99	47.48	11.31	36.17	22.30	8.60	5.27

Table C1. Average streamflow and infiltration for 1966–80, 1981–95, and 1953–68 used for steady-state and transient models of groundwater flow, Idaho National Laboratory and vicinity, Idaho.—Continued

[Values are in cubic feet per second. Map No. Identifier corresponding to a streamflow-gaging station or stream reach on maps located in appendix B figures and as a cross reference with data in other appendixes. Names of stations and reaches are given in table A2 and include the Big Lost River streamflow-gaging stations located near Arco (501), at the inlet to spreading area A (502), at the outlet of spreading area A (503), downstream of the diversion dam used to divert flow into spreading area A (504), at the Lincoln Boulevard bridge near the INTEC (505), and above the Big Lost River Sinks (506). Reaches are located between the model boundary and streamflow-gaging station 501 (600), between streamflow-gaging stations 501 and 504 (601), within spreading area A through D (602–605), between streamflow-gaging stations 504 and 505 (606), between streamflow-gaging station 505 and the Big Lost River Sinks (607), within the Big Lost River Sinks (608), and within Big Lost River Playas 1, 2 (609), and 3 (610). Annual transient stress periods are January through April, May through August, and September through December. **Abbreviations** 1, infiltration; Q, discharge; INL, Idaho National Laboratory]

Year and stress period	Model boundary to INL diversion				INL spreading areas					INL diversion to Lincoln Blvd					Lincoln Blvd to Playa 3		
	I	Q	I	Q	I	Q	I	I	I	Q	I	Q	I	Q	I	I	I
	600	501	601	502	602	503	603	604	605	504	606	505	607	506	608	609	610
					Streamflow and infiltration of 1980-95 for Big Lost River adjusted for evaporation and used in transient model												
1980																	
1	6.37		14.63		10.55		24.73	7.49	0.00		12.64		10.94		20.19	7.94	4.93
1981	6.45	42.51	14.03	1.23	.88	0.31	.30	.00	.00	26.88	7.44	19.27	6.05	13.04	11.68	.00	.00
2	6.51		18.01		1.34		.46	.00	.00		3.42		5.34		.00	.00	.00
3	6.34		12.99		1.31		.46	.00	.00		18.75		12.74		34.66	.00	.00
4	6.49		11.16		.00		.00	.00	.00		.00		.00		.00	.00	.00
1982	6.42	218.12	26.93	64.32	14.80	48.79	37.44	10.69	.00	126.52	13.77	112.55	15.43	97.03	55.25	18.19	16.99
5	6.50		18.06		1.24		1.90	.00	.00		4.36		11.98		12.41	.00	.00
6	6.30		35.05		23.36		95.80	31.73	.00		26.90		19.20		75.49	46.03	50.42
7	6.47		27.47		19.50		13.57	.00	.00		9.78		15.01		76.97	8.02	.00
1983	6.39	477.91	27.04	291.28	33.78	256.21	190.68	63.32	.00	159.26	23.76	135.31	16.19	118.91	63.15	20.72	27.84
8	6.47		19.14		24.48		89.73	29.85	.00		12.02		13.45		47.50	.00	.00
9	6.28		18.57		37.68		250.27	82.95	.00		34.12		20.79		83.81	52.06	82.61
10	6.41		43.36		38.99		229.89	76.44	.00		24.86		14.26		57.73	9.51	.00
1984	6.38	488.42	28.45	381.62	36.45	343.65	255.54	84.82	.00	78.01	16.73	61.08	13.29	47.57	41.23	3.59	.00
11	6.44		54.48		39.17		206.65	68.74	.00		11.76		11.95		11.71	.00	.00
12	6.27		18.58		46.91		399.41	132.32	.00		23.65		13.41		49.24	.00	.00
13	6.44		12.59		23.20		158.98	52.86	.00		14.67		14.51		62.42	10.76	.00
1985	6.43	120.76	20.24	39.38	7.21	32.00	23.94	7.91	.00	60.76	11.41	49.13	12.81	36.09	30.87	3.57	.00
14	6.48		37.29		18.65		72.35	24.07	.00		12.27		12.19		17.39	.00	.00
15	6.34		5.58		.23		.13	.00	.00		6.24		11.73		12.74	.00	.00
16	6.48		18.26		3.00		.34	.00	.00		15.79		14.51		62.41	10.68	.00

Table C1. Average streamflow and infiltration for 1966–80, 1981–95, and 1953–68 used for steady-state and transient models of groundwater flow, Idaho National Laboratory and vicinity, Idaho.—Continued

[Values are in cubic feet per second. Map No. Identifier corresponding to a streamflow-gaging station or stream reach on maps located in appendix B figures and as a cross reference with data in other appendixes. Names of stations and reaches are given in table A2 and include the Big Lost River streamflow-gaging stations located near Arco (501), at the inlet to spreading area A (502), at the outlet of spreading area A (503), downstream of the diversion dam used to divert flow into spreading area A (504), at the Lincoln Boulevard bridge near the INTEC (505), and above the Big Lost River Sinks (506). Reaches are located between the model boundary and streamflow-gaging stations 501 and 504 (601), within spreading area A through D (602–605), between streamflow-gaging stations 501 and 504 (601), between streamflow-gaging stations 504 and 505 (606), between streamflow-gaging stations 505 and the Big Lost River Sinks (607), within the Big Lost River Sinks (608), and within Big Lost River Playas 1, 2 (609), and 3 (610). Annual transient stress periods are January through April, May through August, and September through December. **Abbreviations** I, infiltration; Q, discharge; INL, Idaho National Laboratory]

Year and stress period	Model boundary to INL diversion				INL spreading areas					INL diversion to Lincoln Blvd			Lincoln Blvd to Playa 3				
	I	Q	I	Q	I	Q	I	I	I	Q	I	Q	I	Q	I	I	I
	600	501	601	502	602	503	603	604	605	504	606	505	607	506	608	609	610
Streamflow and infiltration of 1980–95 for Big Lost River adjusted for evaporation and used in transient model—Continued																	
1986	6.43	204.94	23.07	59.09	16.79	41.52	32.14	8.84	0.00	122.42	22.44	99.78	17.35	82.21	48.92	20.17	7.79
17	6.49		26.82		4.59		1.83	.00	.00		28.42		12.48		24.51	.00	.00
18	6.32		21.90		24.68		79.19	26.23	.00		30.21		16.98		68.76	46.11	15.38
19	6.47		20.57		20.83		14.52	.00	.00		8.74		22.52		52.92	13.85	7.79
1987	6.45	29.21	11.49	10.27	7.36	2.76	2.75	.00	.00	7.10	5.52	1.41	1.37	.00	.00	.00	.00
20	6.50		15.65		22.39		8.35	.00	.00		15.02		4.17		.00	.00	.00
21	6.37		6.52		.00		.00	.00	.00		1.74		.00		.00	.00	.00
22	6.49		12.40		.00		.00	.00	.00		.00		.00		.00	.00	.00
1988	4.28	2.91	2.58	.00	.00	.00	.00	.00	.00	.00	.00	.00	.00	.00	.00	.00	.00
23	6.51		7.53		.00		.00	.00	.00		.00		.00		.00	.00	.00
24	6.32		.28		.00		.00	.00	.00		.00		.00		.00	.00	.00
25	.00		.00		.00		.00	.00	.00		.00		.00		.00	.00	.00
1989	.00	0	.00	.00	.00	.00	.00	.00	.00	.00	.00	.00	.00	.00	.00	.00	.00
26	.00		.00		.00		.00	.00	.00		.00		.00		.00	.00	.00
27	.00		.00		.00		.00	.00	.00		.00		.00		.00	.00	.00
28	.00		.00		.00		.00	.00	.00		.00		.00		.00	.00	.00
1990	.00	0	.00	.00	.00	.00	.00	.00	.00	.00	.00	.00	.00	.00	.00	.00	.00
29	.00		.00		.00		.00	.00	.00		.00		.00		.00	.00	.00
30	.00		.00		.00		.00	.00	.00		.00		.00		.00	.00	.00
31	.00		.00		.00		.00	.00	.00		.00		.00		.00	.00	.00
1991	.00	0	.00	.00	.00	.00	.00	.00	.00	.00	.00	.00	.00	.00	.00	.00	.00
32	.00		.00		.00		.00	.00	.00		.00		.00		.00	.00	.00
33	.00		.00		.00		.00	.00	.00		.00		.00		.00	.00	.00
34	.00		.00		.00		.00	.00	.00		.00		.00		.00	.00	.00

Table C1. Average streamflow and infiltration for 1966–80, 1981–95, and 1953–68 used for steady-state and transient models of groundwater flow, Idaho National Laboratory and vicinity, Idaho.—Continued

[Values are in cubic feet per second. Map No. Identifier corresponding to a streamflow-gaging station or stream reach on maps located in appendix B figures and as a cross reference with data in other appendixes. Names of stations and reaches are given in table A2 and include the Big Lost River streamflow-gaging stations located near Arco (501), at the inlet to spreading area A (502), at the outlet of spreading area A (503), downstream of the diversion dam used to divert flow into spreading area A (504), at the Lincoln Boulevard bridge near the INTEC (505), and above the Big Lost River Sinks (506). Reaches are located between the model boundary and streamflow-gaging stations 501 and 504 (601), within spreading area A through D (602–605), between streamflow-gaging stations 501 and 504 (600), between streamflow-gaging stations 504 and 505 (606), between streamflow-gaging station 505 and the Big Lost River Sinks (607), within the Big Lost River Sinks (608), and within Big Lost River Playas 1, 2 (609), and 3 (610). Annual transient stress periods are January through April, May through August, and September through December. **Abbreviations** I, infiltration; Q, discharge; INL, Idaho National Laboratory]

Year and stress period	Model boundary to INL diversion				INL spreading areas					INL diversion to Lincoln Blvd				Lincoln Blvd to Playa 3			
	I	Q	I	Q	I	Q	I	I	I	Q	I	Q	I	Q	I	I	I
	600	501	601	502	602	503	603	604	605	504	606	505	607	506	608	609	610
Streamflow and infiltration of 1980–95 for Big Lost River adjusted for evaporation and used in transient model—Continued																	
1992	0.00	0	0.00	0.00	0.00	0.00	0.00	0.00	0.00	0.00	0.00	0.00	0.00	0.00	0.00	0.00	0.00
35	.00		.00		.00		.00	.00	.00		.00		.00		.00	.00	.00
36	.00		.00		.00		.00	.00	.00		.00		.00		.00	.00	.00
37	.00		.00		.00		.00	.00	.00		.00		.00		.00	.00	.00
1993	2.14	14.69	3.72	.22	.20	.00	.00	.00	.00	10.53	3.35	7.05	3.89	3.03	2.71	.00	.00
38	.00		.00		.00		.00	.00	.00		.00		.00		.00	.00	.00
39	6.35		11.05		.60		.00	.00	.00		9.93		11.53		8.04	.00	.00
40	.00		.00		.00		.00	.00	.00		.00		.00		.00	.00	.00
1994	.00	0	.00	.00	.00	.00	.00	.00	.00	.00	.00	.00	.00	.00	.00	.00	.00
41	.00		.00		.00		.00	.00	.00		.00		.00		.00	.00	.00
42	.00		.00		.00		.00	.00	.00		.00		.00		.00	.00	.00
43	.00		.00		.00		.00	.00	.00		.00		.00		.00	.00	.00
1995	4.29	125.87	11.89	50.14	15.15	33.84	20.02	6.62	6.63	63.54	9.10	54.27	5.74	48.39	22.99	21.08	.00
44	.00		.00		.00		.00	.00	.00		.00		.05		.00	.00	.00
45	6.30		13.59		44.97		59.42	19.64	19.68		17.13		16.98		68.23	62.55	.00
46	6.49		21.86		.00		.00	.00	.00		9.96		.00		.00	.00	.00
Streamflow and infiltration of 1953–68 for Big Lost River unadjusted for evaporation and used in transient particle path analysis																	
1953	6.56	75.44	15.95	0.00	0.00	0.00	0.00	0.00	0.00	59.49	15.88	43.61	18.22	25.39	17.26	8.13	0.00
1	6.56		15.95		.00		.00	.00	.00		10.22		6.10		.70	.00	.00
1954	6.56	32.97	15.95	.00	.00	.00	.00	.00	.00	17.02	15.88	6.80	18.55	.70	2.13	.00	.00
2	6.56		15.95		.00		.00	.00	.00		2.62		.00		.00	.00	.00
3	6.56		15.95		.00		.00	.00	.00		12.32		.00		.00	.00	.00
4	6.56		15.95		.00		.00	.00	.00		.00		.00		.00	.00	.00

Table C1. Average streamflow and infiltration for 1966–80, 1981–95, and 1953–68 used for steady-state and transient models of groundwater flow, Idaho National Laboratory and vicinity, Idaho.—Continued

[Values are in cubic feet per second. Map No. Identifier corresponding to a streamflow-gaging station or stream reach on maps located in appendix B figures and as a cross reference with data in other appendixes. Names of stations and reaches are given in table A2 and include the Big Lost River streamflow-gaging stations located near Arco (501), at the inlet to spreading area A (502), at the outlet of spreading area A (503), downstream of the diversion dam used to divert flow into spreading area A (504), at the Lincoln Boulevard bridge near the INTEC (505), and above the Big Lost River Sinks (506). Reaches are located between the model boundary and streamflow-gaging station 501 (600), between streamflow-gaging stations 501 and 504 (601), within spreading area A through D (602–605), between streamflow-gaging stations 504 and 505 (606), between streamflow-gaging station 505 and the Big Lost River Sinks (607), within the Big Lost River Sinks (608), and within Big Lost River Playas 1, 2 (609), and 3 (610). Annual transient stress periods are January through April, May through August, and September through December. **Abbreviations** 1, infiltration; Q, discharge; INL, Idaho National Laboratory]

Year and stress period	Model boundary to INL diversion				INL spreading areas					INL diversion to Lincoln Blvd				Lincoln Blvd to Playa 3			
	I	Q	I	Q	I	Q	I	I	I	Q	I	Q	I	Q	I	I	I
	600	501	601	502	602	503	603	604	605	504	606	505	607	506	608	609	610
Streamflow and infiltration of 1953–68 for Big Lost River unadjusted for evaporation and used in transient particle path analysis																	
1955	6.56	15.27	13.09	0.00	0.00	0.00	0.00	0.00	0.00	2.19	2.19	0.00	0.00	0.00	0.00	0.00	0.00
5	6.56		15.95		.00		.00	.00	.00		1.98		.00		.00	.00	.00
6	6.56		7.45		.00		.00	.00	.00		.00		.00		.00	.00	.00
7	6.56		15.95		.00		.00	.00	.00		4.60		.00		.00	.00	.00
1956	6.56	93.06	15.95	.00	.00	.00	.00	.00	.00	77.11	15.88	61.23	18.55	42.68	42.68	.00	.00
8	6.56		15.95		.00		.00	.00	.00		15.88		18.55		11.10	.00	.00
9	6.56		15.95		.00		.00	.00	.00		15.88		18.55		115.14	.00	.00
10	6.56		15.95		.00		.00	.00	.00		15.88		18.55		.96	.00	.00
1957	6.56	99.81	15.95	.00	.00	.00	.00	.00	.00	83.86	15.88	67.98	14.55	53.43	53.43	.00	.00
11	6.56		15.95		.00		.00	.00	.00		15.88		6.37		.00	.00	.00
12	6.56		15.95		.00		.00	.00	.00		15.88		18.55		127.85	.00	.00
13	6.56		15.95		.00		.00	.00	.00		15.88		18.55		30.96	.00	.00
1958	6.56	196.4	15.95	.00	.00	.00	.00	.00	.00	180.45	15.88	164.57	18.55	146.02	52.51	15.17	78.33
14	6.56		15.95		.00		.00	.00	.00		15.88		18.55		52.51	4.18	.00
15	6.56		15.95		.00		.00	.00	.00		15.88		18.55		52.51	20.56	214.15
16	6.56		15.95		.00		.00	.00	.00		15.88		18.55		52.51	20.56	18.45
1959	6.56	32.73	12.81	.00	.00	.00	.00	.00	.00	19.91	5.22	14.69	6.10	8.59	8.59	.00	.00
17	6.56		15.95		.00		.00	.00	.00		15.88		18.55		26.14	.00	.00
18	6.56		9.71		.00		.00	.00	.00		.00		.00		.00	.00	.00
19	6.56		12.86		.00		.00	.00	.00		.00		.00		.00	.00	.00
1960	6.58	3.64	3.64	.00	.00	.00	.00	.00	.00	.00	.00	.00	.00	.00	.00	.00	.00
20	6.56		10.65		.00		.00	.00	.00		.00		.00		.00	.00	.00
21	6.56		.30		.00		.00	.00	.00		.00		.00		.00	.00	.00
22	6.56		.02		.00		.00	.00	.00		.00		.00		.00	.00	.00

Table C1. Average streamflow and infiltration for 1966–80, 1981–95, and 1953–68 used for steady-state and transient models of groundwater flow, Idaho National Laboratory and vicinity, Idaho.—Continued

[Values are in cubic feet per second. Map No. Identifier corresponding to a streamflow-gaging station or stream reach on maps located in appendix B figures and as a cross reference with data in other appendixes. Names of stations and reaches are given in table A2 and include the Big Lost River streamflow-gaging stations located near Arco (501), at the inlet to spreading area A (502), at the outlet of spreading area A (503), downstream of the diversion dam used to divert flow into spreading area A (504), at the Lincoln Boulevard bridge near the INTEC (505), and above the Big Lost River Sinks (506). Reaches are located between the model boundary and streamflow-gaging station 501 (600), between streamflow-gaging stations 501 and 504 (601), within spreading area A through D (602–605), between streamflow-gaging stations 504 and 505 (606), between streamflow-gaging station 505 and the Big Lost River Sinks (607), within the Big Lost River Sinks (608), and within Big Lost River Playas 1, 2 (609), and 3 (610). Annual transient stress periods are January through April, May through August, and September through December. **Abbreviations** I, infiltration; Q, discharge; INL, Idaho National Laboratory]

Year and stress period	Model boundary to INL diversion				INL spreading areas					INL diversion to Lincoln Blvd				Lincoln Blvd to Playa 3			
	I	Q	I	Q	I	Q	I	I	I	Q	I	Q	I	Q	I	I	I
	600	501	601	502	602	503	603	604	605	504	606	505	607	506	608	609	610
	Streamflow and infiltration of 1953–68 for Big Lost River unadjusted for evaporation and used in transient particle path analysis—Continued																
1961	2.16	0.01	0.01	0.00	0.00	0.00	0.00	0.00	0.00	0.00	0.00	0.00	0.00	0.00	0.00	0.00	0.00
23	6.56		.03		.00		.00	.00	.00		.00		.00		.00	.00	.00
24	.00		.00		.00		.00	.00	.00		.00		.00		.00	.00	.00
25	.00		.00		.00		.00	.00	.00		.00		.00		.00	.00	.00
1962	4.4	20.11	10.71	.00	.00		.00	.00	.00	9.40	6.67	2.73	2.73	.00	.00	.00	.00
26	.00		.00		.00		.00	.00	.00		.00		.00		.00	.00	.00
27	6.56		15.95		.00		.00	.00	.00		4.05		.00		.00	.00	.00
28	6.56		15.95		.00		.00	.00	.00		15.88		8.17		.00	.00	.00
1963	6.56	91.89	15.95	.00	.00	.00	.00	.00	.00	75.94	15.88	60.06	18.55	41.51	41.51	.00	.00
29	6.56		15.95		.00		.00	.00	.00		15.88		18.55		9.62	.00	.00
30	6.56		15.95		.00		.00	.00	.00		15.88		18.55		59.62	.00	.00
31	6.56		15.95		.00		.00	.00	.00		15.88		18.55		54.62	.00	.00
1964	6.58	95.37	15.99	.00	.00	.00	.00	.00	.00	79.38	15.92	63.45	18.60	44.85	44.85	.00	.00
32	6.56		15.95		.00		.00	.00	.00		15.88		18.55		19.62	.00	.00
33	6.56		15.95		.00		.00	.00	.00		15.88		18.55		59.62	.00	.00
34	6.56		15.95		.00		.00	.00	.00		15.88		18.55		54.62	.00	.00
1965	6.56	343.67	15.95	158.94	13.42	145.52	109.14	36.38	.00	168.78	15.88	152.90	18.55	134.35	29.38	6.93	98.03
35	6.56		15.95		.00		.00	.00	.00		15.88		18.55		20.06	.00	.00
36	6.56		15.95		20.00		177.14	59.05	.00		15.88		18.55		52.51	20.56	290.91
37	6.56		15.95		20.00		147.93	49.31	.00		15.88		18.55		15.24	.00	.00
1966	6.54	70.1	12.89	45.14	6.56	38.58	28.94	9.65	.00	12.07	7.02	5.04	5.04	.00	.00	.00	.00
38	6.56		19.34		20.00		88.26	29.42	.00		12.31		15.38		.00	.00	.00
39	6.56		11.38		.00		.00	.00	.00		8.89		.00		.00	.00	.00
40	6.56		8.17		.00		.00	.00	.00		.00		.00		.00	.00	.00

Table C1. Average streamflow and infiltration for 1966–80, 1981–95, and 1953–68 used for steady-state and transient models of groundwater flow, Idaho National Laboratory and vicinity, Idaho.—Continued

[Values are in cubic feet per second. Map No. Identifier corresponding to a streamflow-gaging station or stream reach on maps located in appendix B figures and as a cross reference with data in other appendixes. Names of stations and reaches are given in table A2 and include the Big Lost River streamflow-gaging stations located near Arco (501), at the inlet to spreading area A (502), at the outlet of spreading area A (503), downstream of the diversion dam used to divert flow into spreading area A (504), at the Lincoln Boulevard bridge near the INTEC (505), and above the Big Lost River Sinks (506). Reaches are located between the model boundary and streamflow-gaging stations 501 and 504 (600), between streamflow-gaging stations 501 and 504 (601), within spreading area A through D (602–605), between streamflow-gaging stations 504 and 505 (606), between streamflow-gaging station 505 and the Big Lost River Sinks (607), within the Big Lost River Sinks (608), and within Big Lost River Playas 1, 2 (609), and 3 (610). Annual transient stress periods are January through April, May through August, and September through December. **Abbreviations** I, infiltration; Q, discharge; INL, Idaho National Laboratory]

Year and stress period	Model boundary to INL diversion				INL spreading areas					INL diversion to Lincoln Blvd			Lincoln Blvd to Playa 3				
	I	Q	I	Q	I	Q	I	Q	I	Q	I	Q	I	Q	I	I	I
	600	501	601	502	602	503	603	604	605	504	606	505	607	506	608	609	610
Streamflow and infiltration of 1953–68 for Big Lost River unadjusted for evaporation and used in transient particle path analysis—Continued																	
1967	6.56	232.55	15.81	108.22	13.42	94.79	71.10	23.70	0.00	108.52	21.81	86.71	15.07	71.64	44.68	26.96	0.00
41	6.56		8.39		.00		.00	.00	.00		.00		.00		.00	.00	.00
42	6.56		19.66		20.00		172.85	57.62	.00		45.44		25.51		78.87	80.01	.00
43	6.56		19.24		20.00		38.44	12.81	.00		19.43		19.36		54.16	.00	.00
1968	6.58	88.96	19.00	34.71	13.32	21.39	17.22	4.18	.00	35.25	12.89	22.35	13.49	8.86	8.86	.00	.00
44	6.56		19.33		20.00		14.13	.00	.00		11.71		16.59		7.03	.00	.00
45	6.56		17.48		.00		.00	.00	.00		8.41		6.53		.00	.00	.00
46	6.56		20.06		20.00		37.49	12.50	.00		18.48		17.33		19.53	.00	.00

Appendix D. Industrial Wastewater Disposal to Model Layer 1 Used for Steady-State and Transient Models of Groundwater Flow, Idaho National Laboratory, Idaho

Table D1.   Industrial wastewater disposal to model layer 1 used for steady-state and transient models of groundwater flow, Idaho National Laboratory, Idaho.

[Values are in cubic feet per second. **Map No.:** Identifier corresponding to locations of production wells on maps shown in appendix B figures and as a cross reference with data in other appendixes. Locations of infiltration from disposal of industrial waste to ponds and ditches or landscape irrigation at facilities and facility names are also shown on figure 18. **Transient stress periods:** (a) January through April, (b) May through August, (c) September through December. **Abbreviations:** ARA, Army Reactor Area; CFA, Central Facilities Area; EBR I, Experimental Breeder Reactor I; INTEC, Idaho Nuclear Technology and Engineering Center; MFC, Materials and Fuels Complex; NRF, Naval Reactors Facility; PBF, Power Burst Facility; RTC, Reactor Technology Center; RWMC, Radioactive Waste Management Complex; TAN, Test Area North]

Stress period	Simulated year	Four month period	29 ARA Waste	51 CFA Waste	80 EBR I Waste	66 INTEC Infiltration ponds	66 INTEC Sewage	66 INTEC Waste to gravel pit	82 MFC Pond	82 MFC Sewage	131 NRF Sewage	131 NRF Waste	205 PBF Waste	246 RTC Chemical waste	246 RTC Cold waste disposal	246 RTC Irrigation	246 RTC Sewage	246 RTC Warm waste	191 RWMC Waste	86 LOFT disposal	207 TAN Waste	207 TAN TAN+LOFT disposal	24 WRRTF sewage
1	[1]1980		0.144	0.245	0.001	0.000	0.059	0.000	0.219	0.028	0.076	0.615	0.072	0.049	0.000	0.107	0.035	0.241	0.016	0.624	0.067	0.001	0.003
2	1981	(a)	.166	.173	.001	.000	.048	.000	.193	.018	.072	.182	.018	.036	.000	.013	.028	.210	.018	.544	.057	.000	.002
3		(b)	.163	.555	.001	.000	.057	.000	.266	.024	.067	.607	.039	.039	.000	.113	.026	.233	.037	.425	.079	.001	.003
4		(c)	.132	.240	.001	.000	.057	.000	.181	.023	.062	.389	.043	.045	.000	.029	.019	.237	.055	.478	.064	.002	.004
5	1982	(a)	.113	.224	.001	.000	.065	.000	.192	.029	.056	.884	.075	.048	.653	.019	.021	.287	.050	.346	.064	.001	.002
6		(b)	.135	.481	.001	.000	.059	.000	.210	.024	.070	.444	.101	.034	.961	.181	.021	.189	.028	2.823	.081	.001	.002
7		(c)	.140	.107	.001	.980	.057	1.215	.093	.022	.077	.352	.072	.030	1.120	.039	.023	.160	.036	.327	.063	.001	.002
8	1983	(a)	.122	.023	.001	.000	.052	.000	.121	.025	.073	.186	.085	.031	1.048	.016	.023	.116	.014	.391	.058	.002	.001
9		(b)	.129	.424	.001	.000	.039	.000	.161	.017	.105	.685	.101	.026	1.004	.124	.033	.122	.012	.555	.096	.001	.001
10		(c)	.111	.286	.001	.000	.029	.000	.221	.016	.075	.476	.066	.028	1.002	.047	.023	.134	.005	.425	.084	.001	.001
11	1984	(a)	.081	.206	.001	1.112	.030	.000	.158	.016	.083	.474	.068	.025	1.015	.019	.020	.087	.009	.314	.076	.001	.001
12		(b)	.059	.409	.001	2.513	.031	.000	.195	.018	.140	.548	.110	.022	1.039	.146	.029	.070	.015	.118	.090	.001	.002
13		(c)	.061	.261	.000	2.028	.029	.000	.245	.022	.103	.611	.069	.027	1.105	.066	.027	.039	.009	.728	.078	.001	.001
14	1985	(a)	.059	.228	.001	1.844	.032	.000	.214	.035	.112	.356	.047	.025	.848	.000	.027	.023	.026	.212	.085	.001	.076
15		(b)	.075	.563	.001	1.587	.028	.000	.290	.028	.086	.724	.054	.026	.963	.132	.036	.094	.034	.807	.079	.001	.047
16		(c)	.068	.154	.001	2.457	.023	.000	.250	.037	.104	.077	.037	.025	1.015	.038	.027	.084	.025	.327	.076	.001	.030
17	1986	(a)	.044	.128	.001	2.054	.034	.000	.177	.063	.154	.120	.049	.029	1.534	.025	.035	.118	.007	.087	.091	.001	.020
18		(b)	.083	.466	.001	2.453	.031	.000	.300	.066	.117	.524	.030	.025	1.120	.118	.042	.108	.001	.098	.149	.000	.023
19		(c)	.129	.173	.000	2.424	.030	.000	.225	.061	.111	.412	.026	.028	.816	.062	.030	.092	.001	.064	.131	.001	.023
20	1987	(a)	.112	.113	.000	2.553	.040	.000	.198	.058	.138	.340	.008	.027	.756	.080	.025	.083	.001	.073	.116	.002	.013
21		(b)	.167	.401	.000	2.619	.036	.000	.284	.014	.118	.315	.056	.023	.788	.215	.034	.088	.001	.069	.155	.000	.009
22		(c)	.169	.142	.000	2.451	.037	.000	.285	.011	.141	.046	.051	.020	.723	.071	.027	.069	.001	.059	.120	.000	.010
23	1988	(a)	.077	.090	.000	2.556	.043	.000	.207	.011	.198	.041	.054	.023	1.190	.052	.026	.080	.001	.017	.120	.000	.009
24		(b)	.050	.593	.000	2.468	.036	.000	.234	.012	.231	.256	.079	.016	.829	.425	.036	.074	.001	.000	.168	.000	.008
25		(c)	.009	.171	.000	2.161	.037	.000	.212	.026	.168	.115	.045	.014	.831	.240	.031	.078	.001	.000	.167	.000	.019

Table D1. Industrial wastewater disposal to model layer 1 used for steady-state and transient models of groundwater flow, Idaho National Laboratory, Idaho.—Continued

[Values are in cubic feet per second. **Map No.:** Identifier corresponding to locations of production wells on maps shown in appendix B figures and as a cross reference with data in other appendixes. Locations of infiltration from disposal of industrial waste to ponds and ditches or landscape irrigation at facilities and facility names are also shown on figure 18. **Transient stress periods:** (a) January through April, (b) May through August, (c) September through December. **Abbreviations:** ARA, Army Reactor Area; CFA, Central Facilities Area; EBR I, Experimental Breeder Reactor I; INTEC, Idaho Nuclear Technology and Engineering Center; MFC, Materials and Fuels Complex; NRF, Naval Reactors Facility; PBF, Power Burst Facility; RTC, Reactor Technology Center; RWMC, Radioactive Waste Management Complex; TAN, Test Area North]

Stress period	Simulated year	Four month period	29 ARA Waste	51 CFA Waste	80 EBR I Waste	66 INTEC Infiltration ponds	66 INTEC Sewage	66 INTEC Waste to gravel pit	82 MFC Pond	82 MFC Sewage	131 NRF Sewage	131 NRF Waste	205 PBF Waste	246 RTC Chemical waste	246 RTC Cold waste	246 RTC Irrigation	246 RTC Sewage	246 RTC Warm waste	191 RWMC Waste	86 LOFT disposal	207 TAN Waste	24 WRRTF TAN+LOFT disposal sewage
26	1989	(a)	0.000	0.138	0.000	1.935	0.044	0.000	0.139	0.021	0.238	0.176	0.013	0.032	1.127	0.100	0.028	0.097	0.001	0.118	0.124	0.020
27		(b)	.000	.647	.000	1.676	.038	.000	.103	.022	.117	.488	.041	.032	1.339	.442	.038	.081	.001	.134	.121	.019
28		(c)	.000	.199	.000	1.977	.042	.000	.112	.019	.068	.462	.010	.034	1.269	.183	.038	.108	.001	.290	.183	.026
29	1990	(a)	.000	.145	.000	2.313	.045	.000	.181	.015	.078	.590	.015	.030	1.006	.012	.034	.076	.001	.098	.087	.016
30		(b)	.000	.704	.000	2.990	.050	.000	.497	.014	.065	.416	.028	.036	1.173	.516	.044	.092	.001	.115	.127	.011
31		(c)	.000	.301	.000	2.606	.049	.000	.213	.011	.047	.227	.031	.029	1.053	.121	.031	.082	.001	.109	.109	.007
32	1991	(a)	.000	.210	.000	2.194	.050	.000	.208	.018	.060	.331	.013	.034	.964	.000	.037	.112	.001	1.367	.134	.043
33		(b)	.000	.598	.000	2.376	.042	.000	.161	.011	.062	.762	.045	.042	1.204	.378	.070	.144	.001	.098	.120	.012
34		(c)	.000	.304	.000	2.501	.046	.000	.182	.014	.047	.530	.045	.033	.452	.575	.060	.108	.003	.038	.077	.016
35	1992	(a)	.000	.247	.000	2.850	.069	.000	.124	.013	.055	.556	.009	.033	.390	.389	.055	.108	.002	.010	.233	.018
36		(b)	.000	.761	.000	2.457	.072	.000	.251	.015	.076	.842	.026	.041	.600	.770	.051	.092	.003	.000	.185	.007
37		(c)	.000	.259	.000	2.307	.096	.000	.179	.016	.057	.774	.015	.035	.831	.431	.053	.097	.001	.000	.081	.010
38	1993	(a)	.000	.179	.000	3.029	.059	.000	.156	.033	.046	.711	.011	.028	.904	.306	.055	.098	.002	.000	.093	.018
39		(b)	.000	.622	.000	2.825	.055	.000	.135	.032	.059	.785	.014	.027	1.107	.464	.055	.081	.004	.000	.078	.013
40		(c)	.000	.250	.000	2.593	.053	.000	.095	.074	.047	.679	.014	.026	1.165	.142	.062	.041	.006	.000	.125	.004
41	1994	(a)	.000	.146	.000	2.551	.081	.000	.067	.053	.049	.547	.014	.027	.922	.085	.068	.029	.002	.000	.111	.006
42		(b)	.000	.576	.000	1.701	.196	.000	.130	.018	.063	.583	.026	.024	1.089	.507	.061	.026	.002	.000	.142	.018
43		(c)	.000	.234	.000	1.991	.077	.000	.077	.017	.057	.423	.015	.026	.704	.007	.090	.031	.002	.000	.090	.005
44	1995	(a)	.000	.500	.000	2.240	.096	.000	.047	.016	.072	.445	.014	.026	.680	.000	.099	.033	.003	.000	.072	.003
45		(b)	.000	.596	.000	2.070	.080	.000	.061	.016	.082	.334	.017	.030	.795	.000	.096	.033	.003	.000	.089	.002
46		(c)	.000	.461	.000	1.570	.067	.000	.060	.013	.074	.208	.014	.025	.678	.000	.089	.028	.005	.000	.071	.002

[1] Steady-state stress period.

Appendix E. Irrigation Well Withdrawals Used for Steady-State and Transient Models of Groundwater Flow, Idaho National Laboratory and Vicinity, Idaho

Table E1.   Irrigation well withdrawals used in the calibration of steady-state and transient models of groundwater flow, Idaho National Laboratory and vicinity, Idaho.

[Values are annual total in cubic feet per second **Map No.:** Identifier used to locate wells on maps located in appendix B figures and as a cross reference with data in other appendixes **Local name:** Local well identifier used in this study **Row, Column,** and **Layer:** Model grid coordinates Negative values are withdrawals]

Map No.	Local name	Row	Column	Layer 1	Layer 2	Layer 3	Layer 4	Layer 5
2	434556	47	178		-0.596	-0.726	-0.356	
3	434647	51	192	-0.217	-.219	-.219	-.099	
4	434650	49	188			-.136	-.619	
121	ML11	51	194		-.268	-.327	-.160	
122	ML12	46	175			-1.199		
123	ML13	46	188				-.755	
	ML21-11	68	266	-.215	-.215	-.022	-.044	
	ML21-9	66	260	-.146	-.146	-.146	-.293	-0.477
	ML23-11	76	264	-.046	-.046	-.002	-.003	
	ML24-10	78	263	-.102	-.102	-.047	-.005	
	ML24-11	80	267	-.003	-.003	-.003	-.005	
	ML25-11a	82	266	-.078	-.078	-.051	-.103	
	ML25-11b	83	265	-.078	-.078	-.051	-.103	
	ML26-10a	86	261	-.143	-.036			
	ML26-10b	86	261	-.143	-.036			
	ML26-10c	85	264	-.143	-.036			
	ML26-11a	87	265	-.180	-.176			
	ML26-11b	88	267	-.180	-.176			
	ML27-10a	90	264	-.132	-.128			
	ML27-10b	91	264	-.430	-.128			
	ML27-11	92	268	-.543	-.529			
	ML27-12a	89	269	-.164	-.137			
	ML27-12b	89	269	-.164	-.137			
	ML27-12c	89	271	-.164	-.137			
	ML27-13	89	272	-.411	-.101			
	ML28-12a	94	271	-.336				
	ML28-12b	95	271	-.336				
	ML28-13	96	275	-.101				
	ML28-15	96	280	-.335	-.082			
	ML28-17	96	292	-.512				
	ML29-14a	98	277	-.263	-.065			
	ML29-14b	98	278	-.263	-.065			
	ML29-15a	100	281	-.132	-.106			
	ML29-15b	98	283	-.430	-.106			
	ML29-15c	97	283	-.430	-.106			
	ML29-16	99	287	-.173	-.043			
	ML29-17a	97	289	-.258	-.038			
	ML29-17b	99	290	-.258	-.038			
	ML29-18a	97	295	-.502	-.019			
	ML29-18b	99	296	-.502	-.019			

Table E1.   Irrigation well withdrawals used in the calibration of steady-state and transient models of groundwater flow, Idaho National Laboratory and vicinity, Idaho.—Continued

[Values are annual total in cubic feet per second **Map No.:** Identifier used to locate wells on maps located in appendix B figures and as a cross reference with data in other appendixes **Local name:** Local well identifier used in this study **Row, Column,** and **Layer:** Model grid coordinates Negative values are withdrawals]

Map No.	Local name	Row	Column	Layer 1	Layer 2	Layer 3	Layer 4	Layer 5
	ML29-19	101	299	-0.553				
	ML30-15a	102	280	-.699	-0.063			
	ML30-15b	103	283	-.699	-.063			
	ML30-16a	101	285	-.669	-.265	-0.033	-0.065	
	ML30-16b	102	286	-.669	-.265	-.033	-.065	
	ML30-16c	103	287	-.669	-.265	-.033	-.065	
	ML30-17a	102	289	-.257	-.084			
	ML30-17b	101	291	-.257	-.084			
	ML30-17c	104	292	-.257	-.084			
	ML30-18a	102	293	-.283	-.042			
	ML30-18b	104	296	-.283	-.042			
	ML30-19a	102	296	-.511				
	ML30-19b	102	297	-.511				
	ML31-16	105	284	-1.201	-.264			
	ML31-17a	106	290	-.374	-.055			
	ML31-17b	106	292	-.374	-.055			
	ML32-17	111	288	-.690				
	ML32-18	110	295	-1.605	-.161			
	MontM	41	265	-.482	-1.124	-.126	-.252	
	MontN	38	265	-.340	-.303			
	MontS	42	263	-.009	-.009	-.123	-.247	
	RenoM	12	273	-.880				
	RenoN	12	279	-.359	-.127			
	RenoPoint	12	275	-.485				
	RenoS	12	267	-.993				

Appendix F. Industrial Well Withdrawals or Injections Used for Steady-State and Transient Models of Groundwater Flow, Idaho National Laboratory, Idaho

Table F1.   Industrial well withdrawals or injections used for steady-state and transient models of groundwater flow, Idaho National Laboratory, Idaho.

[Values are in cubic feet per second. **Map No.** Identifier used to locate wells on maps located in appendix B figures and as a cross reference with data in other appendixes. **Local name:** Local well identifier used in this study. **Transient stress periods:** (a) January through April, (b) May through August, (c) September through December. Negative entries are withdrawals and positive entries are injection or infiltration]

Layer 1

Stress period	Simulated year	Four month period	24 ANP 8	28 ARA 2	29 ARA 3	34 BADGING FACILITY	50 CFA 1	51 CFA 2	66 CPP 1	67 CPP 2	68 CPP 3	69 CPP 4	80 EBR 1	81 EBR 2-1	82 EBR II-2	86 FET1	87 FET2	88 FET DISP3	89 FIRE STA2
1	1980[1]		-0.140	-0.042		0.000	-0.021	-0.169	-0.636	-0.380	1.389	-0.024	-0.003	-0.300	-0.270	-0.324	-0.155	0.091	-0.017
2	1981	(a)	-.123	-.046		.000	.000	-.095	-.840	-.502	1.874	-.019	-.002	-.212	-.191	-.290	-.084	.000	-.028
3		(b)	-.095	-.055		.000	-.058	-.229	-.769	-.459	1.705	-.023	-.003	-.387	-.348	-.129	-.191	.000	-.006
4		(c)	-.168	-.049		.000	-.009	-.131	-.897	-.536	2.412	-.023	-.002	-.317	-.285	-.155	-.155	.000	-.009
5	1982	(a)	-.098	-.040		.000	-.005	-.093	-.930	-.555	2.517	-.026	-.001	-.218	-.196	-.201	-.092	.000	-.010
6		(b)	-.063	-.047		.000	-.171	-.142	-.796	-.475	2.150	-.024	-.002	-.348	-.313	-.237	-1.170	.000	-.005
7		(c)	-.061	-.051		.000	-.188	-.024	-.811	-.484	.000	-.023	-.001	-.188	-.169	-.191	-.139	.000	-.005
8	1983	(a)	-.062	-.046		.000	-.004	-.103	-.729	-.435	1.973	-.021	.000	-.219	-.198	-.076	-.169	.000	-.010
9		(b)	-.062	-.046		.000	-.061	-.223	-.950	-.567	2.598	-.016	.000	-.293	-.264	-.233	-.132	.000	-.003
10		(c)	-.105	-.048		.000	-.059	-.104	-.853	-.509	2.339	-.012	-.001	-.283	-.254	-.050	-.186	.000	-.003
11	1984	(a)	-.150	-.029		.000	-.090	-.055	-.889	-.531	1.363	-.012	.000	-.202	-.181	-.037	-.119	.000	-.003
12		(b)	-.193	-.010		.000	-.206	-.099	-.905	-.541	.003	-.013	-.002	-.339	-.305	-.101	-.009	.000	-.002
13		(c)	-.080	-.016		.000	-.136	-.063	-.730	-.436	.000	-.012	-.001	-.320	-.288	-.500	-.072	.000	-.033
14	1985	(a)	-.076	-.014		.000	-.096	-.058	-.664	-.397	.000	-.013	-.005	-.284	-.255	-.199	-.061	.000	-.034
15		(b)	-.047	-.008		.000	-.278	-.125	-.715	-.427	.000	-.011	-.003	-.400	-.360	-.470	-.128	.000	-.001
16		(c)	-.030	-.014		.000	-.091	-.069	-.888	-.530	.000	-.009	-.001	-.334	-.300	-.207	-.106	.000	-.003
17	1986	(a)	-.020	-.008		.000	-.079	-.054	-.877	-.524	.000	-.014	.000	-.256	-.231	-.131	-.070	.000	-.003
18		(b)	-.023	-.008		.000	-.195	-.130	-.883	-.527	.000	-.012	.000	-.425	-.383	-.084	-.043	.000	-.002
19		(c)	-.023	-.017		.000	-.127	-.063	-.873	-.522	.000	-.012	.000	-.348	-.314	-.083	-.049	.000	-.004
20	1987	(a)	-.013	-.011		.000	-.081	-.047	-.919	-.549	.000	-.016	.000	-.316	-.284	-.066	-.054	.000	-.003
21		(b)	-.009	-.014		.000	-.234	-.079	-.943	-.563	.000	-.015	.000	-.418	-.377	-.084	-.046	.000	-.001
22		(c)	-.010	-.017		.000	-.132	-.058	-.882	-.527	.000	-.015	.000	-.405	-.365	-.091	-.055	.000	-.019
23	1988	(a)	-.009	-.014		.000	-.001	-.111	-.920	-.550	.000	-.017	.000	-.296	-.266	-.070	-.041	.000	-.017
24		(b)	-.008	-.029		.000	-.060	-.324	-.888	-.531	.000	-.015	.000	-.454	-.409	.000	.000	.000	-.008
25		(c)	-.019	-.008		.000	-.018	-.139	-.778	-.465	.000	-.015	.000	-.405	-.364	.000	.000	.000	-.022

Table F1. Industrial well withdrawals or injections used for steady-state and transient models of groundwater flow, Idaho National Laboratory, Idaho.—Continued

[Values are in cubic feet per second. **Map No.** Identifier used to locate wells on maps located in appendix B figures and as a cross reference with data in other appendixes. **Local name:** Local well identifier used in this study. **Transient stress periods:** (a) January through April, (b) May through August, (c) September through December. Negative entries are withdrawals and positive entries are injection or infiltration]

Stress period	Simulated year	Four month period	24 ANP 8	28 ARA 2	29 ARA 3	34 BADGING FACILITY	50 CFA 1	51 CFA 2	66 CPP 1	67 CPP 2	68 CPP 3	69 CPP 4	80 EBR 1	81 EBR 2-1	82 EBR II-2	86 FET1	87 FET2	88 FET DISP3	89 FIRE STA2
									Layer 1—Continued									
26	1989	(a)	-0.020	0.000		0.000	0.000	-0.109	-0.696	-0.416	0.000	-0.018	0.000	-0.282	-0.254	-0.046	-0.025	0.000	-0.027
27		(b)	-.019	.000		.000	-.090	-.309	-.603	-.360	.000	-.015	.000	-.150	-.135	-.052	-.029	.000	-.065
28		(c)	-.026	.000		.000	-.003	-.180	-.712	-.425	.000	-.017	.000	-.183	-.164	-.087	-.075	.000	-.020
29	1990	(a)	-.016	.000		.000	-.001	-.128	-.832	-.497	.000	-.018	.000	-.254	-.229	-.039	-.020	.000	-.001
30		(b)	-.011	.000		.000	-.022	-.321	-1.076	-.643	.000	-.020	.000	-.411	-.370	-.046	-.024	.000	-.063
31		(c)	-.015	.000		.000	-.011	-.216	-.938	-.560	.000	-.020	.000	-.310	-.279	-.050	-.020	.000	-.024
32	1991	(a)	-.012	.000		-.001	-.024	-.161	-1.086	-.649	.000	-.020	.000	-.263	-.236	-.039	-.554	.000	-.002
33		(b)	-.009	.000		-.002	-.102	-.253	-1.239	-.740	.000	-.017	-.001	-.189	-.170	-.031	-.020	.000	-.001
34		(c)	-.013	.000		-.001	-.137	-.129	-1.279	-.764	.000	-.019	.000	-.246	-.222	-.026	-.014	.000	-.002
35	1992	(a)	-.022	.000		-.001	-.001	-.111	-1.385	-.827	.000	-.028	.000	-.130	-.117	-.004	-.002	.000	-.001
36		(b)	-.009	.000		-.010	-.124	-.334	-1.166	-.696	.000	-.029	-.001	-.276	-.249	.000	.000	.000	-.003
37		(c)	-.012	.000		-.001	-.004	-.156	-1.073	-.641	.000	-.038	-.001	-.274	-.247	.000	.000	.000	-.002
38	1993	(a)	-.011	.000		-.002	-.001	-.137	-1.237	-.739	.000	-.023	.000	-.282	-.254	.000	.000	.000	-.002
39		(b)	-.014	.000		-.036	-.024	-.295	-1.183	-.706	.000	-.022	.000	-.277	-.249	.000	.000	.000	-.003
40		(c)	-.005	.000		-.354	-.008	-.144	-.793	-.474	.000	-.021	.000	-.239	-.215	.000	.000	.000	-.002
41	1994	(a)	-.006	.000		-.001	-.005	-.132	-.950	-.567	.000	-.032	-.001	-.170	-.153	.000	.000	.000	-.002
42		(b)	-.017	.000		-.003	-.024	-.294	-.839	-.501	.000	-.079	.000	-.303	-.272	.000	.000	.000	-.001
43		(c)	-.005	.000		-.003	-.330	-.153	-.593	-.354	.000	-.031	-.001	-.153	-.138	.000	.000	.000	-.003
44	1995	(a)	-.005	.000		-.001	-.064	-.168	-.627	-.374	.000	-.038	.000	-.059	-.053	-.007	-.004	.000	-.001
45		(b)	-.006	.000		-.006	-.006	-.235	-.638	-.381	.000	-.032	.000	-.065	-.059	-.010	-.008	.000	-.030
46		(c)	-.004	.000		-.002	-.162	-.092	-.705	-.421	.000	-.027	.000	-.068	-.061	-.011	-.008	.000	-.023

Table F1.   Industrial well withdrawals or injections used for steady-state and transient models of groundwater flow, Idaho National Laboratory, Idaho.—Continued

[Values are in cubic feet per second. **Map No.** Identifier used to locate wells on maps located in appendix B figures and as a cross reference with data in other appendixes. **Local name:** Local well identifier used in this study. **Transient stress periods:** (a) January through April, (b) May through August, (c) September through December. Negative entries are withdrawals and positive entries are injection or infiltration]

Stress period	Simulated year	Four month period	24 ANP 8	28 ARA 2	29 ARA 3	34 BADGING FACILITY	50 CFA 1	51 CFA 2	66 CPP 1	67 CPP 2	68 CPP 3	69 CPP 4	80 EBR 1	81 EBR 2-1	82 EBR II-2	86 FET1	87 FET2	88 FET DISP3	89 FIRE STA2
									Layer 2										
1	1980[1]			-0.030		0.000	-0.013	-0.253	-0.247	-0.503	0.305	-0.024	-0.003		-0.030	-0.086	-0.155		
2	1981	(a)		-.032		.000	.000	-.142	-.327	-.665	.411	-.019	-.002		-.021	-.077	-.084		
3		(b)		-.038		.000	-.035	-.344	-.299	-.609	.374	-.023	-.003		-.039	-.034	-.191		
4		(c)		-.034		.000	-.005	-.197	-.349	-.710	.000	-.023	-.002		-.032	-.041	-.155		
5	1982	(a)		-.028		.000	-.003	-.140	-.362	-.736	.000	-.026	-.001		-.022	-.053	-.092		
6		(b)		-.033		.000	-.105	-.213	-.309	-.630	.000	-.024	-.002		-.035	-.063	-1.170		
7		(c)		-.035		.000	-.115	-.036	-.315	-.642	.000	-.023	-.001		-.019	-.051	-.139		
8	1983	(a)		-.032		.000	-.003	-.156	-.284	-.577	.000	-.021	.000		-.022	-.020	.169		
9		(b)		-.032		.000	-.038	-.335	-.369	-.752	.000	-.016	.000		-.029	-.062	-.132		
10		(c)		-.033		.000	-.036	-.157	-.332	-.675	.000	-.012	-.001		-.028	-.013	-.186		
11	1984	(a)		-.020		.000	-.055	-.082	-.346	-.704	.000	-.012	.000		-.020	-.010	-.119		
12		(b)		-.007		.000	-.126	-.149	-.352	-.717	.000	-.013	-.002		-.034	-.027	-.009		
13		(c)		-.011		.000	-.083	-.095	-.284	-.578	.000	-.012	-.001		-.032	-.133	-.072		
14	1985	(a)		-.010		.000	-.059	-.087	-.258	-.526	.000	-.013	-.005		-.028	-.053	-.061		
15		(b)		-.006		.000	-.170	-.187	-.278	-.566	.000	-.011	-.003		-.040	-.125	-.128		
16		(c)		-.010		.000	-.056	-.104	-.345	-.703	.000	-.009	-.001		-.033	-.055	-.106		
17	1986	(a)		-.006		.000	-.048	-.080	-.341	-.694	.000	-.014	.000		-.026	-.035	-.070		
18		(b)		-.005		.000	-.120	-.195	-.343	-.699	.000	-.012	.000		-.043	-.022	-.043		
19		(c)		-.012		.000	-.078	-.094	-.340	-.691	.000	-.012	.000		-.035	-.022	-.049		
20	1987	(a)		-.008		.000	-.050	-.070	-.357	-.728	.000	-.016	.000		-.032	-.017	-.054		
21		(b)		-.010		.000	-.143	-.119	-.367	-.746	.000	-.015	.000		-.042	-.022	-.046		
22		(c)		-.012		.000	-.081	-.088	-.343	-.698	.000	-.015	.000		-.041	-.024	-.055		
23	1988	(a)		-.010		.000	.000	-.167	-.358	-.728	.000	-.017	.000		-.030	-.019	-.041		
24		(b)		-.020		.000	-.037	-.486	-.346	-.703	.000	-.015	.000		-.045	.000	.000		
25		(c)		-.004		.000	-.011	-.209	-.303	-.616	.000	-.015	.000		-.040	.000	.000		

Table F1. Industrial well withdrawals or injections used for steady-state and transient models of groundwater flow, Idaho National Laboratory, Idaho.—Continued

[Values are in cubic feet per second. **Map No.** Identifier used to locate wells on maps located in appendix B figures and as a cross reference with data in other appendixes. **Local name:** Local well identifier used in this study. **Transient stress periods:** (a) January through April, (b) May through August, (c) September through December. Negative entries are withdrawals and positive entries are injection or infiltration]

Stress period	Simulated year	Four month period	24 ANP 8	28 ARA 2	29 ARA 3	34 BADGING FACILITY	50 CFA 1	51 CFA 2	66 CPP 1	67 CPP 2	68 CPP 3	69 CPP 4	80 EBR 1	81 EBR 2-1	82 EBR II-2	86 FET1	87 FET2	88 FET DISP3	89 FIRE STA2
									Layer 2—Continued										
26	1989	(a)		0.000		0.000	0.000	-0.164	-0.271	-0.551	0.000	-0.018	0.000		-0.028	-0.012	-0.025		
27		(b)		.000		.000	-.055	-.464	-.235	-.478	.000	-.015	.000		-.015	-.014	-.029		
28		(c)		.000		.000	-.002	-.270	-.277	-.563	.000	-.017	.000		-.018	-.023	-.075		
29	1990	(a)		.000		.000	-.001	-.192	-.324	-.659	.000	-.018	.000		-.025	-.010	-.020		
30		(b)		.000		.000	-.014	-.482	-.418	-.852	.000	-.020	.000		-.041	-.012	-.024		
31		(c)		.000		.000	-.006	-.324	-.365	-.743	.000	-.020	.000		-.031	-.013	-.020		
32	1991	(a)		.000		-.001	-.003	-.241	-.422	-.860	.000	-.020	.000		-.026	-.010	-.554		
33		(b)		.000		-.001	-.062	-.379	-.482	-.981	.000	-.017	-.001		-.019	-.008	-.020		
34		(c)		.000		-.001	-.084	-.194	-.497	-1.012	.000	-.019	.000		-.025	-.007	-.014		
35	1992	(a)		.000		.000	-.001	-.166	-.539	-1.096	.000	-.028	.000		-.013	-.001	-.002		
36		(b)		.000		-.006	-.076	-.501	-.543	-.923	.000	-.029	-.001		-.028	.000	.000		
37		(c)		.000		-.001	-.003	-.243	-.417	-.849	.000	-.038	-.001		-.027	.000	.000		
38	1993	(a)		.000		-.001	.000	-.206	-.481	-.979	.000	-.023	.000		-.028	.000	.000		
39		(b)		.000		-.021	-.015	-.443	-.460	-.936	.000	-.022	.000		-.028	.000	.000		
40		(c)		.000		-.208	-.005	-.216	-.308	-.628	.000	-.021	.000		-.024	.000	.000		
41	1994	(a)		.000		-.001	-.003	-.196	-.369	-.752	.000	-.032	-.001		-.017	.000	.000		
42		(b)		.000		-.002	-.015	-.442	-.326	-.664	.000	-.079	.000		-.030	.000	.000		
43		(c)		.000		-.002	-.202	-.229	-.231	-.470	.000	-.031	-.001		-.015	.000	.000		
44	1995	(a)		.000		-.001	-.039	-.252	-.244	-.496	.000	-.038	.000		-.006	-.002	-.004		
45		(b)		.000		-.004	-.004	-.353	-.248	-.505	.000	-.032	.000		-.007	-.003	-.008		
46		(c)		.000		-.001	-.100	-.138	-.274	-.558	.000	-.027	.000		-.007	-.003	-.008		

Table F1. Industrial well withdrawals or injections used for steady-state and transient models of groundwater flow, Idaho National Laboratory, Idaho.—Continued

[Values are in cubic feet per second. **Map No.** Identifier used to locate wells on maps located in appendix B figures and as a cross reference with data in other appendixes. **Local name:** Local well identifier used in this study. **Transient stress periods:** (a) January through April, (b) May through August, (c) September through December. Negative entries are withdrawals and positive entries are injection or infiltration]

Stress period	Simulated year	Four month period	24 ANP 8	28 ARA 2	29 ARA 3	34 BADGING FACILITY	50 CFA 1	51 CFA 2	66 CPP 1	67 CPP 2	68 CPP 3	69 CPP 4	80 EBR 1	81 EBR 2-1	82 EBR II-2	86 FET1	87 FET2	88 FET DISP3	89 FIRE STA2
											Layer 3							
1	1980[1]											-0.012		-0.003			-0.059		
2	1981	(a)										-.010		-.002			-.032		
3		(b)										-.011		-.003			-.073		
4		(c)										-.011		-.002			-.059		
5	1982	(a)										-.013		-.001			-.035		
6		(b)										-.012		-.002			-.446		
7		(c)										-.011		-.001			-.053		
8	1983	(a)										-.010		.000			-.064		
9		(b)										-.008		.000			-.050		
10		(c)										-.006		-.001			-.071		
11	1984	(a)										-.006		.000			-.046		
12		(b)										-.006		-.002			-.003		
13		(c)										-.006		-.001			-.027		
14	1985	(a)										-.006		-.005			-.023		
15		(b)										-.006		-.005			-.049		
16		(c)										-.005		-.001			-.040		
17	1986	(a)										-.007		.000			-.027		
18		(b)										-.006		.000			-.017		
19		(c)										-.006		.000			-.019		
20	1987	(a)										-.008		.000			-.021		
21		(b)										-.007		.000			-.018		
22		(c)										-.007		.000			-.021		
23	1988	(a)										-.009		.000			-.016		
24		(b)										-.007		.000			.000		
25		(c)										-.007		.000			.000		

Map No. and local name corresponding to well

Table F1. Industrial well withdrawals or injections used for steady-state and transient models of groundwater flow, Idaho National Laboratory, Idaho.—Continued

[Values are in cubic feet per second. **Map No.** Identifier used to locate wells on maps located in appendix B figures and as a cross reference with data in other appendixes. **Local name:** Local well identifier used in this study. **Transient stress periods:** (a) January through April, (b) May through August, (c) September through December. Negative entries are withdrawals and positive entries are injection or infiltration]

Stress period	Simulated year	Four month period	Map No. and local name corresponding to well															
			24	28	29	34	50	51	66	67	68	69	80	81	82	86	87	88	89
			ANP 8	ARA 2	ARA 3	BADGING FACILITY	CFA 1	CFA 2	CPP 1	CPP 2	CPP 3	CPP 4	EBR 1	EBR 2-1	EBR II-2	FET1	FET2	FET DISP3	FIRE STA2
									Layer 3—Continued										
26	1989	(a)										-0.009		0.000			-0.010		
27		(b)										-.008		.000			-.011		
28		(c)										-.008		.000			-.029		
29	1990	(a)										-.009		.000			-.008		
30		(b)										-.010		.000			-.009		
31		(c)										-.010		.000			-.007		
32	1991	(a)										-.010		.000			-.211		
33		(b)										-.008		-.001			-.008		
34		(c)										.000		.000			-.005		
35	1992	(a)										-.014		.000			-.001		
36		(b)										-.014		-.001			.000		
37		(c)										-.019		-.001			.000		
38	1993	(a)										-.012		.000			.000		
39		(b)										-.011		.000			.000		
40		(c)										-.011		.000			.000		
41	1994	(a)										-.016		-.001			.000		
42		(b)										-.039		.000			.000		
43		(c)										-.015		-.001			.000		
44	1995	(a)										-.019		.000			-.001		
45		(b)										-.016		.000			-.003		
46		(c)										-.013		.000			-.003		

Table F1.   Industrial well withdrawals or injections used for steady-state and transient models of groundwater flow, Idaho National Laboratory, Idaho.—Continued

[Values are in cubic feet per second. **Map No.** Identifier used to locate wells on maps located in appendix B figures and as a cross reference with data in other appendixes. **Local name:** Local well identifier used in this study. **Transient stress periods:** (a) January through April, (b) May through August, (c) September through December. Negative entries are withdrawals and positive entries are injection or infiltration]

Stress period	Simulated year	Four month period	24 ANP 8	28 ARA 2	29 ARA 3	34 BADGING FACILITY	50 CFA 1	51 CFA 2	66 CPP 1	67 CPP 2	68 CPP 3	69 CPP 4	80 EBR 1	81 EBR 2-1	82 EBR II-2	86 FET1	87 FET2	88 FET DISP3	89 FIRE STA2
									Layer 4										
1	1980[1]				-0.024								-0.005						
2	1981	(a)			-.030								-.004						
3		(b)			-.024								-.006						
4		(c)			-.017								-.003						
5	1982	(a)			-.015								-.002						
6		(b)			-.019								-.004						
7		(c)			-.018								-.001						
8	1983	(a)			-.015								.000						
9		(b)			-.017								.000						
10		(c)			-.015								.000						
11	1984	(a)			-.011								.000						
12		(b)			-.014								.000						
13		(c)			-.012								.000						
14	1985	(a)			-.012								.000						
15		(b)			-.021								.000						
16		(c)			-.015								.000						
17	1986	(a)			-.010								.000						
18		(b)			-.024								.000						
19		(c)			-.034								.000						
20	1987	(a)			-.032								.000						
21		(b)			-.048								.000						
22		(c)			-.048								.000						
23	1988	(a)			-.018								.000						
24		(b)			.000								.000						
25		(c)			.000								.000						

Table F1. Industrial well withdrawals or injections used for steady-state and transient models of groundwater flow, Idaho National Laboratory, Idaho.—Continued

[Values are in cubic feet per second. **Map No.** Identifier used to locate wells on maps located in appendix B figures and as a cross reference with data in other appendixes. **Local name:** Local well identifier used in this study. **Transient stress periods:** (a) January through April, (b) May through August, (c) September through December. Negative entries are withdrawals and positive entries are injection or infiltration]

Stress period	Simulated year	Four month period	\	\	\	\	\	\	\	\	\	\	\	\	\	\	\	\	\
			24	28	29	34	50	51	66	67	68	69	80	81	82	86	87	88	89
			ANP 8	ARA 2	ARA 3	BADGING FACILITY	CFA 1	CFA 2	CPP 1	CPP 2	CPP 3	CPP 4	EBR 1	EBR 2-1	EBR II-2	FET1	FET2	FET DISP3	FIRE STA2
Layer 4—Continued																			
26	1989	(a)			0.000								0.000						
27		(b)			.000								.000						
28		(c)			.000								.000						
29	1990	(a)			.000								.000						
30		(b)			.000								.000						
31		(c)			.000								-.001						
32	1991	(a)			.000								.000						
33		(b)			.000								-.001						
34		(c)			.000								-.001						
35	1992	(a)			.000								-.001						
36		(b)			.000								-.001						
37		(c)			.000								-.001						
38	1993	(a)			.000								.000						
39		(b)			.000								.000						
40		(c)			.000								.000						
41	1994	(a)			.000								-.001						
42		(b)			.000								-.001						
43		(c)			.000								-.001						
44	1995	(a)			.000								-.001						
45		(b)			.000								.000						
46		(c)			.000								.000						

Table F1. Industrial well withdrawals or injections used for steady-state and transient models of groundwater flow, Idaho National Laboratory, Idaho.—Continued

[Values are in cubic feet per second. **Map No.** Identifier used to locate wells on maps located in appendix B figures and as a cross reference with data in other appendixes. **Local name:** Local well identifier used in this study. **Transient stress periods:** (a) January through April, (b) May through August, (c) September through December. Negative entries are withdrawals and positive entries are injection or infiltration]

Stress period	Simulated year	Four month period	24 ANP 8	28 ARA 2	29 ARA 3	34 BADGING FACILITY	50 CFA 1	51 CFA 2	66 CPP 1	67 CPP 2	68 CPP 3	69 CPP 4	80 EBR 1	81 EBR 2-1	82 EBR II-2	86 FET1	87 FET2	88 FET DISP3	89 FIRE STA2
									Layer 5										
1	1980[1]				-0.048														
2	1981	(a)			-.058														
3		(b)			-.046														
4		(c)			-.033														
5	1982	(a)			-.029														
6		(b)			-.036														
7		(c)			-.036														
8	1983	(a)			-.029														
9		(b)			-.033														
10		(c)			-.028														
11	1984	(a)			-.021														
12		(b)			-.027														
13		(c)			-.023														
14	1985	(a)			-.024														
15		(b)			-.040														
16		(c)			-.029														
17	1986	(a)			-.020														
18		(b)			-.046														
19		(c)			-.066														
20	1987	(a)			-.061														
21		(b)			-.094														
22		(c)			-.093														
23	1988	(a)			-.035														
24		(b)			.000														
25		(c)			.000														

Table F1. Industrial well withdrawals or injections used for steady-state and transient models of groundwater flow, Idaho National Laboratory, Idaho.—Continued

[Values are in cubic feet per second. **Map No.** Identifier used to locate wells on maps located in appendix B figures and as a cross reference with data in other appendixes. **Local name:** Local well identifier used in this study. **Transient stress periods:** (a) January through April, (b) May through August, (c) September through December. Negative entries are withdrawals and positive entries are injection or infiltration]

Stress period	Simulated year	Four month period	24 ANP 8	28 ARA 2	29 ARA 3	34 BADGING FACILITY	50 CFA 1	51 CFA 2	66 CPP 1	67 CPP 2	68 CPP 3	69 CPP 4	80 EBR 1	81 EBR 2-1	82 EBR II-2	86 FET1	87 FET2	88 FET DISP3	89 FIRE STA2
								Layer 5—Continued										
26	1989	(a)			0.000														
27		(b)			.000														
28		(c)			.000														
29	1990	(a)			.000														
30		(b)			.000														
31		(c)			.000														
32	1991	(a)			.000														
33		(b)			.000														
34		(c)			.000														
35	1992	(a)			.000														
36		(b)			.000														
37		(c)			.000														
38	1993	(a)			.000														
39		(b)			.000														
40		(c)			.000														
41	1994	(a)			.000														
42		(b)			.000														
43		(c)			.000														
44	1995	(a)			.000														
45		(b)			.000														
46		(c)			.000														

Table F1. Industrial well withdrawals or injections used for steady-state and transient models of groundwater flow, Idaho National Laboratory, Idaho.—Continued

[Values are in cubic feet per second. **Map No.** Identifier used to locate wells on maps located in appendix B figures and as a cross reference with data in other appendixes. **Local name:** Local well identifier used in this study. **Transient stress periods:** (a) January through April, (b) May through August, (c) September through December. Negative entries are withdrawals and positive entries are injection or infiltration]

Stress period	Simulated year	Four month period	118 LPTF DISP	131 NRF 1	132 NRF 2	133 NRF 3	134 NRF 4	191 RWMC	205 SPERT 1	206 SPERT 2	207 TAN 1	208 TAN 2	238 TRA 1	239 TRA 3	240 TRA 4	245 TRA DISP
							Layer 1									
1	1980[1]		0.136	-0.598	0.000		0.000	-0.001	-0.058		-0.046	-0.014	-0.174	-0.016		0.874
2	1981	(a)	.121	-2.198	.000		.000	-.006	-.016		-.020	-.034	-.148	.000		.733
3		(b)	.092	-.205	-.043		.000	-.016	-.033		-.024	-.053	-.184	-.029		.690
4		(c)	.165	-1.112	.000		.000	-.032	-.034		-.252	-3.440	-.125	-.002		.511
5	1982	(a)	.087	-1.464	-.004		.000	-.031	-.056		-.042	-.006	-.082	-.009		.299
6		(b)	.061	-.780	-.001		.000	-.012	-.019		-.027	-.038	-.259	-.008		.000
7		(c)	.059	-.866	-.060		.000	-.023	-.031		-.020	-.031	-.221	-.028		.000
8	1983	(a)	.061	-.300	-.002		.000	-.009	-.053		-.007	-.043	-.094	-.018		.000
9		(b)	.061	-.377	-.001		.000	-.009	-.044		-.011	-.073	-.148	.000		.000
10		(c)	.104	-.453	.000		.000	-.002	-.024		-.011	-.060	-.039	-.111		.000
11	1984	(a)	.149	-.321	-.001		.000	-.005	-.035		-.009	-.055	-1.506	-.040		.000
12		(b)	.191	-.356	-.008		.000	-.004	-.042		-.007	-.072	-.245	-.130		.000
13		(c)	.079	-1.210	-.083		.000	-.002	-.027		-.017	-.048	-1.053	-.338		.000
14	1985	(a)	.000	-1.472	-.005		.000	-.002	-.021		-.015	-.057	-.056	-.647		.000
15		(b)	.000	-2.548	-.009		.000	-.013	-.024		-.024	-.048	-1.276	-.400		.000
16		(c)	.000	-1.349	-.105		.000	-.015	-.030		-.011	-.054	-.005	-.001		.000
17	1986	(a)	.000	-1.553	-.319		.000	-.014	-.036		-.022	-.053	-.040	-.199		.000
18		(b)	.000	-1.212	-.687		.000	-.016	-.018		-.056	-.063	-.089	-.001		.000
19		(c)	.000	-1.443	-.025		.000	-.012	-.009		-.056	-.048	-.269	.000		.000
20	1987	(a)	.000	-1.968	-.493		.000	-.008	-.002		-.043	-.051	-.254	-.003		.000
21		(b)	.000	-.802	-.146		.000	-.009	-.015		-.049	-.077	-.455	-.091		.000
22		(c)	.000	-.061	-.811		.000	-.009	-.047		-.045	-.051	-.029	.000		.000
23	1988	(a)	.000	-.063	-.698		.000	-.007	-.043		-.035	-.070	-1.166	-.712		.000
24		(b)	.000	-.836	-.436		.000	-.014	-.009		-.006	-.141	-2.470	.000		.000
25		(c)	.000	-.902	-.036		.000	-.007	-.022		-.075	-.057	-.802	-.033		.000

Table F1. Industrial well withdrawals or injections used for steady-state and transient models of groundwater flow, Idaho National Laboratory, Idaho.—Continued

[Values are in cubic feet per second. **Map No.** Identifier used to locate wells on maps located in appendix B figures and as a cross reference with data in other appendixes. **Local name:** Local well identifier used in this study. **Transient stress periods:** (a) January through April, (b) May through August, (c) September through December. Negative entries are withdrawals and positive entries are injection or infiltration]

Stress period	Simulated year	Four month period	Map number and Local name corresponding to well														
			118	131	132	133	134	191	205	206	207	208	238	239	240	245
			LPTF DISP	NRF 1	NRF 2	NRF 3	NRF 4	RWMC	SPERT 1	SPERT 2	TAN 1	TAN 2	TRA 1	TRA 3	TRA 4	TRA DISP
							Layer 1—Continued									
26	1989	(a)	0.000	-0.724	-0.006		0.000	-0.008	-0.003		-0.046	-0.054	-0.489	-0.011		0.000
27		(b)	.000	-.992	-.018		.000	-.025	-.014		-.056	-.064	-.938	-.058		.000
28		(c)	.000	-.605	-.266		.000	-.011	-.007		-.067	-.081	-1.008	-.017		.000
29	1990	(a)	.000	-.111	-1.972		-.024	-.004	-.013		-.044	-.024	-.345	.000		.000
30		(b)	.000	-.022	-.566		-.113	-.009	-.015		-.087	-.006	-.640	-.013		.000
31		(c)	.000	-.009	-.371		-.116	-.010	-.017		-.044	-.043	-.142	-.024		.000
32	1991	(a)	.000	-.001	-.595		-.142	-.011	-.009		-.098	.000	-.228	-.004		.000
33		(b)	.000	-.040	-.982		-.256	-.009	-.023		-.048	-.061	-.689	.000		.000
34		(c)	.000	-.002	-.672		-.234	-.005	-.016		-.004	-.079	-.952	-.016		.000
35	1992	(a)	.000	-.012	-.617		-.210	-.174	-.005		-.013	-.072	-.097	.000		.000
36		(b)	.000	-.104	-.829		-.278	-.182	-.011		-.035	-.076	-.456	-.157		.000
37		(c)	.000	-.001	-.794		-.307	-.008	-.007		-.039	-.026	-.153	.000		.000
38	1993	(a)	.000	-.001	-.521		-.203	-.007	-.003		-.044	-.027	-.799	-.086		.000
39		(b)	.000	-.116	-.617		-.254	-.016	-.015		-.041	-.052	-.156	.000		.000
40		(c)	.000	-.024	-.503		-.226	-.022	-.014		-.043	-.057	.000	.000		.000
41	1994	(a)	.000	-.065	-.284		-.207	-.010	-.013		-.044	-.045	-.012	.000		.000
42		(b)	.000	-.135	-.129		-.266	-.058	-.019		-.072	-.038	.000	.000		.000
43		(c)	.000	-.029	-.126		-.229	-.016	-.007		-.021	-.054	-.041	.000		.000
44	1995	(a)	.000	-.012	-.119		-.247	-.007	-.006		-.047	-.027	-.768	-.297		.000
45		(b)	.000	-.083	-.106		-.134	-.016	-.009		-.032	-.052	-.684	-.496		.000
46		(c)	.000	-.044	-.101		-.083	-.027	-.007		-.031	-.046	-.496	-.257		.000

Table F1.   Industrial well withdrawals or injections used for steady-state and transient models of groundwater flow, Idaho National Laboratory, Idaho.—Continued

[Values are in cubic feet per second. **Map No.** Identifier used to locate wells on maps located in appendix B figures and as a cross reference with data in other appendixes. **Local name:** Local well identifier used in this study. **Transient stress periods:** (a) January through April, (b) May through August, (c) September through December. Negative entries are withdrawals and positive entries are injection or infiltration]

Stress period	Simulated year	Four month period	118 LPTF DISP	131 NRF 1	132 NRF 2	133 NRF 3	134 NRF 4	191 RWMC	205 SPERT 1	206 SPERT 2	207 TAN 1	208 TAN 2	238 TRA 1	239 TRA 3	240 TRA 4	245 TRA DISP
							Layer 2—Continued									
26	1989	(a)		-0.080		-0.001	0.000	-0.004	-0.001		-0.017	-0.007	-0.138	-0.005		0.000
27		(b)		-.110		-.008	.000	-.011	-.003		-.021	-.009	-.265	-.029		.000
28		(c)		-.067		-.007	.000	-.005	-.002		-.025	-.011	-.284	-.009		.000
29	1990	(a)		-.012		-.006	-.006	-.002	-.003		-.016	-.003	-.097	.000		.000
30		(b)		-.002		-.007	-.029	-.004	-.003		-.032	-.001	-.180	-.006		.000
31		(c)		-.001		-.010	-.030	-.005	-.004		-.016	-.006	-.040	-.012		.000
32	1991	(a)		.000		-.015	-.036	-.005	-.002		-.036	.000	-.064	-.002		.000
33		(b)		-.004		.000	-.065	-.004	-.005		-.018	-.008	-.194	.000		.000
34		(c)		.000		-.004	-.059	-.002	-.005		-.002	-.011	-.268	-.008		.000
35	1992	(a)		-.001		-.006	-.054	-.078	-.001		-.005	-.010	-.027	.000		.000
36		(b)		-.012		-.002	-.071	-.082	-.003		-.013	-.010	-.129	-.077		.000
37		(c)		.000		-.002	-.078	-.003	-.002		-.014	-.004	-.043	.000		.000
38	1993	(a)		.000		-.004	-.052	-.003	-.001		-.016	-.004	-.225	-.043		.000
39		(b)		-.013		-.006	-.064	-.007	-.004		-.015	-.007	-.044	.000		.000
40		(c)		-.003		-.001	-.057	-.010	-.003		-.016	-.008	.000	.000		.000
41	1994	(a)		-.007		-.005	-.053	-.005	-.003		-.016	-.006	-.003	.000		.000
42		(b)		-.015		-.001	-.068	-.026	-.005		-.027	-.005	.000	.000		.000
43		(c)		-.003		-.001	-.058	-.007	-.002		-.008	-.007	-.012	.000		.000
44	1995	(a)		-.001		-.001	-.063	-.003	-.001		-.018	-.004	-.217	-.146		.000
45		(b)		-.009		-.001	-.034	-.007	-.002		-.012	-.007	-.193	-.244		.000
46		(c)		-.005		-.001	-.021	-.012	-.002		-.011	-.006	-.140	-.127		.000

Table F1. Industrial well withdrawals or injections used for steady-state and transient models of groundwater flow, Idaho National Laboratory, Idaho.—Continued

[Values are in cubic feet per second. **Map No.** Identifier used to locate wells on maps located in appendix B figures and as a cross reference with data in other appendixes. **Local name:** Local well identifier used in this study. **Transient stress periods:** (a) January through April, (b) May through August, (c) September through December. Negative entries are withdrawals and positive entries are injection or infiltration]

Stress period	Simulated year	Four month period	118 LPTF DISP	131 NRF 1	132 NRF 2	133 NRF 3	134 NRF 4	191 RWMC	205 SPERT 1	206 SPERT 2	207 TAN 1	208 TAN 2	238 TRA 1	239 TRA 3	240 TRA 4	245 TRA DISP
							Layer 3									
1	1980[1]						0.000									
2	1981	(a)					.000									
3		(b)					.000									
4		(c)					.000									
5	1982	(a)					.000									
6		(b)					.000									
7		(c)					.000									
8	1983	(a)					.000									
9		(b)					.000									
10		(c)					.000									
11	1984	(a)					.000									
12		(b)					.000									
13		(c)					.000									
14	1985	(a)					.000									
15		(b)					.000									
16		(c)					.000									
17	1986	(a)					.000									
18		(b)					.000									
19		(c)					.000									
20	1987	(a)					.000									
21		(b)					.000									
22		(c)					.000									
23	1988	(a)					.000									
24		(b)					.000									
25		(c)					.000									

Table F1.   Industrial well withdrawals or injections used for steady-state and transient models of groundwater flow, Idaho National Laboratory, Idaho.—Continued

[Values are in cubic feet per second. **Map No.** Identifier used to locate wells on maps located in appendix B figures and as a cross reference with data in other appendixes. **Local name:** Local well identifier used in this study. **Transient stress periods:** (a) January through April, (b) May through August, (c) September through December. Negative entries are withdrawals and positive entries are injection or infiltration]

Stress period	Simulated year	Four month period	118 LPTF DISP	131 NRF 1	132 NRF 2	133 NRF 3	134 NRF 4	191 RWMC	205 SPERT 1	206 SPERT 2	207 TAN 1	208 TAN 2	238 TRA 1	239 TRA 3	240 TRA 4	245 TRA DISP
							Layer 3—Continued									
26	1989	(a)					0.000									
27		(b)					.000									
28		(c)					.000									
29	1990	(a)					-.011									
30		(b)					-.050									
31		(c)					-.051									
32	1991	(a)					-.062									
33		(b)					-.113									
34		(c)					-.103									
35	1992	(a)					-.093									
36		(b)					-.123									
37		(c)					-.135									
38	1993	(a)					-.089									
39		(b)					-.112									
40		(c)					-.100									
41	1994	(a)					-.091									
42		(b)					-.117									
43		(c)					-.101									
44	1995	(a)					-.109									
45		(b)					-.059									
46		(c)					-.036									

Table F1. Industrial well withdrawals or injections used for steady-state and transient models of groundwater flow, Idaho National Laboratory, Idaho.—Continued

[Values are in cubic feet per second. **Map No.** Identifier used to locate wells on maps located in appendix B figures and as a cross reference with data in other appendixes. **Local name:** Local well identifier used in this study. **Transient stress periods:** (a) January through April, (b) May through August, (c) September through December. Negative entries are withdrawals and positive entries are injection or infiltration]

Stress period	Simulated year	Four month period	Map number and Local name corresponding to well													
			118 LPTF DISP	131 NRF 1	132 NRF 2	133 NRF 3	134 NRF 4	191 RWMC	205 SPERT 1	206 SPERT 2	207 TAN 1	208 TAN 2	238 TRA 1	239 TRA 3	240 TRA 4	245 TRA DISP
							Layer 4									
1	1980[1]									-0.001					-2.504	0.069
2	1981	(a)								.000					-2.250	.058
3		(b)								.000					-2.629	.055
4		(c)								.000					-2.197	.041
5	1982	(a)								-.001					-2.481	.024
6		(b)								-.006					-2.370	.000
7		(c)								-.003					-2.404	.000
8	1983	(a)								-.002					-2.506	.000
9		(b)								-.004					-2.737	.000
10		(c)								-.003					-2.646	.000
11	1984	(a)								-.002					-.688	.000
12		(b)								-.006					-2.421	.000
13		(c)								-.002					-.759	.000
14	1985	(a)								-.003					-1.318	.000
15		(b)								-.002					-.686	.000
16		(c)								.000					-2.504	.000
17	1986	(a)								-.001					-1.876	.000
18		(b)								-.001					-2.568	.000
19		(c)								-.001					-2.410	.000
20	1987	(a)								-.001					-2.426	.000
21		(b)								-.003					-2.361	.000
22		(c)								.000					-2.823	.000
23	1988	(a)								-.001					-.496	.000
24		(b)								-.006					.000	.000
25		(c)								-.002					-2.066	.000

Table F1.   Industrial well withdrawals or injections used for steady-state and transient models of groundwater flow, Idaho National Laboratory, Idaho.—Continued

[Values are in cubic feet per second. **Map No.** Identifier used to locate wells on maps located in appendix B figures and as a cross reference with data in other appendixes. **Local name:** Local well identifier used in this study. **Transient stress periods:** (a) January through April, (b) May through August, (c) September through December. Negative entries are withdrawals and positive entries are injection or infiltration]

Stress period	Simulated year	Four month period	Map number and Local name corresponding to well													
			118	131	132	133	134	191	205	206	207	208	238	239	240	245
			LPTF DISP	NRF 1	NRF 2	NRF 3	NRF 4	RWMC	SPERT 1	SPERT 2	TAN 1	TAN 2	TRA 1	TRA 3	TRA 4	TRA DISP
			Layer 4—Continued													
26	1989	(a)								-0.001					-1.740	0.000
27		(b)								-.002					-1.555	.000
28		(c)								.000					-1.614	.000
29	1990	(a)								-.001					-2.215	.000
30		(b)								-.001					-2.724	.000
31		(c)								-.001					-2.642	.000
32	1991	(a)								.000					-2.242	.000
33		(b)								-.002					-1.996	.000
34		(c)								-.003					-1.323	.000
35	1992	(a)								.000					-2.249	.000
36		(b)								-.001					-1.772	.000
37		(c)								-.001					-2.308	.000
38	1993	(a)								.000					-1.334	.000
39		(b)								.000					-1.589	.000
40		(c)								.000					-2.313	.000
41	1994	(a)								.000					-1.751	.000
42		(b)								-.001					-1.808	.000
43		(c)								-.001					-2.295	.000
44	1995	(a)								-.001					-.706	.000
45		(b)								-.001					-.941	.000
46		(c)								.000					-1.462	.000

Table F1. Industrial well withdrawals or injections used for steady-state and transient models of groundwater flow, Idaho National Laboratory, Idaho.—Continued

[Values are in cubic feet per second. **Map No.** Identifier used to locate wells on maps located in appendix B figures and as a cross reference with data in other appendixes. **Local name:** Local well identifier used in this study. **Transient stress periods:** (a) January through April, (b) May through August, (c) September through December. Negative entries are withdrawals and positive entries are injection or infiltration]

Stress period	Simulated year	Four month period	Map number and Local name corresponding to well													
			118	131	132	133	134	191	205	206	207	208	238	239	240	245
			LPTF DISP	NRF 1	NRF 2	NRF 3	NRF 4	RWMC	SPERT 1	SPERT 2	TAN 1	TAN 2	TRA 1	TRA 3	TRA 4	TRA DISP
								Layer 5								
1	1980[1]									-0.007						
2	1981	(a)								.000						
3		(b)								.000						
4		(c)								-.002						
5	1982	(a)								-.015						
6		(b)								-.078						
7		(c)								-.044						
8	1983	(a)								-.029						
9		(b)								-.055						
10		(c)								-.037						
11	1984	(a)								-.027						
12		(b)								-.079						
13		(c)								-.029						
14	1985	(a)								-.037						
15		(b)								-.030						
16		(c)								-.004						
17	1986	(a)								-.015						
18		(b)								-.013						
19		(c)								-.019						
20	1987	(a)								-.007						
21		(b)								-.040						
22		(c)								-.003						
23	1988	(a)								-.010						
24		(b)								-.077						
25		(c)								-.026						

Table F1. Industrial well withdrawals or injections used for steady-state and transient models of groundwater flow, Idaho National Laboratory, Idaho.—Continued

[Values are in cubic feet per second. **Map No.** Identifier used to locate wells on maps located in appendix B figures and as a cross reference with data in other appendixes. **Local name:** Local well identifier used in this study. **Transient stress periods:** (a) January through April, (b) May through August, (c) September through December. Negative entries are withdrawals and positive entries are injection or infiltration]

Stress period	Simulated year	Four month period	118 LPTF DISP	131 NRF 1	132 NRF 2	133 NRF 3	134 NRF 4	191 RWMC	205 SPERT 1	206 SPERT 2	207 TAN 1	208 TAN 2	238 TRA 1	239 TRA 3	240 TRA 4	245 TRA DISP
										Layer 5—Continued						
26	1989	(a)								-0.011						
27		(b)								-.029						
28		(c)								-.003						
29	1990	(a)								-.016						
30		(b)								-.014						
31		(c)								-.014						
32	1991	(a)								-.004						
33		(b)								-.023						
34		(c)								-.040						
35	1992	(a)								-.005						
36		(b)								-.017						
37		(c)								-.009						
38	1993	(a)								-.005						
39		(b)								-.001						
40		(c)								.000						
41	1994	(a)								.000						
42		(b)								-.008						
43		(c)								-.008						
44	1995	(a)								-.008						
45		(b)								-.009						
46		(c)								-.005						

[1] Steady-state stress period.

Appendix G. Water-Level Data Used to Represent 1980 Steady-State Head for Calibration of Steady-State and Transient Models of Groundwater Flow, Idaho National Laboratory and Vicinity, Idaho

Table G1.   Water-level data used to represent 1980 steady-state head for the calibration of steady-state and transient models of groundwater flow, Idaho National Laboratory and vicinity, Idaho.

[**Map No.:** Identifier used to locate wells on maps located in <u>appendix B</u> figures and as a cross reference with data in other appendixes **Local name:** Local well identifier used in this study **Observation name:** Model observation name used in this study **Weighting statistic:** Variance of the measurement error of head **Abbreviation**: ft, foot]

Map No.	Site identifier	Local name	Observation name	Date measured	Observed head (ft)	Weighting statistic (1/ft)
1	434819112380501	2ND OWSLEY	2dOwsly_780	July 1980	4,561.62	1.50
2	434556112575601	434556	434556_491	Apr. 1991	4,551.63	9.20
3	434647112534101	434647	434647_491	Apr. 1991	4,530.43	9.20
4	434650112545501	434650	434650_880	Aug. 1980	4,513.64	3.30
5	434714112175801	434714	434714_1180	Nov. 1980	4,554.20	9.20
6	434726112244101	434726	434726_889	Aug. 1989	4,533.95	9.20
7	434756112212101	434756	434756_1291	Dec. 1991	4,541.57	9.20
8	435026112253101	435026	435026_1180	Nov. 1980	4,562.86	9.20
9	435100112271601	435100	435100_1180	Nov. 1980	4,580.29	9.20
10	440109112391301	440109	440109_989	Sept. 1989	4,584.40	37.00
12	433534112392901	ANL MW 11	ANLMW11_1292	Dec. 1992	4,482.74	3.30
13	433545112394102	ANL MW 13	ANLMW13_1104	Nov. 2004	4,485.00	9.20
14	433545112394101	ANL OBS A 001	ANLOBSA_301	Mar. 2001	4,485.21	37.00
21	435308112454101	ANP 5	ANP5_780	July 1980	4,579.82	1.50
22	435152112443101	ANP 6	ANP6_780	July 1980	4,579.82	1.50
23	435522112444201	ANP 7	ANP7_780	July 1980	4,581.18	1.50
25	434856112400001	ANP 9	ANP9_1280	Dec. 1980	4,563.58	1.50
26	434909112400401	ANP 10	ANP10_780	July 1980	4,565.80	1.50
27	435053112423201	ANP DISP 3	ANPdsp3_784	July 1984	4,580.49	9.20
30	433509112384801	ARBOR TEST	ArbrTst_1280	Dec. 1980	4,485.52	.37
31	433223112470201	AREA 2	Area2_780	July 1980	4,458.78	3.30
49	432618112555501	CERRO GRANDE	CroGrnd_1280	Dec. 1980	4,426.37	.37
51	433144112563501	CFA 2	CFA2_189	Jan. 1989	4,457.15	3.30
55	433216112563301	CFA LF 2-10	CFAL210_691	June 1991	4,453.47	3.30
56	433230112561701	CFA LF 2-11	CFAL211_1289	Dec. 1989	4,459.32	9.20
59	433216112571001	CFA LF 3-9	CFALF39_1190	Nov. 1990	4,455.98	3.30
63	432927112410101	COREHOLE 1	CrHole1_1280	Dec. 1980	4,435.74	9.20
65	431519113112901	COX WELL	CoxWell_849	Aug. 1949	4,319.47	37.00
69	433440112554401	CPP 4	CPP4_1083	Oct. 1983	4,466.00	9.20
69	433440112554402	CPP 5	CPP5_1193	Nov. 1993	4,457.50	9.20
74	434611112504301	DH 1B	DH1B_1292	Dec. 1992	4,527.08	9.20
75	434547112512801	DH 2A	DH2A_293	Feb. 1993	4,526.58	9.20
83	431831113312901	ELLSWORTH	Elswrth_567	May 1967	4,107.30	37.00
88	435124112433701	FET DISP 3	FETdsp3_592	May 1992	4,582.05	3.30
90	432424113165301	FNGR BUTTE	FngrBtt_1072	Oct. 1972	4,370.00	37.00
91	434947112414301	GIN 1 (old)	GIN1old_592	May 1992	4,579.01	9.20
92	434949112413401	GIN 2	GIN2old_780	July 1980	4,577.43	3.30
93	434945112413101	GIN 3 (old)	GIN3old_493	Apr. 1993	4,576.41	9.20
94	434949112413601	GIN 4	GIN4old_493	Apr. 1993	4,576.50	3.30

Table G1. Water-level data used to represent 1980 steady-state head for the calibration of steady-state and transient models of groundwater flow, Idaho National Laboratory and vicinity, Idaho.—Continued

[**Map No.:** Identifier used to locate wells on maps located in appendix B figures and as a cross reference with data in other appendixes **Local name:** Local well identifier used in this study **Observation name:** Model observation name used in this study **Weighting statistic:** Variance of the measurement error of head **Abbreviation**: ft, foot]

Map No.	Site identifier	Local name	Observation name	Date measured	Observed head (ft)	Weighting statistic (1/ft)
95	434953112413301	GIN 5 (old)	GIN5old_493	Apr. 1993	4,576.61	9.20
114	431439113071401	Houghland	Hghlnd_380	Mar. 1980	4,379.85	37.00
115	435153112420501	IET 1 DISP	IETdisp_780	July 1980	4,580.32	3.30
121	434700112530401	ML 11	ML11_491	Apr. 1991	4,534.45	9.20
122	434558112585301	ML 12	ML12_880	Aug. 1980	4,534.71	3.30
123	434723112552701	ML 13	ML13_380 [1]	Mar. 1980	4,535.40	9.20
124	433520112572601	MTR TEST	MTRtest_1280	Dec. 1980	4,457.62	1.50
128	435038112453401	NO NAME 1	NoName1_780	July 1980	4,577.82	9.20
129	433449112523101	NPR TEST	NPRtest_1191	Nov. 1991	4,470.70	3.30
136	433910112550101	NRF 6	NRF6_991	Sept. 1991	4,479.04	3.30
138	433920112543601	NRF 7	NRF7_991	Sept. 1991	4,479.16	3.30
140	433843112550901	NRF 8	NRF8_898	Aug. 1998	4,479.23	3.30
141	433840112550201	NRF 9	NRF9_898	Aug. 1998	4,478.27	3.30
142	433841112545201	NRF 10	NRF10_898	Aug. 1998	4,479.15	3.30
143	433847112544201	NRF 11	NRF11_898	Aug. 1998	4,479.66	3.30
144	433855112543201	NRF 12	NRF12_898	Aug. 1998	4,478.38	3.30
145	433928112545401	NRF 13	NRF13_898	Aug. 1998	4,478.35	3.30
150	435416112460401	P&W 1	P&W1_780	July 1980	4,580.07	1.50
151	435419112453101	P&W 2	P&W2_780	July 1980	4,579.83	1.50
152	435443112435801	P&W 3	P&W3_780	July 1980	4,580.32	1.50
155	434941112454201	PSTF TEST	PSTFtst_780	July 1980	4,578.32	1.50
170	432632113095901	QAB	QAB_282	Feb. 1982	4,424.48	37.00
184	432956113030901	RWMC M1SA	WMCM1SA_1192	Nov. 1992	4,425.56	3.30
185	433008113021801	RWMC M3S	RWMCM3S_1192	Nov. 1992	4,426.38	3.30
186	432939113030101	RWMC M4D	RWMCM4D_1192	Nov. 1992	4,427.19	3.30
187	432931113015001	RWMC M6S	RWMCM6S_1192	Nov. 1992	4,425.51	3.30
188	433023113014801	RWMC M7S	RWMCM7S_1192	Nov. 1992	4,426.97	3.30
196	432854113201002	SITE 1	Site1_1280	Dec. 1980	4,378.26	37.00
197	431946113161401	SITE 2	Site2_1280	Dec. 1980	4,158.31	37.00
199	433826112510701	SITE 6	Site6_780	July 1980	4,478.20	1.50
200	433123112530101	SITE 9	Site9_1280	Dec. 1980	4,453.49	3.30
201	434334112463101	SITE 14	Site14_1280	Dec. 1980	4,524.51	1.50
202	433545112391501	SITE 16	Site16_780	July 1980	4,484.84	3.30
203	434027112575701	SITE 17	Site17_780	July 1980	4,483.84	1.50
204	433522112582101	SITE 19	Site19_780	July 1980	4,457.50	9.20
207	435056112420001	TAN 1	TAN1_1187	Nov. 1987	4,586.28	9.20
208	435100112420701	TAN 2	TAN2_1187 [1]	Nov. 1987	4,585.68	9.20
209	435104112420301	TAN 3	TAN3_190	Jan. 1990	4,582.74	9.20
210	435055112421302	TAN 5	TAN5_190	Jan. 1990	4,582.33	9.20

Table G1. Water-level data used to represent 1980 steady-state head for the calibration of steady-state and transient models of groundwater flow, Idaho National Laboratory and vicinity, Idaho.—Continued

[**Map No.:** Identifier used to locate wells on maps located in appendix B figures and as a cross reference with data in other appendixes **Local name:** Local well identifier used in this study **Observation name:** Model observation name used in this study **Weighting statistic:** Variance of the measurement error of head **Abbreviation**: ft, foot]

Map No.	Site identifier	Local name	Observation name	Date measured	Observed head (ft)	Weighting statistic (1/ft)
211	435039112412601	TAN 6	TAN6_990	Sept. 1990	4,582.27	9.20
212	435038112412601	TAN 7	TAN7_990	Sept. 1990	4,582.19	9.20
213	435034112421701	TAN 8	TAN8_80e [2]	Dec. 1980e	4,578.53	9.20
214	435053112423202	TAN 9	TAN9_1289	Dec. 1989	4,584.85	9.20
216	435051112423201	TAN 10 A	TAN10A_490	Apr. 1990	4,586.51	9.20
217	435050112423202	TAN 11	TAN11_1289	Dec. 1989	4,584.96	9.20
219	435040112423701	TAN 13	TAN13_1090	Oct. 1990	4,580.57	9.20
220	435040112423801	TAN 13A	TAN13A_1190	Nov. 1990	4,578.82	9.20
221	435039112423701	TAN 14	TAN14_1090	Oct. 1990	4,580.89	9.20
222	435021112412701	TAN 15	TAN15_991	Sept. 1991	4,580.99	3.30
224	435034112421601	TAN 17	TAN17_80e [2]	Dec. 1980e	4,578.98	9.20
225	435051112421401	TAN 18	TAN18_1192	Nov. 1992	4,576.95	3.30
226	435051112421501	TAN 19	TAN19_1092	Oct. 1992	4,577.69	3.30
227	435046112425001	TAN 20	TAN20_1092	Oct. 1992	4,577.86	3.30
228	435009112420001	TAN 21	TAN21_1092	Oct. 1992	4,575.98	3.30
229	435020112412702	TAN 22	TAN22_892	Aug. 1992	4,579.05	9.20
230	435019112412701	TAN 22A	TAN22A_1092	Oct. 1992	4,577.09	9.20
231	435020112412703	TAN 23	TAN23_792	July 1992	4,578.76	9.20
231	435020112412704	TAN 23A	TAN23A_1092	Oct. 1992	4,576.96	3.30
232	434942112411101	TAN 24	TAN24_992	Sept. 1992	4,576.61	9.20
232	434942112411001	TAN 24A	TAN24A_1092	Oct. 1992	4,575.69	3.30
233	435058112423401	TAN CH 1	TANCH1_190	Jan. 1990	4,584.14	9.20
234	435033112421702	TAN CH 2B	TANCH2b_401	Apr. 2001	4,577.41	9.20
242	433446112574701	TRA 6	TRA6_1290	Dec. 1990	4,455.60	9.20
243	433449112575901	TRA 7	TRA7_990	Sept. 1990	4,457.10	9.20
244	433431112580101	TRA 8	TRA8_1290	Dec. 1990	4,457.10	9.20
245	433506112572301	TRA DISP	TRAdisp_791	July 1991	4,458.39	3.30
247	432700112470801	USGS 1	USGS1_1280	Dec. 1980	4,435.21	.37
248	433320112432301	USGS 2	USGS2_1280	Dec. 1980	4,466.83	1.50
250	434657112282201	USGS 4	USGS4_1280	Dec. 1980	4,531.61	3.30
251	433543112493801	USGS 5	USGS5_1280	Dec. 1980	4,469.86	.37
252	434031112453701	USGS 6	USGS6_1280	Dec. 1980	4,485.14	3.30
253	434915112443901	USGS 7	USGS7_1280	Dec. 1980	4,575.85	1.50
254	433121113115801	USGS 8	USGS8_1280	Dec. 1980	4,429.93	37.00
255	432740113044501	USGS 9	USGS9_1280	Dec. 1980	4,423.77	.37
256	432336113064201	USGS 11	USGS11_1280	Dec. 1980	4,415.68	1.50
257	434126112550701	USGS 12	USGS12_1280	Dec. 1980	4,490.67	9.20
258	432731113143902	USGS 13	USGS13_781	July 1981	4,388.64	9.20
259	432019112563201	USGS 14	USGS14_1280	Dec. 1980	4,418.12	1.50

Table G1.   Water-level data used to represent 1980 steady-state head for the calibration of steady-state and transient models of groundwater flow, Idaho National Laboratory and vicinity, Idaho.—Continued

[**Map No.:** Identifier used to locate wells on maps located in <u>appendix B</u> figures and as a cross reference with data in other appendixes **Local name:** Local well identifier used in this study **Observation name:** Model observation name used in this study **Weighting statistic:** Variance of the measurement error of head **Abbreviation**: ft, foot]

Map No.	Site identifier	Local name	Observation name	Date measured	Observed head (ft)	Weighting statistic (1/ft)
260	434234112551701	USGS 15	USGS15_780	July 1980	4,492.56	1.50
262	433937112515401	USGS 17	USGS17_1280	Dec. 1980	4,478.38	1.50
263	434540112440901	USGS 18	USGS18_1280	Dec. 1980	4,533.62	1.50
264	434426112575701	USGS 19	USGS19_1280	Dec. 1980	4,527.93	9.20
265	433253112545901	USGS 20	USGS20_1280	Dec. 1980	4,453.86	.37
266	434307112382601	USGS 21	USGS21_1280	Dec. 1980	4,506.06	1.50
267	433422113031701	USGS 22	USGS22_1280	Dec. 1980	4,438.13	.37
268	434055112595901	USGS 23	USGS23_1280	Dec. 1980	4,484.31	1.50
269	435053112420801	USGS 24	USGS24_780	July 1980	4,579.49	1.50
270	435339112444601	USGS 25	USGS25_1280	Dec. 1980	4,579.07	1.50
271	435212112394001	USGS 26	USGS26_1280	Dec. 1980	4,579.21	1.50
272	434851112321801	USGS 27	USGS27_380	Mar. 1980	4,560.28	1.50
273	434600112360101	USGS 28	USGS28_780	July 1980	4,541.02	1.50
274	434407112285101	USGS 29	USGS29_780	July 1980	4,522.48	1.50
275	434601112315403	USGS 30A	USGS30A_1280	Dec. 1980	4,537.16	1.50
275	434601112315402	USGS 30B	USGS30B_1280	Dec. 1980	4,528.16	1.50
275	434601112315401	USGS 30C	USGS30C_1280	Dec. 1980	4,526.07	1.50
276	434625112342101	USGS 31	USGS31_780	July 1980	4,529.69	3.30
277	434444112322101	USGS 32	USGS32_780	July 1980	4,522.28	3.30
279	433334112565501	USGS 34	USGS34_780	July 1980	4,456.43	3.30
280	433339112565801	USGS 35	USGS35_780	July 1980	4,456.25	.37
281	433330112565201	USGS 36	USGS36_780	July 1980	4,456.53	.37
282	433326112564801	USGS 37	USGS37_1280	Dec. 1980	4,456.26	.37
283	433322112564301	USGS 38	GS38new_780	July 1980	4,456.59	.37
284	433343112570001	USGS 39	USGS39_780	July 1980	4,456.45	.37
285	433411112561101	USGS 40	GS40new_1280	Dec. 1980	4,457.25	.37
286	433409112561301	USGS 41	USGS41_780	July 1980	4,457.68	3.30
287	433404112561301	USGS 42	USGS42_780	July 1980	4,457.64	3.30
288	433415112561501	USGS 43	USGS43_780	July 1980	4,457.64	3.30
289	433409112562101	USGS 44	USGS44_780	July 1980	4,457.76	1.50
290	433402112561801	USGS 45	USGS45_780	July 1980	4,456.81	1.50
291	433407112561501	USGS 46	USGS46_1280	Dec. 1980	4,457.62	1.50
292	433407112560301	USGS 47	USGS47_780	July 1980	4,458.45	1.50
293	433401112560301	USGS 48	USGS48_780	July 1980	4,457.81	3.30
294	433403112555401	USGS 49	GS49new_1191	Nov. 1991	4,456.99	3.30
296	433350112560601	USGS 51	USGS51_780	July 1980	4,457.79	1.50
297	433414112554201	USGS 52	USGS52_780	July 1980	4,457.81	1.50
302	433344112562601	USGS 57	USGS57_780	July 1980	4,456.42	9.20
303	433500112572502	USGS 58	USGS58_1280	Dec. 1980	4,457.61	.37

Table G1.   Water-level data used to represent 1980 steady-state head for the calibration of steady-state and transient models of groundwater flow, Idaho National Laboratory and vicinity, Idaho.—Continued

[**Map No.:** Identifier used to locate wells on maps located in <u>appendix B</u> figures and as a cross reference with data in other appendixes **Local name:** Local well identifier used in this study **Observation name:** Model observation name used in this study **Weighting statistic:** Variance of the measurement error of head **Abbreviation**: ft, foot]

Map No.	Site identifier	Local name	Observation name	Date measured	Observed head (ft)	Weighting statistic (1/ft)
304	433354112554701	USGS 59	USGS59_780	July 1980	4,457.49	1.50
310	433447112574501	USGS 65	USGS65_780	July 1980	4,458.28	.37
312	433344112554101	USGS 67	USGS67_780	July 1980	4,457.19	3.30
321	433425112573201	USGS 76	USGS76_780	July 1980	4,456.79	3.30
322	433315112560301	USGS 77	USGS77_380	Mar. 1980	4,456.95	1.50
324	433505112581901	USGS 79	USGS79_780	July 1980	4,457.40	3.30
327	433401112551001	USGS 82	USGS82_1280	Dec. 1980	4,457.37	3.30
328	433023112561501	USGS 83	USGS83_1280	Dec. 1980	4,442.82	3.30
329	433356112574201	USGS 84	USGS84_1280	Dec. 1980	4,456.35	.37
330	433246112571201	USGS 85 (old)	USGS85_1280	Dec. 1980	4,456.00	1.50
331	432935113080001	USGS 86	USGS86_1280	Dec. 1980	4,428.72	1.50
332	433013113024201	USGS 87	USGS87_1080	Oct. 1980	4,430.13	.37
333	432940113030201	USGS 88	USGS88_1080 [3]	Oct. 1980	4,428.56	9.20
334	433005113032801	USGS 89	USGS89_1080	Oct. 1980	4,429.24	.37
335	432954113020501	USGS 90	USGS90_1080	Oct. 1980	4,427.83	.37
343	433807112551501	USGS 97	USGS97_1280	Dec. 1980	4,477.95	1.50
344	433657112563601	USGS 98	USGS98_780	July 1980	4,472.15	3.30
345	433705112552101	USGS 99	USGS99_780	July 1980	4,477.43	1.50
346	433503112400701	USGS 100	USGS100_780	July 1980	4,482.09	.37
347	433255112381801	USGS 101	USGS101_780	July 1980	4,481.62	3.30
348	433853112551601	USGS 102	USGS102_1291	Dec. 1991	4,478.61	3.30
349	432714112560701	USGS 103	USGS103_1280	Dec. 1980	4,426.71	1.50
350	432856112560801	USGS 104	USGS104_1280	Dec. 1980	4,431.94	1.50
351	432703113001801	USGS 105	USGS105_1180	Nov. 1980	4,426.29	1.50
352	432959112593101	USGS 106	USGS106_1180	Nov. 1980	4,428.19	1.50
353	432942112532801	USGS 107	USGS107_1180	Nov. 1980	4,438.48	3.30
354	432659112582601	USGS 108	USGS108_1280	Dec. 1980	4,426.00	1.50
355	432701113025601	USGS 109	USGS109_1180	Nov. 1980	4,423.92	1.50
356	432717112501501	USGS 110	USGS110_1080	Oct. 1980	4,429.91	9.20
357	433331112560501	USGS 111	USGS111_1091	Oct. 1991	4,455.51	9.20
358	433314112563001	USGS 112	USGS112_1091	Oct. 1991	4,456.21	3.30
359	433314112561801	USGS 113	USGS113_1091	Oct. 1991	4,456.88	9.20
360	433318112555001	USGS 114	USGS114_1091	Oct. 1991	4,456.50	9.20
361	433320112554101	USGS 115	USGS115_1091	Oct. 1991	4,456.31	3.30
362	433331112553201	USGS 116	USGS116_1091	Oct. 1991	4,456.76	3.30
363	432955113025901	USGS 117	USGS117_1091	Oct. 1991	4,427.62	3.30
364	432947113023001	USGS 118	USGS118_1192	Nov. 1992	4,425.98	3.30
365	432945113023401	USGS 119	USGS119_1091	Oct. 1991	4,426.89	3.30
366	432919113031501	USGS 120	USGS120_1291	Dec. 1991	4,425.72	3.30

Table G1. Water-level data used to represent 1980 steady-state head for the calibration of steady-state and transient models of groundwater flow, Idaho National Laboratory and vicinity, Idaho.—Continued

[**Map No.:** Identifier used to locate wells on maps located in appendix B figures and as a cross reference with data in other appendixes **Local name:** Local well identifier used in this study **Observation name:** Model observation name used in this study **Weighting statistic:** Variance of the measurement error of head **Abbreviation:** ft, foot]

Map No.	Site identifier	Local name	Observation name	Date measured	Observed head (ft)	Weighting statistic (1/ft)
367	433450112560301	USGS 121	USGS121_1091	Oct. 1991	4,457.28	3.30
368	433353112555201	USGS 122	USGS122_1091	Oct. 1991	4,456.88	3.30
369	433352112561401	USGS 123	USGS123_1091	Oct. 1991	4,457.94	3.30
370	432307112583101	USGS 124	USGS124_499	Apr. 1999	4,420.85	9.20
371	432602113052801	USGS 125	USGS125_499	Apr. 1999	4,423.27	9.20
375	433445113202801	Weaver and Lowe	WeavLow_585	May 1985	4,475.89	37.00
376	433716112563601	WS INEL 1	WSINEL1_1280	Dec. 1980	4,476.72	3.30

[1] Transient model only

[2] Head estimated by extrapolation from March 2004

[3] Steady-state model only

Appendix H. Water-Level Data Used for Calibration of Transient Model of Groundwater Flow, Idaho National Laboratory and Vicinity, Idaho

Table H1. Water-level data used for calibration of transient model of groundwater flow, Idaho National Laboratory and vicinity, Idaho.

[**Map No.:** Identifier used to locate wells on maps located in appendix B figures and as a cross reference with data in other appendixes **Local name:** Local well used in this study **Observation name:** Model observation name used in this study]

Map No.	Local name	Observation name	Date measured	Observed head	Number of observations	Initial simulated date	Final observed and simulated date
1	2ND OWSLEY	2dOwsly_780	July 1980	4,561.62	74	12-31-1980	05-30-1995
2	434556	434556_491	04-08-1991	4,551.63	1	04-08-1991	
3	434647	434647_491	04-03-1991	4,530.43	1	04-03-1991	
4	434650	434650_880 [1]	Aug. 1980	4,513.64	2	12-31-1980	04-03-1991
5	434714	434714_1180	Nov. 1980	4,554.20	4	12-31-1980	04-13-1989
6	434726	434726_489	04-12-1989	4,539.22	2	04-12-1989	08-31-1989
7	434756	434756_686	06-03-1986	4,547.37	107	06-03-1986	11-08-1995
8	435026	435026_1180	Nov. 1980	4,562.86	3	12-31-1980	10-20-1981
9	435100	435100_1180	Nov. 1980	4,580.29	4	12-31-1980	08-31-1989
10	440109	440109_489 [1]	04-17-1989	4,587.49	2	04-17-1989	09-01-1989
12	ANL MW 11	ANLMW11_1292	12-17-1992	4,482.74	1	12-17-1992	
13	ANL MW 13	ANLMW13_1196	Nov. 1996	4,485.00	1	12-31-1980	
14	ANL OBS A 001	ANLOBSA_694	06-24-1994	4,482.84	1	06-24-1994	
21	ANP 5	ANP5_780	July 1980	4,579.82	16	12-31-1980	05-22-1995
22	ANP 6	ANP6_780	July 1980	4,579.82	23	12-31-1980	06-15-1995
23	ANP 7	ANP7_780	July 1980	4,581.18	16	12-31-1980	05-22-1995
25	ANP 9	ANP9_1280	Dec. 1980	4,563.58	101	12-31-1980	10-10-1995
26	ANP 10	ANP10_780	July 1980	4,565.80	12	12-31-1980	05-23-1995
27	ANP DISP 3	ANPdsp3_784	07-19-1984	4,580.49	7	07-19-1984	12-08-1988
30	ARBOR TEST	ArbTst_1280	Dec. 1980	4,485.52	107	12-31-1980	10-19-1995
31	AREA 2	Area2_780	July 1980	4,458.78	21	12-31-1980	07-14-1995
49	CERRO GRANDE	CrGrnd_1280	Dec. 1980	4,426.37	98	12-31-1980	11-22-1995
51	CFA 2	CFA2_189	01-25-1989	4,457.15	1	01-25-1989	
55	CFA LF 2-10	CFAL210_189	01-26-1989	4,456.53	8	01-26-1989	10-25-1995
56	CFA LF 2-11	CFAL211_1289	12-06-1989	4,459.32	2	12-06-1989	06-03-1994
59	CFA LF 3-9	CFALF39_1190	11-13-1990	4,455.98	5	11-13-1990	04-27-1995
63	COREHOLE 1	CrHle1_1280	Dec. 1980	4,435.74	90	12-31-1980	05-18-1995
65	COX WELL	CoxWell_849	Aug. 1949	4,319.47	1	12-31-1980	
69	CPP 4	CPP4_1083	10-17-1983	4,466.00	1	10-17-1983	
69	CPP 5	CPP5_1193	11-30-1993	4,457.50	1	11-30-1993	
74	DH 1B	DH1B_785 [1]	07-19-1985	4,524.04	72	07-19-1985	11-16-1995
75	DH 2A	DH2A_785 [1]	07-19-1985	4,522.57	25	07-19-1985	09-18-1995
83	ELLSWORTH	Elswrth_567	May 1967	4,107.30	1	12-31-1980	
88	FET DISP 3	FETdsp3_784	07-19-1984	4,581.26	11	07-19-1984	05-31-1995
90	FNGR BUTTE	FngrBtt_1072	Oct. 1972	4,376.00	1	12-31-1980	
91	GIN 1 (old)	GIN1old_784	07-12-1984	4,578.75	3	07-12-1984	04-23-1993
91	GIN 1 (new)	GIN1new_593	05-21-1993	4,576.21	3	05-21-1993	05-19-1995
92	GIN 2	GIN2old_780	01-02-1981	4,577.43	16	01-02-1981	05-21-1993
93	GIN 3 (old)	GIN3old_784	07-12-1984	4,579.21	5	07-12-1984	04-23-1993
93	GIN 3 (new)	GIN3new_594	05-26-1994	4,575.53	2	05-26-1994	05-19-1995

Table H1.   Water-level data used for calibration of transient model of groundwater flow, Idaho National Laboratory and vicinity, Idaho.—Continued

[**Map No.:** Identifier used to locate wells on maps located in appendix B figures and as a cross reference with data in other appendixes **Local name:** Local well used in this study **Observation name:** Model observation name used in this study]

Map No.	Local name	Observation name	Date measured	Observed head	Number of observations	Initial simulated date	Final observed and simulated date
94	GIN 4	GIN4_784	07-12-1984	4,579.20	6	07-12-1984	05-19-1995
95	GIN 5 (old)	GIN5old_784	07-12-1984	4,579.28	5	07-12-1984	05-26-1994
95	GIN 5 (new)	GIN5new_595	05-19-1995	4,574.33	1	05-19-1995	
114	Houghland	Hghlnd_380	Mar. 1980	4,379.85	1	12-31-1980	
115	IET 1 DISP	IETdisp_780	July 1980	4,580.32	24	12-31-1980	07-07-1995
121	ML 11	ML11_491 [1]	04-03-1991	4,534.45	2	04-03-1991	08-23-1992
122	ML 12	ML12_880	Dec. 1980	4,534.71	1	12-31-1980	
123	ML 13	ML13_380	Mar. 1980	4,535.40	2	12-31-1980	04-04-1991
124	MTR TEST	MTRtst_1280	Dec. 1980	4,457.62	167	12-31-1980	12-20-1995
128	NO NAME 1	NoName1_780	July 1980	4,577.82	25	12-31-1980	10-03-1995
129	NPR TEST	NPRtst_784	07-26-1984	4,473.78	57	07-26-1984	10-23-1995
136	NRF 6	NRF6_991	09-09-1991	4,479.04	20	09-09-1991	11-07-1995
138	NRF 7	NRF7_991	09-10-1991	4,479.16	20	09-10-1991	11-08-1995
140	NRF 8	NRF8_895	08-04-1995	4,473.30	1	08-04-1995	
141	NRF 9	NRF9_895	08-04-1995	4,472.11	1	08-04-1995	
142	NRF 10	NRF10_895	08-04-1995	4,472.82	1	08-04-1995	
143	NRF 11	NRF11_895	08-04-1995	4,472.23	1	08-04-1995	
144	NRF 12	NRF12_895	08-04-1995	4,471.89	1	08-04-1995	
145	NRF 13	NRF13_895	08-04-1995	4,477.80	1	08-04-1995	
150	P&W 1	P&W1_780	July 1980	4,580.07	16	12-31-1980	05-19-1995
151	P&W 2	P&W2_780	July 1980	4,579.83	44	12-31-1980	10-18-1995
152	P&W 3	P&W3_780	July 1980	4,580.32	16	12-31-1980	05-31-1995
155	PSTF TEST	PSTFtst_780 [1]	July 1980	4,578.32	25	12-31-1980	10-04-1995
170	QAB	QAB_282	02-13-1982	4,424.48	8	02-13-1982	04-05-1988
184	RWMC M1SA	WMCM1SA_1192	11-30-1992	4,425.56	7	11-30-1992	08-08-1995
185	RWMC M3S	RWMCM3S_1192	11-30-1992	4,426.38	11	11-30-1992	09-05-1995
186	RWMC M4D	RWMCM4D_1192	11-30-1992	4,427.19	8	11-30-1992	09-05-1995
187	RWMC M6S	RWMCM6S_1192	11-30-1992	4,425.51	8	11-30-1992	09-05-1995
188	RWMC M7S	RWMCM7S_1192	11-30-1992	4,426.97	11	11-30-1992	09-05-1995
196	SITE 1	Site1_1280	Dec. 1980	4,378.26	35	12-31-1980	09-21-1995
197	SITE 2	Site2_1280	Dec. 1980	4,158.31	35	12-31-1980	09-21-1995
199	SITE 6	Site6_780	July 1980	4,478.20	16	12-31-1980	06-01-1995
200	SITE 9	Site9_1280	Dec. 1980	4,453.49	95	12-31-1980	10-24-1995
201	SITE 14	Site14_1280	Dec. 1980	4,524.51	112	12-31-1980	10-18-1995
202	SITE 16	Site16_780	July 1980	4,484.84	24	12-31-1980	03-15-1995
203	SITE 17	Site17_780	July 1980	4,483.84	23	12-31-1980	06-16-1995
204	SITE 19	Site19_780	Dec. 1980	4,457.50	34	12-31-1980	07-06-1995
207	TAN 1	TAN1_1187a	11-06-1987	4,586.28	2	11-06-1987	11-17-1987
208	TAN 2	TAN2_1187a	11-06-1987	4,585.68	2	11-06-1987	11-17-1987
209	TAN 3	TAN3_190	01-24-1990	4,582.74	1	01-24-1990	

Table H1.   Water-level data used for calibration of transient model of groundwater flow, Idaho National Laboratory and vicinity, Idaho.—Continued

[**Map No.:** Identifier used to locate wells on maps located in <u>appendix B</u> figures and as a cross reference with data in other appendixes **Local name:** Local well used in this study **Observation name:** Model observation name used in this study]

Map No.	Local name	Observation name	Date measured	Observed head	Number of observations	Initial simulated date	Final observed and simulated date
210	TAN 5	TAN5_190	01-24-1990	4,582.33	1	01-24-1990	
211	TAN 6	TAN6_990	09-09-1990	4,582.27	1	09-09-1990	
212	TAN 7	TAN7_990	09-09-1990	4,582.19	1	09-09-1990	
213	TAN 8	TAN8_190	01-24-1990	4,581.58	1	01-24-1990	
214	TAN 9	TAN9_1289	12-11-1989	4,584.85	1	12-11-1989	
216	TAN 10 A	TAN10A_490	04-25-1990	4,586.51	1	04-25-1990	
217	TAN 11	TAN11_1289	12-14-1989	4,584.96	1	12-14-1989	
219	TAN 13	TAN13_1090	10-24-1990	4,580.57	1	10-24-1990	
220	TAN 13A	TAN13A_1190	11-29-1990	4,578.82	1	11-29-1990	
221	TAN 14	TAN14_1090	10-24-1990	4,580.89	1	10-24-1990	
222	TAN 15	TAN15_991	09-09-1991	4,580.99	1	09-09-1991	
224	TAN 17	TAN17_80e [2]	Dec. 1980	4,578.98	1	12-31-1980	
225	TAN 18	TAN18_1192	11-03-1992	4,576.95	1	11-03-1992	
226	TAN 19	TAN19_1092	10-14-1992	4,577.69	1	10-14-1992	
227	TAN 20	TAN20_1092	10-14-1992	4,577.86	1	10-14-1992	
228	TAN 21	TAN21_1092	10-14-1992	4,575.98	1	10-14-1992	
229	TAN 22	TAN22_892	08-10-1992	4,579.05	1	08-10-1992	
230	TAN 22A	TAN22A_1092	10-14-1992	4,577.09	1	10-14-1992	
231	TAN 23	TAN23_792	07-07-1992	4,578.76	1	07-07-1992	
231	TAN 23A	TAN23A_1092	10-14-1992	4,576.96	1	10-14-1992	
232	TAN 24	TAN24_992	09-25-1992	4,576.61	1	09-25-1992	
232	TAN 24A	TAN24A_1092	10-14-1992	4,575.69	1	10-14-1992	
233	TAN CH 1	TANCH1_190	01-24-1990	4,584.14	1	01-24-1990	
234	TAN CH 2B	TANCH2B_401	April 2001	4,577.41	1	12-31-1980	
242	TRA 6	TRA6_1290	12-07-1990	4,455.60	1	12-07-1990	
243	TRA 7	TRA7_990	09-28-1990	4,457.10	1	09-28-1990	
244	TRA 8	TRA8_1290	12-19-1990	4,457.10	1	12-19-1990	
245	TRA DISP	TRAdisp_784	07-25-1984	4,463.36	30	07-25-1984	10-16-1995
247	USGS 1	USGS1_1280	Dec. 1980	4,435.21	190	12-31-1980	12-20-1995
248	USGS 2	USGS2_1280	Dec. 1980	4,466.83	96	12-31-1980	10-24-1995
250	USGS 4	USGS4_1280	Dec. 1980	4,531.61	100	12-31-1980	10-10-1995
251	USGS 5	USGS5_1280	July 1980	4,469.86	101	12-31-1980	10-25-1995
252	USGS 6	USGS6_1280	Dec. 1980	4,485.14	95	12-31-1980	10-24-1995
253	USGS 7	USGS7_1280	Dec. 1980	4,575.85	104	12-31-1980	10-04-1995
254	USGS 8	USGS8_1280	Dec. 1980	4,429.93	93	12-31-1980	10-12-1995
255	USGS 9	USGS9_1280	Dec. 1980	4,423.77	178	12-31-1980	12-20-1995
256	USGS 11	USGS11_1280	Dec. 1980	4,415.68	96	12-31-1980	10-26-1995
257	USGS 12	USGS12_1280	Dec. 1980	4,490.67	186	12-31-1980	12-28-1995
258	USGS 13	USGS13_781	07-11-1981	4,388.64	28	07-11-1981	04-10-1995
259	USGS 14	USGS14_1280	Dec. 1980	4,418.12	96	12-31-1980	10-26-1995

Table H1. Water-level data used for calibration of transient model of groundwater flow, Idaho National Laboratory and vicinity, Idaho.—Continued

[**Map No.:** Identifier used to locate wells on maps located in <u>appendix B</u> figures and as a cross reference with data in other appendixes **Local name:** Local well used in this study **Observation name:** Model observation name used in this study]

Map No.	Local name	Observation name	Date measured	Observed head	Number of observations	Initial simulated date	Final observed and simulated date
260	USGS 15	USGS15_780	July 1980	4,492.56	43	12-31-1980	11-02-1995
262	USGS 17	USGS17_1280	Dec. 1980	4,478.38	110	12-31-1980	11-07-1995
263	USGS 18	USGS18_1280	Dec. 1980	4,533.62	103	12-31-1980	10-20-1995
264	USGS 19	USGS19_1280	Dec. 1980	4,527.93	183	12-31-1980	12-18-1995
265	USGS 20	USGS20_1280	Dec. 1980	4,453.86	98	12-31-1980	10-16-1995
266	USGS 21	USGS21_1280	Dec. 1980	4,506.06	179	12-31-1980	12-15-1995
267	USGS 22	USGS22_1280	Dec. 1980	4,438.13	97	12-31-1980	10-13-1995
268	USGS 23	USGS23_1280	Dec. 1980	4,484.31	98	12-31-1980	10-10-1995
269	USGS 24	USGS24_780	July 1980	4,579.49	104	12-31-1980	12-19-1995
270	USGS 25	USGS25_1280	Dec. 1980	4,579.07	184	12-31-1980	12-15-1995
271	USGS 26	USGS26_1280	Dec. 1980	4,579.21	99	12-31-1980	10-04-1995
272	USGS 27	USGS27_380	Mar. 1980	4,560.28	183	12-31-1980	12-21-1995
273	USGS 28	USGS28_780	July 1980	4,541.02	24	12-31-1980	04-24-1995
274	USGS 29	USGS29_780	July 1980	4,522.48	29	12-31-1980	06-15-1995
275	USGS 30A	USGS30A_1280	Dec. 1980	4,537.16	96	12-31-1980	11-21-1995
275	USGS 30B	USGS30B_1280	Dec. 1980	4,528.16	96	12-31-1980	11-21-1995
275	USGS 30C	GS30C_1280	Dec. 1980	4,526.07	97	12-31-1980	11-21-1995
276	USGS 31	USGS31_780	July 1980	4,529.69	27	12-31-1980	06-15-1995
277	USGS 32	USGS32_780	July 1980	4,522.28	28	12-31-1980	06-15-1995
279	USGS 34	USGS34_780	July 1980	4,456.43	34	12-31-1980	10-11-1995
280	USGS 35	USGS35_780	July 1980	4,456.25	34	12-31-1980	10-16-1995
281	USGS 36	USGS36_780	July 1980	4,456.53	39	12-31-1980	10-11-1995
282	USGS 37	USGS37_1280	Dec. 1980	4,456.26	92	12-31-1980	10-11-1995
283	USGS 38	GS38new_780	July 1980	4,456.59	33	12-31-1980	10-12-1995
284	USGS 39	USGS39_780	July 1980	4,456.45	39	12-31-1980	10-10-1995
285	USGS 40	GS40nw_1280	Dec. 1980	4,457.25	141	12-31-1980	10-19-1995
286	USGS 41	USGS41_780	July 1980	4,457.68	35	12-31-1980	10-19-1995
287	USGS 42	USGS42_780	July 1980	4,457.64	35	12-31-1980	10-19-1995
288	USGS 43	USGS43_780	July 1980	4,457.64	35	12-31-1980	11-13-1995
289	USGS 44	USGS44_780	July 1980	4,457.76	34	12-31-1980	10-16-1995
290	USGS 45	USGS45_780	July 1980	4,456.81	34	12-31-1980	10-11-1995
291	USGS 46	USGS46_1280	Dec. 1980	4,457.62	77	12-31-1980	10-16-1995
292	USGS 47	USGS47_780	July 1980	4,458.45	48	12-31-1980	10-16-1995
293	USGS 48	USGS48_780	July 1980	4,457.81	32	12-31-1980	10-16-1995
294	USGS 49	GS49new_1186	11-14-1986	4,460.00	11	11-14-1986	04-27-1992
296	USGS 51	USGS51_780	July 1980	4,457.79	34	12-31-1980	10-24-1995
297	USGS 52	USGS52_780	July 1980	4,457.81	33	12-31-1980	10-19-1995
302	USGS 57	USGS57_780	July 1980	4,456.42	51	12-31-1980	10-11-1995
303	USGS 58	USGS58_1280	Dec. 1980	4,457.61	135	12-31-1980	10-30-1995
304	USGS 59	USGS59_780	July 1980	4,457.49	32	12-31-1980	10-23-1995

Table H1.   Water-level data used for calibration of transient model of groundwater flow, Idaho National Laboratory and vicinity, Idaho.—Continued

[**Map No.:** Identifier used to locate wells on maps located in <u>appendix B</u> figures and as a cross reference with data in other appendixes   **Local name:** Local well used in this study   **Observation name:** Model observation name used in this study]

Map No.	Local name	Observation name	Date measured	Observed head	Number of observations	Initial simulated date	Final observed and simulated date
310	USGS 65	USGS65_780	July 1980	4,458.28	50	12-31-1980	10-11-1995
312	USGS 67	USGS67_780	Dec. 1980	4,457.19	27	12-31-1980	10-17-1995
321	USGS 76	USGS76_780	July 1980	4,456.79	40	12-31-1980	10-30-1995
322	USGS 77	USGS77_380	July 1980	4,456.95	32	12-31-1980	10-24-1995
324	USGS 79	USGS79_780	July 1980	4,457.40	69	12-31-1980	10-30-1995
327	USGS 82	USGS82_1280	Dec. 1980	4,457.37	88	12-31-1980	10-17-1995
328	USGS 83	USGS83_1280	Dec. 1980	4,442.82	98	12-31-1980	10-26-1995
329	USGS 84	USGS84_1280	Dec. 1980	4,456.35	86	12-31-1980	10-18-1995
330	USGS 85 (old)	GS85old_1280	Dec. 1980	4,456.00	89	12-31-1980	04-19-1993
330	USGS 85 (new)	GS85new_793	07-28-1993	4,453.39	10	07-28-1993	10-23-1995
331	USGS 86	USGS86_1280	Dec. 1980	4,428.72	98	12-31-1980	10-12-1995
332	USGS 87	USGS87_1080	Oct. 1980	4,430.13	107	12-31-1980	10-16-1995
334	USGS 89	USGS89_1080	Oct. 1980	4,429.24	110	12-31-1980	10-25-1995
335	USGS 90	USGS90_1080	Oct. 1980	4,427.83	98	12-31-1980	10-25-1995
343	USGS 97	USGS97_1280	Dec. 1980	4,477.95	162	12-31-1980	12-26-1995
344	USGS 98	USGS98_780	July 1980	4,472.15	62	12-31-1980	11-06-1995
345	USGS 99	USGS99_780	July 1980	4,477.43	58	12-31-1980	11-06-1995
346	USGS 100	USGS100_780	July 1980	4,482.09	52	12-31-1980	10-19-1995
347	USGS 101	USGS101_780	July 1980	4,481.62	24	12-31-1980	10-19-1995
348	USGS 102	GS102_1289	12-20-1989	4,482.56	35	12-20-1989	11-07-1995
349	USGS 103	USGS103_1280	Dec. 1980	4,426.71	74	12-31-1980	10-10-1995
350	USGS 104	USGS104_1280	Dec. 1980	4,431.94	54	12-31-1980	10-10-1995
351	USGS 105	USGS105_1180	Nov. 1980	4,426.29	41	12-31-1980	10-10-1995
352	USGS 106	GS106_1180	Nov. 1980	4,428.19	90	12-31-1980	10-10-1995
353	USGS 107	USGS107_1180	Nov. 1980	4,438.48	82	12-31-1980	10-26-1995
354	USGS 108	USGS108_1280	Dec. 1980	4,426.00	41	12-31-1980	10-10-1995
355	USGS 109	GS109_1180	Nov. 1980	4,423.92	65	12-31-1980	10-12-1995
356	USGS 110	USGS110_1080	Oct. 1980	4,429.91	70	12-31-1980	04-01-1992
356	USGS 110A	GS110A_1095	10-25-1995	4,433.80	1	10-25-1995	
357	USGS 111	USGS111_884	08-31-1984	4,464.26	38	08-31-1984	10-26-1995
358	USGS 112	USGS112_785	07-06-1985	4,463.87	43	07-06-1985	10-11-1995
359	USGS 113	USGS113_785	07-06-1985	4,464.87	40	07-06-1985	10-17-1995
360	USGS 114	USGS114_785	07-06-1985	4,463.58	40	07-06-1985	10-25-1995
361	USGS 115	USGS115_785	07-06-1985	4,463.41	42	07-06-1985	10-23-1995
362	USGS 116	USGS116_785	07-06-1985	4,464.33	39	07-06-1985	10-17-1995
363	USGS 117	USGS117_1287	12-08-1987	4,432.23	45	12-08-1987	10-24-1995
364	USGS 118	USGS118_1192	11-30-1992	4,425.98	13	11-30-1992	11-22-1995
365	USGS 119	USGS119_1287	12-08-1987	4,431.89	37	12-08-1987	10-25-1995
366	USGS 120	GS120_1287	12-08-1987	4,429.81	94	12-08-1987	12-20-1995
367	USGS 121	USGS121_191	01-29-1991	4,458.67	14	01-29-1991	10-30-1995

Table H1. Water-level data used for calibration of transient model of groundwater flow, Idaho National Laboratory and vicinity, Idaho.—Continued

[**Map No.:** Identifier used to locate wells on maps located in appendix B figures and as a cross reference with data in other appendixes **Local name:** Local well used in this study **Observation name:** Model observation name used in this study]

Map No.	Local name	Observation name	Date measured	Observed head	Number of observations	Initial simulated date	Final observed and simulated date
368	USGS 122	USGS122_191	01-29-1991	4,458.30	12	01-29-1991	10-30-1995
369	USGS 123	USGS123_191	01-29-1991	4,458.15	13	01-29-1991	10-30-1995
370	USGS 124	USGS124_194	01-27-1994	4,418.84	7	01-27-1994	10-26-1995
371	USGS 125	USGS125_495	04-27-1995	4,419.84	3	04-27-1995	10-26-1995
375	Weaver and Lowe	WeavLow_485	04-15-1985	4,475.23	2	04-15-1985	05-23-1985
376	WS INEL 1	WSINEL1_1280	Dec. 1980	4,476.72	46	12-31-1980	11-06-1995

[1] Head observations only

[2] Head estimated by extrapolation from March 2004

Appendix I. Tritium Disposal and Production Well Pumpage at Idaho Nuclear Technology and Engineering Center and Reactor Technology Complex, Idaho National Laboratory, Idaho

Table I1.   Tritium disposal and production well pumpage at the Idaho Nuclear Technology and Engineering Center and the Reactor Technology Complex, 1952 through 1968, Idaho National Laboratory, Idaho.

[Data from references as indicated at end of table **Abbreviations**: INTEC, Idaho Nuclear Technology and Engineering Center; RTC, Reactor Technology Complex; ft³/s, cubic foot per second; na, data not available; –, no waste disposed]

	Tritium disposal at INTEC and RTC, 1952 through 1968					
Year	INTEC disposal well (CPP 3)		INTEC disposal pit		RTC warm-waste ponds	
	Discharge rates (ft³/s)	Total tritium (Curies)	Discharge rates (ft³/s)	Total tritium (Curies)	Discharge rates (ft³/s)	Total tritium (Curies)
1952	–	–	–	–	[1]0.021	[1,2]7
1953	[3]1.679	[4]456	–	–	[1].064	[2]20
1954	[3].971	[4]608	[3]0.021	na	[1].403	[2]105
1955	[3]1.679	[4]808	[3].021	na	[5].414	[2]114
1956	[3]1.488	[4]1,543	[3].025	na	[5].402	[2]77
1957	[3].979	[4]969	[3].021	na	[5].437	[2]89
1958	[3]1.157	[4]3,504	[3].030	na	[5]1.056	[2]278
1959	[3]1.390	[4]2,565	[3].034	na	[3].837	[2]444
1960	[3].805	[4]679	[3].008	na	[3].932	[2]308
1961	[3].797	[6]578	[3].000	[7,8]12	[3].986	[8]303
1962	[3]1.111	[8]146	[3].020	[8]194.4	[3]1.195	[8]278
1963	[3]1.089	[8]1,021	[3].011	[8]58.2	[3].841	[8]386
1964	[3]1.458	[8]1,696	[3].006	[8]72.5	[3].713	[8]399
1965	[3]1.763	[9]58	[3].003	[9]37.9	[3].713	[9]328
1966	[3]1.543	[10]234	[3].002	[10]15.5	[3].546	[10]389
1967	[3]1.276	[11]857	–	–	[3].754	[11]402
1968	[3]1.149	[12]510	–	–	[3].784	[12]499

	Production well pumpages at INTEC and RTC, 1952 through 1968					
	INTEC well pumpages (ft³/s)		RTC well pumpages (ft³/s)			
Year	CPP 1	CPP 2	TRA 1	TRA 2	TRA 3	TRA 4
1952	–	–	na	na	–	–
1953	na	na	na	na	–	–
1954	na	na	na	na	–	–
1955	na	na	na	na	–	–
1956	na	na	na	na	–	–
1957	na	na	na	na	na	–
1958	na	na	na	na	na	–
1959	[13]0.444	[13]0.931	[13]0.584	[13]0.520	[13]1.599	–
1960	[13].504	[13].696	[13]1.123	[13].383	[13]1.237	–
1961	[13].659	[13].644	[13]2.311	[13].128	[13].492	–
1962	[13].919	[13].726	[13]1.962	[13].064	[13].980	–
1963	[13].751	[13].637	[13]1.221	[13].001	[13]1.570	na
1964	[13].726	[13].800	[13]1.134	[13].000	[13]1.719	na
1965	na	na	na	–	na	na
1966	na	na	na	–	na	na
1967	na	na	na	–	na	na
1968	na	na	na	–	na	na

Table I1.   Tritium disposal and production well pumpage at the Idaho Nuclear Technology and Engineering Center and the Reactor Technology Complex, 1952 through 1968, Idaho National Laboratory, Idaho.—Continued

[Data from references as indicated at end of table **Abbreviations**: IDO, Idaho Operations Office; INTEC, Idaho Nuclear Technology and Engineering Center; RTC, Reactor Technology Complex; ft³/s, cubic foot per second; na, data not available; –, no waste disposed]

References	
1	Schmalz, B.L., 1972, Radionuclide distribution in soil mantle of the lithosphere as a consequence of waste disposal at the National Reactor Testing station: U.S. Atomic Energy Commission, Idaho Operations Office Publication IDO-10049, 80 p.
2	Barraclough, J.T., Teasdale, W.E., Robertson, J.B., and Jensen, R.G., 1967, Hydrology of the National Reactor Testing Station, Idaho, 1966: U.S. Geological Survey Open-File Report, (IDO-22049), 95. p.
3	Robertson, J.B., Schoen, R., and Barraclough, J.T., 1974, The influence of liquid waste disposal on the geochemistry of water at the National Reactor Testing Station, Idaho: 1952-1970, U.S. Geological Survey Open-File Report, (IDO-22053) 231 p.
4	DOE-ID, retrieved May 4, 1989, Radioactive waste management information system nondecay summary database list (P61SH050) list for 1954-1974: FOIA Document No. 0007, RWM-167, 1 p.
5	U.S. Atomic Energy Commission, 1972, U.S. Atomic Energy Commission, Idaho Operations Office, National Reactor Testing Station Radioactive Waste Management Information, 1972, Summary and Record-to-Date: U.S. Atomic Energy Comission, Idaho Operations Office Publication IDO-10054(72), 13 p.
6	Schmalz, B.L., and Keys, W.S., 1962, Retention and migration of radioactive isotopes in the lithosphere at the National Reactor Testing Station – Idaho: U.S. Atomic Energy Commission, Waste Management Section, Idaho Operations Office Publication IDO-12026, 26 p.
7	Lewis, B.D., Eagleton, J.M., and Jensen, R.G., 1985, Aqueous radioactive- and industrial-waste disposal at the Idaho National Engineering Laboratory through 1982: U.S. Geological Open-File Report 85-636, (DOE/ID-22069), 77 p.
8	Osloond, J.H., 1965, Radioactive waste disposal data for the National Reactor Testing Station: U.S. Atomic Energy Commission, Idaho Operations Office Publication IDO-12040, 29 p.
9	Osloond, J.H., and Schmalz, B.L., 1966, Radioactive waste disposal data for the National Reactor Testing Station, Idaho: U.S. Atomic Energy Commission, Idaho Operations Office Publication IDO-12040-S1, 12 p.
10	Osloond, J.H., 1967, Radioactive waste disposal data for the National Reactor Testing Station, Idaho: U.S. Atomic Energy Commission, Idaho Operations Office Publication IDO-12040-S2, 18 p.
11	Osloond, J.H., and Newcomb, D.L., 1968, Radioactive waste disposal data for the National Reactor Testing Station, Idaho: U.S. Atomic Energy Commission, Idaho Operations Office Publication IDO-12040-S3, 15 p.
12	Osloond, J.H., and Newcomb, D.L., 1969, Radioactive waste disposal data for the National Reactor Testing Station, Idaho: U.S. Atomic Energy Commission, Idaho Falls, Idaho, (IDO-12040-S4), 15 p.
13	Morris, D.A., Barraclough, J.T., Chase, G.H., Teasdale, W.E., and Jensen, R.G., 1965, Hydrology of subsurface waste disposal National Reactor Testing Station Idaho annual progress report: U.S. Atomic Energy Commission, Idaho Falls, Idaho, (IDO-22047), [variously paged].

www.ingramcontent.com/pod-product-compliance
Lightning Source LLC
Chambersburg PA
CBHW081439170526
45166CB00008B/2252